HIGH POLYMERS

HIGH POLYMERS

A SERIES OF MONOGRAPHS ON THE CHEMISTRY, PHYSICS, AND TECHNOLOGY OF HIGH POLYMERIC SUBSTANCES

VOLUME XXIII

POLYMER CHEMISTRY OF SYNTHETIC ELASTOMERS

Part I

EDITED BY

JOSEPH P. KENNEDY

Esso Research and Engineering Company
Linden, New Jersey

AND

ERIK G. M. TÖRNQVIST

Esso Research and Engineering Company
Linden, New Jersey

INTERSCIENCE PUBLISHERS

A DIVISION OF JOHN WILEY & SONS, NEW YORK · LONDON · SYDNEY

To Robert McKay Thomas
Inspirer, Inquirer, Inventor

——THE EDITORS

PREFACE

This book, as its title indicates, is a collection of treatises of various aspects of the chemistry of synthetic elastomers. It emphasizes polymer chemistry, and in particular the chemistry involved in polymer formation, as contrasted to polymer physics. The reasons for limiting the book to chemical aspects are manifold but are largely the same as those which prompted the publication of this book in the first place.

To begin with, the discovery of the complex organometallic catalysts by Karl Ziegler and coworkers in 1953 and the subsequent development of numerous stereospecific catalysts of this type has led to the practical synthesis of several novel and highly useful elastomers, including *cis*-1,4 polyisoprene or "synthetic natural rubber"—the original and ultimate goal of chemists interested in synthetic rubber.

These developments alone could very well have justified the publication of this book, especially since they have taken place after the last preceding comprehensive treatise of the same general subject, namely, *Synthetic Rubber*, edited by G. S. Whitby, published in 1954. However, this book could have been justified equally well on the basis of developments in other areas of elastomer chemistry. The tremendous progress made in the area of alkyllithium initiated polymerization leading among others to another method of making *cis*-1,4 polyisoprene may be mentioned as a typical example, but many developments involving more conventional polymerization methods (cationic, free radical, condensation, etc.) could also be used to illustrate the rapid progress made in elastomer synthesis during the past 10–15 years.

Since even before the very rapid and in many respects revolutionary progress in elastomer synthesis began some 15 years ago, it had barely been possible to treat all important aspects of synthetic elastomers in one volume. It was obvious that a satisfactory presentation of the chemical aspects in one volume would require the deletion of many if not most other aspects of synthetic elastomers. Inasmuch as many of the latter and notably the relationship between structure and physical properties in elastomers had been treated recently in great depth and considerable detail in other books and numerous review articles, it was decided to limit the scope of this book to the chemistry of synthetic elastomers and to treat other related subjects only to the extent necessary for a clear presentation of the chemical aspects. This general approach led to the inclusion of an introductory chapter on "Structure–Property Relationships for Elastomeric Materials" and two immediately following chapters dealing primarily with the history and the status and future of synthetic elastomers.

Even with these limitations, the subject matter of the book remained so diverse and extensive that it could hardly be treated in a satisfactory manner by one author or even by a few closely cooperating authors, at least not within a reasonable period of time. In view of the rapid progress still being made in this field, it was obvious, therefore, that a truly satisfactory and up to date book could be written only by a large number of specialists. At first, this seemed to pose some problems, since many of the foremost experts on elastomer synthesis were associated with industrial firms having considerable economic interests in this area. However, it turned out that not only did most of the scientists invited to contribute to our book respond enthusiastically but the various firms with which many of the prospective authors were associated also readily gave them permission to contribute.

While offering the only satisfactory solution, the publication of a book written by numerous authors also poses certain problems. For instance, every author has his personal style and preferences. This leads automatically to a considerable variation in the method of presentation of the various subjects. It will also lead to some repetition that could be avoided in a book written by a single author. It is one of the duties of the editors to minimize these negative aspects of multiple authorship; however, there is a limit to what can and should be done in this respect. It has been the policy of the editors to give the individual authors as much freedom as possible not only in selecting their methods of presentation but also in deciding what should be included for a proper presentation of their subject matter within the guidelines offered by the editors.

Some alert readers will undoubtedly notice that certain elastomeric materials, e.g., elastomeric polyesters, polycarbonates, polyalkylene sulfides, nitroso rubbers, and certain recently developed A-B-A type block copolymers and related graft copolymers, etc., have not been treated or could not be treated in detail. The decision not to discuss these materials is somewhat arbitrary, but is based primarily on the fact that they are presently of limited importance. Even with these omissions, it turned out, ironically, that our hope of publishing this book in one

volume was unrealistic.   This, of course, is a further indication of the
great expansion of our knowledge of elastomer synthesis that has occurred
in recent years.

The publication of this book would not have been possible had it not
been for the positive and encouraging attitude of a large number of
industrial firms and scientific institutions which gave the various authors
permission and opportunity to contribute.   We, the editors, should like
in particular to express our gratitude to the Esso Research and Engineering Company and its management for the encouraging attitude which
was shown during the preparation of this book.   We should also like to
express our gratitude for many valuable suggestions and much good
advice offered by Drs. John Rehner, Jr. and F. P. Baldwin as well as by
many other colleagues.

J. P. Kennedy
E. G. M. Törnqvist
March, 1968

CONTRIBUTORS TO VOLUME XXIII

GLEN ALLIGER, *The Firestone Tire and Rubber Company, Akron, Ohio*

M. B. BERENBAUM, *Thiokol Chemical Corporation, Trenton, New Jersey*

J. R. COOPER, *E. I. du Pont de Nemours and Company, Wilmington, Delaware*

GINO DALL'ASTA, *Montecatini S.p.A., Milano, Italy*

J. V. DAWKINS, *Duke University, Durham, North Carolina*

C. FITZSIMMONDS, *University of Liverpool, Liverpool, England*

L. E. FORMAN, *The Firestone Tire and Rubber Company, Akron, Ohio*

J. FURUKAWA, *Kyoto University, Sakyo, Kyoto, Japan*

R. H. GOBRAN, *Thiokol Chemical Corporation, Trenton, New Jersey*

M. A. GOLUB, *Stanford Research Institute, Menlo Park, California*

CHESTER HARGREAVES, II, *E. I. du Pont de Nemours and Company, Wilmington, Delaware*

W. HOFMANN, *Farbenfabriken Bayer A.G., Leverkusen-Bayerwerk, Germany*

S. IDO, *Japan Synthetic Rubber Company, Limited, Tokyo, Japan*

A. H. JORGENSEN, *The B. F. Goodrich Chemical Company, Avon Lake, Ohio*

J. P. KENNEDY, *Esso Research and Engineering Company, Linden, New Jersey*

W. R. KRIGBAUM, *Duke University, Durham, North Carolina*

JOGINDER LAL, *The Goodyear Tire and Rubber Company, Akron, Ohio*

A. LEDWITH, *University of Liverpool, Liverpool, England*

F. M. LEWIS, *General Electric Company, Waterford, New York*

H. S. MAKOWSKI, *Esso Research and Engineering Company, Linden, New Jersey*

GIULIO NATTA, *Istituto di Chimica Industriale del Politecnico, Milano, Italy*

LIDO PORRI, *Istituto di Chimica Industriale del Politecnico, Milano, Italy*

R. E. RINEHART, *Uniroyal-U.S. Rubber Company, Wayne, New Jersey*

GUIDO SARTORI, *Esso Research S.A., Brussels, Belgium*

J. H. SAUNDERS, *Mobay Chemical Company, Pittsburgh, Pennsylvania*

W. H. SHARKEY, *E. I. du Pont de Nemours and Company, Wilmington, Delaware*

ERIK G. M. TÖRNQVIST, *Esso Research and Engineering Company, Linden, New Jersey*

H. A. TUCKER, *The B. F. Goodrich Company Research Center, Brecksville, Ohio*

CARL A. URANECK, *Phillips Petroleum Company, Bartlesville, Oklahoma*

ALBERTO VALVASSORI, *Montecatini, S.p.A., Milano, Italy*

OTTO VOGL, *E. I. du Pont de Nemours and Company, Wilmington, Delaware*

F. C. WEISSERT, *The Firestone Tire and Rubber Company, Akron, Ohio*

S. YAMASHITA, *Kyoto University, Kyoto, Japan*

CONTENTS FOR PART I

CONTENTS FOR PART II

(Tentative)

CHAPTER 1

STRUCTURE–PROPERTY RELATIONSHIPS FOR ELASTOMERIC MATERIALS

W. R. Krigbaum and J. V. Dawkins

Department of Chemistry
Duke University, Durham, North Carolina

Contents

The rubber chemist must be vitally concerned with a wide range of properties exhibited by the materials he produces in the plant or laboratory. Among the foremost of these are the mechanical properties. Some of these are connected with the stress–strain behavior, and may at least approximate equilibrium properties, while others are concerned with viscoelastic behavior, and hence have a kinetic origin. Aside from mechanical behavior, other properties of interest may include the dielectric constant, vapor barrier, and swelling or solubility behaviors. Whether he is interested in developing a new product, or improving an existing one, his efforts must be guided by some rationale. This chapter will present a brief exploration of the guidelines which can be developed from existing theory concerning the structure–property relationships for elastomers.

In the interest of brevity, we shall refer the reader elsewhere for the

1

background and derivation of the theoretical relationships cited. Furthermore, we will assume at the outset the fulfillment of the essential requirements for long range elasticity. These are, of course, that the material is predominantly amorphous, is above its glass transition temperature, and consists of flexible chains of reasonable length interconnected in some fashion to form a permanent network structure. Our ideal objective is to predict some range, or area, of anticipated property values from a minimum of input data, such as the chemical formula of the chain repeating unit and some parameters characterizing the network structure.

I. STRESS–STRAIN BEHAVIOR

The elastic equation of state derived by statistical mechanics may be used to predict the stress-strain behavior of an elastomer. The reader is referred to a recent review (1) for the historical development of this subject, and for further details of the various treatments.

A. Treatments Assuming No Energy Change on Deformation

Let us consider a vulcanized specimen undergoing simple elongation, and ignore the dilation which accompanies this process. According to the early gaussian treatments, the tension, or retractive force per unit unstrained cross section, is given by

$$f_u = (G/3)(\alpha - 1/\alpha^2) \tag{1}$$

where α is the extension ratio (or ratio of the final and initial sample lengths), and the initial modulus, $G = (\delta f_u/\delta \alpha)_{T,V}$, is

$$G = 3(\nu_e/V)kT \tag{2}$$

Here ν_e/V is the number of elastically effective network chains per unit volume, k is Boltzmann's constant, and T is the absolute temperature. For vulcanized samples ν_e/V clearly depends in some manner on the crosslink density. However, those crosslinks required to form a continuous network from the N initial chains must be discounted, and others may be wasted by the formation of dangling chains and elastically inactive loops. On the other hand, even in an unvulcanized sample some chain entanglements have an appreciable lifetime, and these will increase the number of effective network chains. If the latter two factors are ignored, then ν_e is given approximately by (2)

$$\nu_e = \nu(1 - 2N/\nu) = \nu(1 - 2M_c/M) \tag{3}$$

where ν is the number of crosslinks in the sample and M and M_c are the average molecular weights of the N initial molecules, and of the network chains, respectively.

For our purposes it is important to observe the close analogy between this result and that obtained in the treatment of ideal gases. One sees that ν_e plays the same role as the number of gas particles, and that it can be controlled by varying the initial polymer molecular weight and the crosslinking density. No parameter characteristic of the particular chemical nature of the elastomer appears in this simplest equation of state.

The gaussian model fails to account for the finite extensibility of real network chains. This feature may be incorporated by replacing the gaussian by the inverse Langevin displacement distribution. The result obtained by Treloar (3), for the simplest three-chain model, is:

$$f_u = \frac{G}{3}\left(\alpha - \frac{1}{\alpha^2}\right)\left[1 + \frac{1}{5n}\left(\alpha + \frac{1}{\alpha^2}\right) + \ldots\right] \tag{4}$$

Here the portion of a molecule between crosslinks is represented by a flexibly linked chain having n links of fixed length l. Thus n is given by M_c/bM_0, where b is the number of repeating units, having molecular weight M_0, in a statistical link. This treatment predicts, in general, a somewhat larger force at any extension and, more importantly, an upturn of the stress–strain curve at high extension. This equation is of interest here since it introduces implicitly one parameter, b, characteristic of the flexibility of the particular molecule at hand. The upturn appears at lower extensions as n is made smaller which, for a given network geometry, will occur as the molecular flexibility is decreased.

Rivlin (4) has shown that the stored energy may be generally expressed in terms of a set of strain invariants. A particular example of this general approach, and the simplest one, is the Mooney equation (5):

$$f_u = (2C_1 + 2C_2/\alpha)(\alpha - 1/\alpha^2) \tag{5}$$

This may be regarded as a semiempirical equation of state. It affords a better representation of experimental data, but is of little predictive value, since the molecular significance of the parameter C_2 remains obscure. Since the C_1 term has the same dependence upon α as the result of the gaussian treatment, one is tempted to regard the Mooney equation in the same light as the virial expansion for nonideal gases.

This analogy, however, is incorrect, since the C_1 term does not represent the limiting behavior as $\alpha \to 1$.

B. Energy Effects

All conformations of the gaussian or inverse Langevin chains mentioned above are assumed to correspond to the same energy. Thus, the equations of state deduced from these models yield a retractive force which is purely entropic in nature, i.e., dependent upon the number of distinguishable arrangements of the molecules. However, a number of experimental observations have clearly established that an appreciable fraction of the retractive force of real networks arises from intramolecular energy effects. Due to the barrier hindering rotation about the chain bonds, all conformations cannot, in fact, correspond to the same energy. Hence, in general one must expect that the unperturbed dimensions of the uncrosslinked molecules will be temperature dependent, and that a network formed from these molecules will exhibit a retractive force having an energy component. The fact that the existence of one of these latter conditions necessarily implies the other was first pointed out by Volkenstein (6).

The simplest approach to this problem considers the rotations about the individual bonds of the chain to occur independently, and neglects all interactions other than those involving the groups or atoms directly bonded to the two carbon atoms undergoing rotation. Let two neighboring chain bonds specify a coordinate system, and designate by ϕ the angle by which the third bond is rotated out of the *trans* conformation, as shown in Fig. 1. The potential energy will be a smooth and continuous function of the rotational angle ϕ, as illustrated by the potential diagram for the ethane molecule shown in the lower right-hand side of Fig. 1. Nevertheless, an ethane molecule will spend most of the time in one of the three rotational positions corresponding to the energy minima, which suggests a treatment of hindered rotation based upon a small number of discrete rotational "states." This basic idea has been developed in detail by Volkenstein (7), and the reader is referred there for details.

Figure 2 illustrates schematically the type of diagrams obtained for a number of hydrocarbons. Figure 2a represents the potential diagram for rotation about the central bond of butane. It is assumed that a similar diagram would apply for polyethylene. At absolute zero one might expect to find all the bonds of polyethylene frozen in the lowest energy *trans* state, which corresponds to the highly extended planar zigzag conformation. This, in fact, is the stable conformation in the crystalline

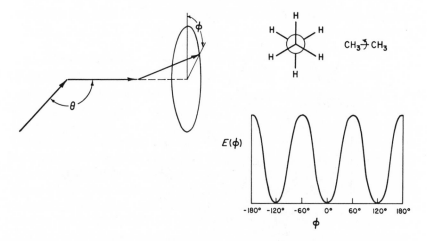

Fig. 1. Model for the rotation about single bonds, and the rotational potential for ethane.

state. As the temperature is raised, the population of bonds in the two higher energy *gauche* states should be augmented. Clearly it is the energy difference designated ΔE in Fig. 2a which will determine, for any temperature, the partition of bonds between the *trans* and *gauche* states. As the fraction of *trans* bonds diminishes with rising temperature, the mean-square displacement length $\langle r^2 \rangle_0$ of the chain should decrease. If such a chain were held at a fixed displacement length, the force required should increase with temperature. Since the energy component of the retractive force, f_e, is given by

$$f_e = f - T(\partial f / \partial T)_{V,L}, \tag{6}$$

this leads to the conclusion that a network of amorphous polyethylene chains should exhibit a negative value of f_e/f, a conclusion verified experimentally by Ciferri, Hoeve, and Flory (8). We mention in passing that the barrier height E^* shown in Fig. 2a will determine the temperature at which exchange between the rotational states becomes possible, and hence is connected with the glass transition phenomena to be considered below.

The barrier to rotation about the central chain bond of 2-methyl butane is shown in Fig. 2b. This might serve as an approximation for

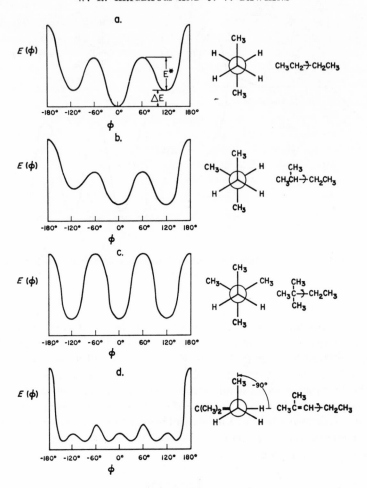

Fig. 2. Schematic potentials for rotation about the central bond of four hydrocarbons (simplest approximation).

the polypropylene molecule. One observes that the two *gauche* states now have different energies, which might suggest the possibility of a helical conformation for the isotactic molecule in the crystalline state, as is actually observed. Figure 2c illustrates the potential diagram for 2,2-dimethyl butane, a possible model compound for polyisobutylene. This potential resembles that of ethane in having three minima of equal energy. Volkenstein has suggested that elastomers are characterized by rotational potentials having minima of nearly equal magnitude, so that $d \ln \langle r^2 \rangle_0 / dT$ and f_e/f are small. Polyisolbutylene is elastomeric, and the same consideration would predict elastomeric behavior for polyisoprene and polybutadiene, as can be inferred from the potential diagram for 2-methyl pentene-2 shown in Fig. 2d.

This appears to be a remarkably simple concept having some predictive value; however, one does not have to look far to find exceptions. For example, it is difficult to predict from these arguments the observed elastomeric behavior of atactic polypropylene or the ethylene-propylene copolymers. Secondly, even the basic premise that rubberlike behavior is associated with small values of f_e/f is apparently not always borne out by experience. One might hope that the basic philosophy is sound, but that some further elaboration is required. Indeed, experience has shown that the magnitude and temperature dependence of $\langle r^2 \rangle_0$ can not be quantitatively predicted if rotations about successive bonds are treated as being independent. Rather, one must consider the rotation of successive bonds to be correlated, and all interactions with neighboring groups along the chain must be taken into account. If one considers only the correlated rotations of *pairs* of bonds, then the potential diagram must be represented as a contour map. Figure 3 illustrates such a diagram for *n*-pentane from the recent work of Abe, Jernigan, and Flory (9). Here the terminal bonds are assumed to be fixed in the *trans* position, and energy minima are marked by crosses. Comparison may be made with Fig. 2a. The more elaborate matrix calculations are capable of predicting $\langle r^2 \rangle_0$ and its temperature dependence. Furthermore, one can view Fig. 3 from the standpoint of a small number of rotational states of low energy; however, much of the simplicity of the early rotational isomer theory has vanished.

We now return to the elastic equation of state, and ask how energy effects will modify the stress-strain curve. If the distribution of displacement lengths is assumed to remain gaussian in shape, but $\langle r^2 \rangle_0$ is taken as temperature dependent, the result (10) is simply to introduce into Eq. 2 the additional factor $\langle r_N^2 \rangle / \langle r^2 \rangle_0$, where $\langle r_N^2 \rangle$ is the mean-square displacement length of the network chains and $\langle r^2 \rangle_0$ is the unperturbed mean-square displacement length which the same chains would exhibit if severed at the network tie-points. One might suspect that a gaussian displacement distribution is not appropriate in this case, since it does not fully account for the fact that different conformations correspond to different energies. Attempts to treat this problem more

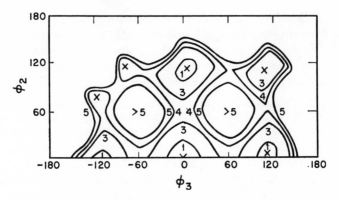

Fig. 3. Computed rotational diagram for the two internal bonds of pentane.   (From Abe, Jernigan, and Flory (9).)

rigorously meet with difficulties owing to the fact that adjacent rotations are correlated, which leads to the familiar Ising lattice problem.   Examination of one three-dimensional model amenable to treatment—the cubic lattice chain—revealed (11) that the shape of the stress–strain curve can be somewhat modified by energy effects.   Nevertheless, it remains true that energy effects mainly modify the temperature dependence of the retractive force, while having a much less pronounced effect upon the shape of the stress–strain curve.   This rather surprising circumstance arises from a partial cancellation of the energy and entropy contributions associated with rotation about the backbone chain bonds, as pointed out by Volkenstein and Ptitsyn (6).

In summary, the elastic equations of state provided by current theory yield the qualitative shape of the elastomeric stress–strain curve, and predict a dependence of the initial modulus upon the crosslink density. These are essentially gaslike treatments, which explains in part why the equations contain no other parameters characteristic of the particular material under investigation.   Attempts have been made (12–14), and are still progressing, to produce a liquidlike elastic equation of state. These may lead to a better understanding of the departures which are currently lumped in the empirical C_2 term.   However, the mere fact that the gaussian relationships afford a reasonable representation of the observed behavior implies that the particular chemical nature of the elastomer is of secondary importance so far as the stress–strain properties of amorphous elastomers are concerned.

II. VISCOELASTIC BEHAVIOR

A. Principle of Corresponding States

Perhaps the foremost development resulting from the study of the many types of viscoelastic phenomena is the discovery of the time–temperature superposition principle. This concept was first applied to viscoelastic behavior by Leaderman (15), and its application was extensively investigated by Tobolsky (16) and Ferry (17) and their respective coworkers. The principle states that if, at some standard temperature, T_S, the time dependence of a viscoelastic property is represented by the function (t/τ), then the time dependence at another temperature T will be given by $(t/a_T\tau)$, where a_T is a temperature shift factor. The reader is referred elsewhere for details concerning the application of time–temperature superposition (16,17). We will merely point out that it permits one to examine, in effect, a very broad frequency range by performing measurements at different temperatures.

Williams, Landel, and Ferry (18) found that, for a wide variety of both polymeric and nonpolymeric materials, the shift factor could be represented in the following way:

$$\ln a_T = -\frac{c_1(T - T_S)}{c_2 + (T - T_S)} \tag{7}$$

where c_1 and c_2 are numerical constants (whose values depend only upon the particular choice of the reference state) and T_S is the temperature which reduces the material under investigation to the selected reference state. Hence the viscoelastic behavior of all materials can be represented by a single master curve if these materials are in corresponding states.

One can establish a relationship between the foregoing empirical equation and the free volume through a second empirical relation, that of Doolittle for the temperature dependence of the viscosity (19):

$$\ln \eta = \ln A + B(v - v_f)/v_f \tag{8}$$

Here v and v_f are the specific volume and the free volume per gram, respectively, and A and B are treated as adjustable parameters. Thus the ratio of the viscosities at temperature T, and at a standard temperature T_S, may be represented as

$$\ln\left(\frac{\eta_T}{\eta_S}\right) = B\left(\frac{1}{\phi_T} - \frac{1}{\phi_S}\right) \tag{9}$$

where $\phi = v_f/v$ is the fractional free volume. The free volume is, unfortunately, an elusive quantity; however, if we assume that its temperature dependence is given by

$$\phi_T = \phi_S + \alpha(T - T_S) \tag{10}$$

where α is the cubical coefficient of thermal expansion, then Eq. 9 can be transformed into a relation having the same form as the WLF Eq. 7:

$$a_T \cong \ln\left(\frac{\eta_T}{\eta_S}\right) = -\frac{(B/\phi_S)(T - T_S)}{(\phi_S/\alpha) + (T - T_S)} \tag{11}$$

We are left with the question: is there a recognizable corresponding state which can be established for a given material without resorting to viscoelastic measurements?

It has long been suggested that different substances are in corresponding states at their glass transition temperatures, T_g. One way to test this suggestion is to determine whether universal viscoelastic behavior is obtained when T_g is chosen as the reference state. Since this effectively replaces the only adjustable parameter T_S in Eq. 7 by T_g, the reduction to a single master curve is not as good; nevertheless, the viscoelastic behavior of most materials may be fairly well represented by a single master curve if the WLF equation is written in the universal form:

$$\ln a_T = -\frac{40.0(T - T_g)}{52 + (T - T_g)} \tag{12}$$

The sucess of this approach is a powerful argument for accepting the glass transition as at least an approximate condition of corresponding states. Comparison with Eq. 11 indicates $B/\phi = 40$ and, if B is assigned a value near unity, it is predicted that the fractional free volume of all materials at the glass transition is 0.025.

We have arrived at Eq. 11 through combination of two empirical relationships, and the operational definition of free volume furnished by Eq. 10. F. Bueche (20) has derived a relation having the form of Eq. 11 from considerations based upon jump frequency and packets of free volume. He suggested that α appearing in Eq. 10 should be replaced by the difference $\alpha_l - \alpha_g$, between the coefficients of thermal expansion in the liquid and glassy states. If we accept $\phi_g = 0.025$, then from the factor 52 appearing in the denominator of Eq. 12 we obtain the prediction that $\alpha_l - \alpha_g = 5 \times 10^{-4}$ for all materials near their glass transition temperatures.

In summary, the viscoelastic behavior of a given material at any temperature of interest can be predicted from the appropriate master curve and, a knowledge of the glass transition temperature of the material. We have already observed that T_g is related to the energy barrier, E^*, to rotation illustrated in Fig. 2a. This explains why polyisobutylene has a higher T_g than polyisoprene, and hence is less resilient at room temperature. However, since E^* values are not generally available, it is appropriate to examine other relationships for the prediction of T_g.

B. The Glass Transition Temperature

If materials are in corresponding states at T_g, and also at the melting point T_m, one might anticipate some universal relationship between T_g and T_m. Indeed, Boyer and Beaman (21) have proposed the following:

$$T_g/T_m = \text{constant} \tag{13}$$

It turns out that this relation is only approximately obeyed, the ratio varying from 0.5 for symmetrical polymers, such as polyethylene, to 0.7 for unsymmetrical ones, such as polystyrene.

The concept of the glass transition as an iso-free volume state has furnished the basis for additional relationships. The first of these, due to Boyer and Spencer (22), was based upon the assumption that the occupied volume (in the "hard spheres" sense) is independent of temperature:

$$\alpha_l T_g \cong 0.16 \tag{14}$$

Simha and Boyer (23) tested this relation using data for fourteen polymers. They found the indicated product to vary from 0.13 to 0.23. They also reported that a more reliable prediction was furnished by

$$(\alpha_l - \alpha_g)T_g = 0.11 \tag{15}$$

which may be deduced from the assumption that the occupied volume has the same temperature coefficient as the glass (see Fig. 4). In this case the free volume will be constant at all temperatures below T_g. They appear to have used the expansivity $e = (dv/dT)_p$ in place of α. For the same fourteen polymers mentioned above, the product varied between 0.08 and 0.13, and thus Eq. 15 has considerable value for the purposes of prediction. An example of the application of this relation to five polymers appears in Table IA. The last column shows that the product of the glass temperature and the difference in expansivities in the liquid and glassy states is nearly constant. We also call attention to column

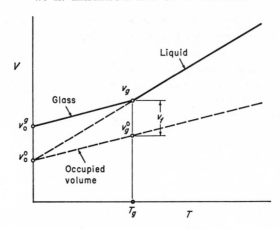

Fig. 4. Thermal expansions of the liquid and glass as assumed by Simha and Boyer
(see Eq. 15).

five of Table IA, which indicates that $(e_l - e_g)$ varies considerably from
one polymer to another. It will be recalled that Bueche's treatment
led to the prediction $(\alpha_l - \alpha_g) = 5 \times 10^{-4}$. While the order of magnitude is correct, this difference in coefficients of thermal expansion actually
varies from 2.8×10^{-4} to 5.2×10^{-4} for the five polymers appearing
in Table I.

Interpretation of the Simha-Boyer relation in terms of free volume
involves a long extrapolation from T_g to $0°K$. However, to the extent
that the expansivities are independent of temperature, one can conclude
from Eq. 15 that $\phi_g = 0.11$, or about four times the value predicted from
viscoelastic data according to Eq. 12, with the assignment $B = 1$. In
view of this difference, it is of interest to estimate ϕ_g in another way
using the tabulated values given by Bondi (24) for the van der Waals
volumes of atoms and groups at $0°K$. The results of these calculations
are given in Table IB. The symbols for the various volumes may be
understood by reference to Fig. 4. The van der Waals specific volumes,
v_{VDW}, were computed directly from Bondi's tables. In order to obtain
the occupied volume at $0°K$, $v_0°$, some assumption must be made concerning the packing density. We have assumed hexagonal close packing
of spheres, so that $v_0° = v_{VDW}/0.74$. As seen from column seven, the
fractional free volume is again nearly constant, but its magnitude is
nearer 0.05. Furthermore, this number will be further reduced if a

TABLE IA

The Fractional Free Volume at T_g: Test of Equation 15

Polymer	$T_g(^\circ K)$	$10^4 e_l{}^a$	$10^4 e_g$	$10^4(e_l - e_g)$	$T_g(e_l - e_g)$
Polyisobutylene (25)	210	6.0	1.5	4.5	0.094
cis-Polyisoprene (26)	201	7.42	2.0	5.4	0.108
Polystyrene (27)	373	5.5	2.5	3.0	0.112
Poly(methyl methacrylate) (28)	378	4.6	2.15	2.45	0.093
Poly(vinyl acetate) (29,30)	300	6.0	2.2	3.8	0.114

TABLE IB

Calculations Based on Bondi's van der Waals Volumes $(cm^3\ g^{-1})$

Polymer	From Bondi's tables (24)						Simha-Boyer (23)			
	v_g	v_{VDW}	$v_0{}^\circ$	$V_g{}^\circ$	v_f	v_f/v_g	$v_0{}^\circ$	$v_g{}^\circ$	v_f	v_f/v_g
Polyisobutylene	1.075	0.730	0.987	1.019	0.056	0.052	0.950	0.981	0.094	0.088
cis-Polyisoprene	1.035	0.700	0.946	0.986	0.049	0.047	0.886	0.926	0.109	0.105
Polystyrene	0.960	0.604	0.817	0.910	0.050	0.052	0.755	0.848	0.112	0.117
Poly(methyl methacrylate)	0.870	0.563	0.761	0.842	0.028	0.032	0.696	0.777	0.093	0.107
Poly(vinyl acetate)	0.846	0.536	0.724	0.790	0.056	0.066	0.666	0.732	0.114	0.135

$^a e = (dv/dT)_p$ $(cm^3\ g^{-1}\ deg^{-1})$.

lower packing density is assumed. For comparison, the last column of Table IB gives the fractional free volume obtained according to the linear extrapolation procedure of Simha and Boyer (23).

For exploratory work it may sometimes be necessary to estimate T_g for a case in which the expansion coefficients are unknown. For this purpose we may utilize a relation derivable from van der Waals' equation:

$$\alpha\Delta E_v = \text{constant} \qquad (16)$$

where ΔE_v is the molar energy of vaporization $(cal/mole)$. One sees from Eqs. 14 and 16 that strong intermolecular attractions, such as hydrogen bonding, will raise T_g. For 63 organic liquids the constant in Eq. 16 was found (31) to be 9.4 ± 0.9. This is, of course, not useful for polymers; however, Eq. 16 may be recast in terms of the internal pressure $P_i \cong \Delta E_v/V$ (or cohesive energy density):

$$\alpha P_i V = \text{constant} \tag{17}$$

This relation, in conjunction with Eq. 14, suggests T_g/P_iV should be approximately constant. For this purpose the internal pressure may be estimated for any repeating unit using tabulated values of the molar attraction constant, $(EV)^{\frac{1}{2}}$, calculated for various atoms and groups by P. A. Small (32). A more recent compilation has been given by Burrell and Immergut (33). If V in Eq. 17 is taken as the van der Waals volume per repeating unit, the calculation can be based entirely upon tabulated values. If a crude estimate is acceptable, convenience can be gained by ignoring the density differences and assuming constancy of T_g/P_iM_0, where M_0 is the molecular weight per repeating unit.

These relations are tested in Table II using data for six polymers. From column seven

$$T_g/P_iM_0 \cong 0.047 \tag{18}$$

with an average error of $\pm 13\%$. Somewhat better results are obtained upon replacement of M_0 by the van der Waals volume per repeating unit, $V_{\text{VDW}} = M_0 v_{\text{VDW}}$:

$$T_g/P_iV_{\text{VDW}} = 0.074 \tag{19}$$

with an average error of $\pm 7\%$, as shown in column eight. The final column provides values for the ratio appearing in Eq. 18 times the expansivity in the glassy state:

$$T_g e_g/P_iM_0 = 9.6 \times 10^{-6} \tag{20}$$

This is probably of little use for the estimation of T_g, but it might find use for the estimation of e_g for cases in which T_g is known. This relation may be of interest in another connection. Attempts to improve oil resistance generally result in increasing the internal pressure P_i, which may lead to a higher T_g and poorer low temperature properties. However, Eq. 20 suggests that this difficulty might be partially alleviated by looking for materials exhibiting larger values of e_g.

Perhaps it is appropriate to conclude with the warning that the relationships appearing in Eqs. 18–20 may be expected to fail badly if there are strong dipolar interactions along the chain.

TABLE II
Relations for Estimation of T_g

Polymer	T_g(°K)	v_{VDW}	$10^4 e_g$	P_i	$P_i M_0$	$10^2(T_g/P_i M_0)$	$10^2(T_g/P_i M_0 v_{VDW})$	$10^6(T_g e_g/P_i M_0)$
cis-Polybutadiene	188	0.693	2.0	70.8	3825	4.9	7.1	9.8
cis-Polyisoprene	201	0.700	2.0	66.4	4515	4.5	6.4	9.0
Polyisobutylene	210	0.730	1.5	59.3	3320	6.3	8.7	9.5
Poly(vinyl acetate)	300	0.536	2.2	88.4	7602	4.0	7.4	8.8
Polystyrene	373	0.604	2.5	82.8	8600	4.3	7.2	10.8
Poly(methyl methacrylate)	378	0.563	2.15	85.6	8560	4.4	7.8	9.5
						4.7 ± 0.6	7.4 ± 0.5	9.6 ± 0.5

III. CRYSTALLINITY

As mentioned at the outset, one of the essential requirements for long range elasticity is that the material be predominantly amorphous. It is evident that if most of the chains are in crystalline domains (e.g., polyethylene), the sample cannot exhibit rubberlike behavior. On the other hand, the ultimate tensile strength is improved if the elastomer crystallizes at high elongations. If the melting point is appropriately placed, both of these conditions can be fulfilled, since the melting point is elevated by strain. Therefore, the rubber chemist must exercise careful control of the factors which influence the melting point. These may be summarized most succinctly by reference to the general thermodynamic relation for the melting temperature, T_m:

$$T_m = \Delta H_f / \Delta S_f \qquad (21)$$

This indicates that the heat of fusion, ΔH_f, must be kept small (weak intermolecular attractions), while the entropy of fusion, ΔS_f, should be large. Since ΔS_f represents the difference in entropies of the liquid and solid states,

$$\Delta S_f = S_l - S_s \qquad (22)$$

we see that the entropy (or number of distinguishable configurations) of the liquid must be large. This will be true if the intramolecular rotational potential allows the molecule to assume many conformations (if there are several minima of nearly equal energy). Secondly, the entropy of the solid, S_s, should be small. Among low molecular weight compounds, those having low symmetry (cf. toluene vs. benzene) can fit into the crystal in few ways, and hence have low melting points. For macromolecular compounds, regularity of the repeating units is important. If a long translation of the chain is required to bring two similar units into register, S_s will be small, and the melting point will be low. Random copolymerization is, of course, one way to lower the entropy of the solid, and hence to depress the melting point. Atactic sequences have a similar effect. However, too large a proportion of noncrystallizable units will eliminate the possibility of crystallization upon extension. The ideal elastomer should have a melting point below its lowest use temperature, but should be capable of crystallizing when the vulcanizate approaches full extension.

IV. CONCLUSIONS

We summarize here the conclusions from the foregoing discussion. The equilibrium stress–strain behavior may be predicted from the elastic equation of state. The gaussian treatment correctly predicts the general shape of the stress–strain curve, and indicates that the initial modulus is only a function of the degree of crosslinking, but not of the particular nature of the network chains. Again, in the gaussian approximation internal energy effects are predicted to alter the temperature dependence of the retractive force, but not the general shape of the stress–strain curve. Deviations from gaussian behavior are represented empirically, at least up to moderate elongations, by incorporating the Mooney-Rivlin C_2 term. Several authors have attempted to obtain a liquidlike elastic equation of state, and this approach may eventually lead to a better theoretical grasp of the contributing factors. Some of the differences observed between real vulcanizates may be due to finite chain extensibility, and this may be investigated using the inverse Langevin distribution. For example, it has been possible to demonstrate (34) in this way the anisotropy of networks formed by crosslinking in the strained state. Also, one should be able to demonstrate that the stress–strain curve is affected by the distribution of network chain lengths. One stumbling block here is the fact that the convenient assumption of an affine transformation on the molecular scale must be abandoned. However, it is difficult to avoid the conclusion that these effects are likely to be of secondary importance, so that really different stress–strain behavior can only be achieved through the use of active fillers, or by two-phase systems, as might be found in block polymers composed of dissimilar units, or having one crystallizable unit.

The principle of corresponding states is of primary significance in systematizing the various types of viscoelastic behavior for amorphous polymers. If the postulate is accepted that all materials at their glass transition temperatures are in corresponding states, then application of the WLF equation permits one to predict the viscoelastic behavior of any material at a temperature of interest from the appropriate master curve, if the glass transition temperature is known or can be estimated. Some of the simpler relationships for predicting T_g are given. The ability to predict viscoelastic behavior in this manner is certainly a major achievement. Nevertheless, it sometimes happens that the small deviations from the generalized behavior assume major importance. As an example, one might cite the barrier properties of elastomeric materials toward gases. The permeability coefficient of oxygen near room temperature is 15 times larger for natural rubber than for polyisobutylene, which would appear to imply a significant difference in the efficiency of molecular packing for these two polymers. Yet they should have the same free volume at the glass transition temperature, and there is only a minor difference in the T_g values of these polymers. As another example, we observe that the expansivities, e_l and e_g, exhibit considerable variation, even among the hydrocarbon polymers, but the relationship of these differences to molecular structure is not well understood.

Current theory finds its major usefulness in explaining elastomeric behavior in broad terms, while the underlying individual differences, which sometimes assume great importance, must still be investigated on an empirical basis. One may hope that theory will be further refined to include all of these details. Along these lines one might mention the developments concerning the thermodynamic equation of state for liquid polymers. Simha and co-workers (35) have investigated the applicability of the principle of corresponding states to polymers in the liquid and glassy states, and have tested relations derived from cell and hole model liquid treatments, while Flory, Orowoll, and Vrij (36) have applied a chain liquid partition function to the thermodynamics of polymer solutions with considerable success.

The foregoing treatments of the stress–strain and viscoelastic behavior assume that the polymer is, at least predominantly, amorphous. Indeed, this is one of the essential criteria for a material to exhibit rubberlike elasticity. Nevertheless, better ultimate tensile properties are found in those elastomers which are capable of crystallizing at high extension, and hence the relationship of melting point to molecular structure is of importance.

References

1. W. R. Krigbaum and R.-J. Roe, *Rubber Chem. Technol.*, **38**, 1039 (1965).
2. P. J. Flory, *Principles of Polymer Chemistry*, Cornell Univ. Press, Ithaca, New York, 1953, Ch. 11.
3. L. R. G. Treloar, *The Physics of Rubber Elasticity*, 2nd ed., Oxford Univ. Press, London, 1958, Ch. 6.
4. R. S. Rivlin, in *Rheology*, F. R. Eirich, Ed., Academic Press, Inc., New York, 1956, Vol. I, Ch. 10.
5. M. Mooney, *J. Appl. Phys.*, **11**, 582 (1940).
6. M. V. Volkenstein and O. B. Ptitsyn, *Dokl. Akad. Nauk SSSR*, **91**, 1313 (1953); *Zh. Tekh. Fiz.*, **25**, 649, 662 (1955).
7. M. V. Volkenstein, *Configurational Statistics of Polymeric Chains*, Wiley, New York, 1963.

8. A. Ciferri, C. A. J. Hoeve, and P. J. Flory, *J. Am. Chem. Soc.*, **83**, 1015 (1961).
9. A. Abe, R. L. Jernigan and P. J. Flory, *J. Am. Chem. Soc.*, **88**, 631 (1966).
10. P. J. Flory, C. A. J. Hoeve, and A. Ciferri, *J. Polymer Sci.*, **34**, 337 (1959).
11. W. R. Krigbaum and M. Kaneko, *J. Chem. Phys.*, **36**, 99 (1962).
12. M. V. Volkenstein, Yu. Ya. Gotlieb, and O. B. Ptitsyn, *Vysokomolekul. Soedin.*, **1**, 1056 (1959).
13. E. A. DiMarzio, *J. Chem. Phys.*, **36**, 1563 (1962).
14. J. L. Jackson, M. C. Shen, and D. A. McQuarrie, *J. Chem. Phys.*, **44**, 2388 (1966).
15. H. Leaderman, *Elastic and Creep Properties of Filamentous Materials*, Textile Foundation, Washington, D.C., 1943.
16. A. V. Tobolsky, *Properties and Structure of Polymers*, Wiley, New York, 1960, Ch. 4.
17. J. D. Ferry, *Viscoelastic Properties of Polymers*, Wiley, New York, 1961, Ch. 11.
18. M. L. Williams, R. F. Landel, and J. D. Ferry, *J. Am. Chem. Soc.*, **77**, 3701 (1955).
19. A. K. Doolittle, *J. Appl. Phys.*, **22**, 1471 (1951).
20. F. Bueche, *Physical Properties of Polymers*, Interscience, 1962, Ch. 4.
21. R. G. Beaman, *J. Polymer Sci.*, **9**, 470 (1952); R. F. Boyer, *J. Appl. Phys.*, **25**, 825 (1954).
22. R. F. Boyer and R. S. Spencer, *J. Appl. Phys.*, **15**, 398 (1944).
23. R. Simha and R. F. Boyer, *J. Chem. Phys.*, **37**, 1003 (1962).
24. A. Bondi, *J. Phys. Chem.*, **68**, 441 (1964).
25. A. A. Miller, *J. Polymer Sci.*, **A1**, 1865 (1963).
26. N. Bekkedahl, *J. Res. Natl. Bur. Std.*, **13**, 411 (1934).
27. A. A. Miller, *J. Polymer Sci.*, **A1**, 1857 (1963).
28. S. S. Rogers and L. Mandelkern, *J. Phys. Chem.*, **61**, 985 (1957).
29. R. H. Wiley and G. M. Brauer, *J. Polymer Sci.*, **4**, 351 (1949).
30. A. Kovacs, *J. Polymer Sci.*, **30**, 131 (1958).
31. F. T. Wall and W. R. Krigbaum, *J. Chem. Phys.*, **12**, 1274 (1949).
32. P. A. Small, *J. Appl. Chem. (London)*, **3**, 71 (1953).
33. H. Burrell and B. Immergut, "Solubility Parameter Values" in *Polymer Handbook*, J. Brandrup and E. H. Immergut, Eds., Interscience, New York, 1966.
34. K. J. Smith, Jr., A. Ciferri, and J. J. Hermans, *J. Polymer Sci.*, **A2**, 1025 (1964).
35. V. S. Nanda and R. Simha, *J. Chem. Phys.*, **41**, 1884, 3870 (1964); *J. Phys. Chem.*, **68**, 3158 (1964); R. Simha and A. J. Havlik, *J. Am. Chem. Soc.*, **86**, 197 (1964).
36. P. J. Flory, R. A. Orowoll, and A. Vrij, *J. Am. Chem. Soc.*, **86**, 3507, 3515 (1964).

CHAPTER 2

THE HISTORICAL BACKGROUND OF SYNTHETIC ELASTOMERS WITH PARTICULAR EMPHASIS ON THE EARLY PERIOD

ERIK G. M. TÖRNQVIST

Esso Research and Engineering Co., Linden, New Jersey

Contents

I. THE NATURAL RUBBER PERIOD

A. The Discovery of Natural Rubber

It is not strange that a material having the properties of natural rubber should arouse the curiosity of man and tease his imagination. We do not know the earliest date at which man became aware of the elastic substance formed by coagulation from the milky juice or latex occurring in the inner bark of certain trees. Samples of fossilized rubber discovered in 1924 in lignite deposits in Germany are believed to stem from the Eocene period, i.e., about 50 million years ago. Man may thus have noticed substances of this type at a very early stage in his development. However, modern man appears to have become aware of natural rubber only with the discovery of the Western Hemisphere.

In his work describing the wanderings and conquests of the Castillians, Herrera y Tordesilla mentions that Columbus, during his second voyage to America, learned of a game played by the natives of Haiti in which balls of an elastic "tree-resin" were used. Columbus supposedly brought some samples of these balls with him upon his return to Europe in 1496 (1). The same author also mentions a festival arranged by the powerful King Montezuma in 1519 in honor of Cortes during which a ball game called *batos* was played in a building specifically designed for this purpose. The balls used in this game were made from a tree resin and, although larger than the Castillian leather balls, were lighter and exhibited much better rebound.

In his historical work on the Indian monarchy published in 1615, de Torquemada mentions that the natives of Mexico prepared a substance *ulli* from the juice of a tree called "Ulcuahuitl" (*Castilla elastica*) growing in that region and that they ascribed great healing powers to it (2). They also used it for making shoes, headgear, clothing, and other articles. Rubber was mentioned even earlier by d'Anghiera (3), Oviedo y Valdéz (4), and others, but it was left to de la Condamine to write the first report of scientific significance on this subject.

By commission of the Academy of Science in Paris, this explorer and mathematician, to whom we are also indebted for the discovery of the cinchona tree and the arrow poison curare, left France for Ecuador in 1735 together with two other scientists. Their primary purpose was to make measurements for testing the correctness of certain hypotheses regarding the shape of the earth which had been based upon the variation in the motion of pendulums at different latitudes. During the journey, which lasted almost eight years, de la Condamine first traveled via difficult over-land trails from Guayaquil to Quito in the surroundings of which he made the desired measurements over a period of seven years. By that time he had become so interested in the animal and plant life of the region that he decided to study it further by traveling eastward across the continent on his return trip to Europe. He left Quito in May of 1743 and traveled via Loja and Jaén to the river Marañon, which he reached somewhere below the fifth degree of latitude south. Traveling first along this river, which is in part very difficult to navigate, and then along the Amazon River, he reached Pará (now Belém) after several months. During the journey he collected numerous interesting specimens of various origin. Some of these were sent to the Academy from Pará, and they included some pieces of a black substance which were accompanied by the following description:

In the forests of the Province Esmeraldas (Ecuador) there grows a tree which the natives call Hhevé. When the bark is cut, a white, milky fluid runs out which slowly hardens and turns black in the air. The natives make candles from it which burn very well without a wick and shine with a clear flame. In the province of Quito linen is coated with this gum and can then be used like wax cloth among us. The tree also grows along the banks of the Amazon River, and the Mainas call the resin they obtain from it "cahutchu." They make water-tight shoes from one piece of this material which looks just like real leather when smoked. They also spread this sap over a flask-shaped clay form, and when the liquid has solidified they break the form and remove the fragments through the neck of the flask and thus obtain a light unbreak-able bottle suitable for holding all sorts of liquids. Still more remarkable use of the "gum milk" is made by the Omaguas, a tribe which is living in the interior of the American Continent along the banks of the Amazon River. They use it for making pear-shaped flasks which end in a wooden pipe. When the bottle is squeezed, a jet of water spurts forth from the pipe. These flasks are thus veritable squirts.

In 1743, shortly before returning to Europe, de la Condamine met François Fresneau, an engineer living in Cayenne who had also got inter-ested in the rubber-producing trees and had discovered several species, including members of the genus *Hevea* (from Hhevé). It is actually believed that Fresneau rather than Condamine was the first European to discover *Hevea brasiliensis,* by far the most important of the rubber pro-ducing trees. De la Condamine formally reported both his own and Fresneau's findings to the Academy in Paris in 1751 (5). The new mate-rial almost immediately created interest in most countries of Western Europe. It was named "Caoutchouc" from the Tupi word *cahutchu,* which literally means "weeping tree."*

* The spelling of the word varies, of course, from language to language, e.g., French: caoutchouc; German: Kautschuk; Italian: caucciu; Spanish: caucho.

B. The Beginning of Rubber Goods Manufacture in Europe and the USA

One problem immediately facing Europeans interested in manufacturing caoutchouc or rubber* articles was that they had only solid rubber and no fresh latex available for the manufacturing processes. This created the first real research problem in the rubber field.

The French chemists L. A. P. Hérissant and P. J. Macquer found around 1763 that rubber could be dissolved in oil of turpentine and ether, and they thus prepared the first rubber *cements* (8). The latter solvent was particularly useful, since the rubber obtained upon evaporation of the ether was firmer and less sticky than that recovered from other solvents. Macquer used ether solutions of rubber for making, among other things, rubber tubing.

In spite of these advances, rubber consumption did not increase very rapidly in Europe and the United States, primarily because of the disappointing properties which rubber articles exhibited under certain conditions. The tendency of rubber to become very sticky or even melt at slightly elevated temperatures, and to become brittle at lower temperatures, was particularly disadvantageous in most applications. Its poor aging properties further aggravated the situation, and though it had earlier been considered almost a miracle material, it now became a material of rather bad reputation.

Great but essentially unsuccessful efforts were made during the following 60–70 years to overcome the undesirable properties. As a consequence, the world consumption of rubber increased only slowly, reaching about 100 long tons in 1824. An interesting improvement in the manufacture of waterproof fabrics was made by Charles Macintosh in 1823. He found that naphtha (primarily benzene), obtained as a by-product in the manufacture of coal gas, was particularly suitable for dissolving rubber. The new solvent evaporated quickly without yielding a highly sticky rubber, and it was considerably cheaper than ether. Macintosh coated fabrics on one side with the help of such a solution, and after allowing the naphtha to evaporate, pressed two such fabrics together with the rubber-coated sides against each other. Garments made from

---

* Joseph Priestley (6), the famous British chemist, reported in 1770 that the material was particularly useful for rubbing out pencil marks on paper. It is believed, however, that an English instrument maker, Edward Nairne, first discovered this property, which gave the name "West India *rubber*" or "India *rubber*" to the material (7).

the two-ply fabric were truly waterproof and achieved much popularity; they became known first as Macintoshes and later on, through a corruption of the spelling, as Mackintoshes, a name familiar even today. However, the less desirable properties of rubber had by no means been eliminated through this manufacturing procedure, even though they became less noticeable, provided the rubber remained between the two layers of fabric. The search for a method of eliminating the undesirable properties therefore continued in many countries.

In 1832, a German chemist, F. Lüdersdorff, published an article which disclosed that the rubber recovered from a solution in oil of turpentine became less tacky if the solvent had been boiled with 3% sulfur until the evolution of hydrogen sulfide had ceased (9). Lüdersdorff does not seem to have fully appreciated the true significance of his observation, and it remained for Charles Nelson Goodyear to make the extremely important discovery which is now known as vulcanization.

C. The Discovery of Vulcanization by Goodyear and the Following Rapid Rise in Rubber Consumption

Goodyear, who was born in 1800 at New Haven, Connecticut, became interested in rubber in the early 1830's. During the following years, he spent much time and effort to find a method of improving the high and low temperature properties of rubber and eliminating its stickiness. In 1838, at which time his efforts had not yet been particularly successful, Goodyear met another fellow American, Nathaniel Hayward, who had discovered that rubber could be made less sticky by mixing it with sulfur. Goodyear bought the right to use this process and started impregnating various fabrics, mail sacks among others, with rubber–sulfur mixtures. The results obtained were not particularly encouraging until he discovered quite accidentally the following year that rubber and sulfur interacted at considerably elevated temperatures, i.e., above the melting point of sulfur, to yield an entirely different product. The discovery was made after Goodyear had coated pieces of fabric with a mixture of rubber, sulfur, white lead, and oil of turpentine and hung them to dry close to a hot stove. Some of the pieces which had come into direct contact with the stove had greatly changed their character and looked as if the rubber had charred. Goodyear immediately recognized the importance of the observation and further experiments demonstrated that an elastic, nontacky product of good high and low temperature properties could be obtained if the heating was carried out under controlled conditions. However, he did not apply for a patent on his discovery until 1841 (10).

In the same year, Goodyear gave some samples of vulcanized rubber to Stephen Moulton, a young Englishman visiting the United States. Moulton took these samples back to England the following year and showed them to William Brockedon, a well-known inventor. Brockedon in turn showed the samples to Thomas Hancock, who had been actively engaged in the field of rubber processing for over twenty years, originally as a coworker of Macintosh's, and who had made very extensive and valuable contributions to this field, including the invention of the masticating process. Hancock immediately recognized the value of the samples he had been shown and began investigating how they had been prepared. After some time he, too, discovered the vulcanization process and obtained a British patent in 1843, one year before Goodyear received a U.S. patent for the same invention. This has sometimes led to a controversy as to who is the inventor of vulcanization. However, there is no doubt that Goodyear is the true inventor, and Hancock never denied this (11). The importance of Hancock's contributions to rubber technology can hardly be overestimated, however, and it was probably fortunate for the development of the rubber industry that Goodyear's discovery, which was given the name vulcanization by Brockedon, came to Hancock's attention at such an early date.

D. The Development of a Critical Rubber Supply Situation and Resulting Drastic Actions

The profound improvement in the properties of various rubber articles brought about by the vulcanization process immediately increased the demand for such articles and within a few years the demand began to outstrip the availability of crude natural rubber. As the imbalance between supply and demand grew successively worse, the price of natural rubber began to rise rapidly. This had some important consequences.

To begin with, the possibility of making huge profits caused the rubber-producing countries to try to increase rubber production by any possible means. This led to a ruthless utilization of native laborers, first in the Amazon Region and then in the Congo (12–14).

Secondly, England decided to try to make itself less dependent on the wild rubber from South America by establishing plantations in parts of the British Empire which have a climate similar to that of the Amazon Basin. The British desire to establish rubber plantations was prompted not only by the rising rubber price, which was partly caused by a considerable Brazilian export duty, but also by the fact that the wild rubber trees were often tapped in such a careless manner that they died. Thus, there was a real threat that the supply of rubber would start decreasing at a time when the demand for it was rapidly increasing.

After a number of unsuccessful attempts had been made during 1873–76 to bring young rubber plants to Calcutta, Sir Joseph Hooker, Director of the Royal Botanical Gardens at Kew near London asked Henry A. Wickham, an Englishman living at Santarém at the junction of the Amazon and Tapajós Rivers, to collect a large number of Hevea seeds and bring them to England. With the help of some Tapuyo Indians, Wickham immediately set about collecting seeds from the best stands of Hevea trees on the plateaus between the Tapajós and Madeira Rivers, and after a couple of weeks about 70,000 seeds were loaded on board the British steamer "Amazonas," which had traveled up the Tapajós River to Itaituba, about 200 miles from Santarém. The ship was successful in passing the Brazilian customs inspection at Pará (Belém) without any objection and she arrived in Liverpool on June 14, 1876. The seeds were rapidly transferred to Kew Gardens, where about 4% of them eventually germinated. About 2000 of the seedlings were then sent to Ceylon, where they were planted primarily in the Botanical Garden at Peradeniya in October of the same year. Small trees were transferred a few years later from here to various places in Malaya (Malaysia) and the Dutch East Indies (Indonesia), and the first rubber plantations were thus started (12–14). However, it was to take many more years before plantation rubber in significant quantities appeared on the world market. By that time, the rubber price had risen to such a level that it had prompted one of the major German chemical companies to start research toward finding a method of preparing synthetic rubber on an industrial scale.

The decision to start such research was undoubtedly a risky one, because only very limited knowledge about the chemical nature of natural rubber existed in the world at that time. However, it was entirely justified by the rubber supply situation, as is illustrated by the following statistics.

E. Consumption Statistics

The rapid increase in world production of natural rubber between 1830 and 1940 is shown in Table I. Since the consumption up until 1910 was essentially governed by the availability of crude rubber, the production and consumption figures are almost identical up to that time. Although the figures for the earliest period are not based upon as carefully collected statistical data as are those from 1890 on, they are believed to be reasonably correct for comparative purposes.

TABLE I

World Production of Natural Rubber 1830–1940[a]

Year	Production, long tons
1830	25
1840	150
1850	750
1860	6,000
1875	9,000
1890	30,750
1900	44,000
1910	94,000
1920	341,000
1930	825,000
1940	1,395,000

[a] Sources: S. Boström, K. Lange, H. Schmidt, and P. Stöcklin, *Kautschuk und verwandte Stoffe*, Union Deutsche Verlagsgesellschaft, Berlin, 1940, p. 21; K. E. Knorr, *World Rubber and its Regulation*, Stanford University Press, Stanford, California, 1945, pp. 247–248; H. Barron, *Modern Rubber Chemistry*, D. van Nostrand, New York, 1946, p. 4.

The particularly rapid increase in rubber production (and consumption) during 1840–1860 reflects the importance of the discovery of vulcanization, which made rubber an excellent material for manufacturing waterproof fabrics and garments, shoes, combs, etc. After a more moderate rise in consumption during the following twenty years, another rapid rise started around 1880 as a consequence of newly developed applications for which rubber was essential. These included insulation of electric cable and wire and manufacture of tires for bicycles, carriages, and finally automobiles.

The price of crude rubber began to rise rapidly as the gap between supply and demand widened because of the new applications. This is illustrated in Fig. 1, where the price of rubber is given in German gold marks per kilogram. The important feature of the curve is the fluctuation in price that occurred after 1880.

The first major price rise occurred around 1882, and this was too early for efforts to prepare synthetic rubber. The second major price rise, which started in 1902 and reached a peak in 1910, was directly responsible for the decision by the Farbenfabriken vorm. Friedrich Bayer & Co. in Leverkusen to start work on synthetic rubber. The price rise also caused

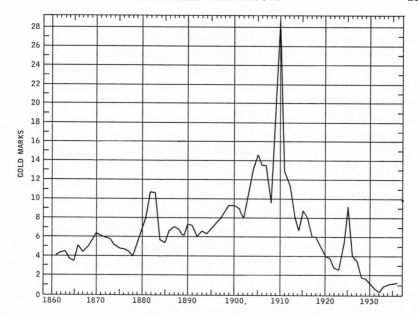

Fig. 1. Price of crude natural rubber in Germany 1861–1936 in gold marks per kilogram. 1 gold mark/kg = 10.8¢/lb before 1934 and 18.3¢/lb thereafter. (After S. Boström et al., *Kautschuk und verwandte Stoffe* (see Table I).)

the search for natural rubber to intensify, resulting in the natives of the rubber-producing regions being treated even more ruthlessly than before. It is difficult to say what would have happened if plantation rubber in large quantities had not begun to appear on the world market at that time.

As recently as 1900, the annual production of plantation rubber had been only about 5 tons, and though it had grown rapidly on a relative basis, it did not reach 1000 tons until 1907, at which time it was still less than 2% of the total world rubber production. However, the growth rate continued strong and production passed 10,000 tons in 1910, reaching 17,100 tons in 1911 when it was high enough to reverse the price trend. The importance of plantation rubber to the total natural rubber production after 1910 is also indicated by the figures in Table I.

The price drop caused by the appearance of plantation rubber in large quantities on the world market actually threatened to put an early end to the synthetic rubber effort. However, World War I broke out before this could happen and the first German synthetic rubber effort was continued until the end of the war, when it was terminated.

Figure 1 shows a third rapid rise in the price of natural rubber beginning in 1923 and reaching a peak in 1925. This increase in price, which caused synthetic rubber research to be restarted in Germany, was the result of the so-called Stevenson plan for restricting the production of plantation rubber, which had been agreed upon by the major rubber producing nations. The sharp price drop occurring during the following three years indicates the failure of the Stevenson plan, and the subsequent drastic decrease to an extremely low price level in 1932 indicates the severe effect of the Great Depression. This price drop caused a drastic curtailment of the synthetic rubber effort in Germany and almost stopped it as far as general purpose rubber is concerned. However, the political changes taking place in Germany at that time brought to power a regime which considered self-sufficiency more important than economic production, and the rubber research effort was soon expanded to a level even higher than that existing before the curtailment. At about the same time and for similar reasons a serious effort was started in the Soviet Union to establish a synthetic rubber industry, but rather little is known about that matter in the outside world even today.

The wide fluctuation in the price of natural rubber, not only from year to year but within almost every year from 1906 to 1965, is shown in Table II which lists New York spot prices for No. 1 Ribbed Smoked Sheet. The two-peak quotations in 1910 and 1925 are also easily recognizable in this table, as is the bottom price of 2.5¢/lb in 1932. It should be noted that the top price in 1910 at New York, $2.88/lb, is close to the price in Germany, $3.08/lb, which was calculated from Fig. 1. It is also apparent from the table that the natural rubber price has become quite stable after 1960 and has dropped to its lowest level since World War II. This is primarily the result of the remarkable developments in the synthetic rubber field in recent years, which have brought to the market elastomers equivalent or even superior to the natural product as general purpose rubbers. The onus is now on natural rubber to stay competitive in price, and also in quality and properties. Thus the historic situation has been reversed, and it is justifiable to say that synthetic elastomers have truly come of age. But behind this development lies a tremendous amount of work which will now be reviewed.

F. Early Work on the Composition and Structure of Natural Rubber

Long before man started thinking of preparing synthetic rubber, his curiosity had caused him to investigate the composition of natural rubber.

TABLE II

New York Annual High, Low, and Average Spot Rubber Prices #1
Ribbed Smoked Sheet[a]

	Price, cents per pound				Price, cents per pound		
Year	High	Low	Average	Year	High	Low	Average
1906	150	86	—	1928	41.3	17	22.5
7	138	93	—	29	26.9	15.5	20.6
8	130	75	—	1930	16.5	7.5	12.0
9	208	128	—	31	8.6	4.3	6.2
1910	288	141	206.6	32	4.8	2.5	3.5
11	184	114	141.3	33	9.9	2.9	6.0
12	140	108	121.6	34	15.9	8.8	12.9
13	113	59	82.0	35	13.8	10.5	12.4
14	93	56	65.3	36	23	13.5	16.4
15	79	58.5	65.9	37	26.9	14	19.4
16	102	55	72.5	38	17.2	10.3	14.6
17	90	52	72.2	39	24	14.9	17.6
18	70	40	60.2	40	24	18.3	20.1
19	57	38.5	48.7	41[b]	24.9	19.1	22.4
1920	56.5	16	36.3	1950	86	18	41.1
21	21.3	11.5	16.4	1955	52	29.4	39.2
22	28.4	13.6	17.5	1960	47.5	28.4	38.1
23	37.1	24.8	29.5	61	32.8	27	29.6
24	40.1	17.6	26.2	62	31	27.1	28.6
25	121	34.4	72.5	63	29.3	23	26.3
26	88.5	36.8	48.5	64	28	23.1	25.3
27	41.8	33	37.7	65	28.2	23.8	

[a] Sources: *Rubber Industry Facts*, The Rubber Manufacturers Association, Inc., New York, 1964, p. 5; *Rubber Age*, various issues.

[b] The Government fixed prices from August 1941 through December 1946 at 22.5¢/lb and from January through April 1947 at 25.8¢/lb.

As a matter of fact, such investigations were started only a few years after de la Condamine's report had been published in 1751 (5). The analytical tools available at that time were, of course, very crude, especially for a material as complex as natural rubber, and it was to take almost 200 years before the structural composition of natural rubber became known with reasonable accuracy.

Although such knowledge may seem a reasonable prerequisite for any attempt to prepare a synthetic elastomer, very little was known about the composition of natural rubber when the first rubber synthesis program was started in Germany. However, as will be discussed below, few if any of the rubber chemists living at that time seem to have realized how little they really knew.

A favorite method of studying the composition of natural products during the early days of chemistry was destructive distillation. This method was also applied to rubber, without its *destructive* aspects being clearly recognized for more than a century. The first recorded distillation of natural rubber seems to have been carried out by Berniard, who obtained an oily product in 1781 (15). Subsequently, several other workers applied destructive distillation to rubber* before the first fairly thorough study along these lines was carried out by Himly (17). In a dissertation published in 1835 he states that 9 parts of an oily, ethereal, brown distillate of a disagreeable, pungent odor and a density of 0.870 can be obtained from 12 parts of dry rubber. By fractionating the distillate, Himly obtained two major products, a light fraction with a density of 0.654 and a boiling point varying from 33 to 44°C, which he called *faradyine*, and a higher boiling (168–171°C) fraction which he called *caoutchine*. The latter was found to have an elemental composition corresponding closely to C_5H_8. The lower boiling fraction obtained according to the procedure of Himly was later shown by Ipatieff and Wittorf (18) to consist primarily of isoprene contaminated with 2-methyl-2-butene. It was named in honor of Faraday, who had also been investigating the composition of rubber and who had established its weight composition as early as 1826 (19). The results obtained by Faraday corresponded fairly closely to C_5H_8 as expressed in modern terms.† At about the same time as Himly, A. Bouchardat (20) carried out a similar investigation and obtained a fraction boiling at 250°C, which he called *hévéène*, and another fraction boiling at 18.2°C which he called *caoutchène*.

In 1860 Williams (21) distilled rubber at as low a temperature as possible to prevent secondary decompositions. He obtained a crude distillate of a very unpleasant odor which contained traces of volatile organic bases derived from the small quantity of protein present in the crude rubber. After these had been removed, the distillate was purified by first being shaken with dilute sulfuric acid, then washed with water and finally digested with caustic soda. Fractionation of the product yielded two major cuts, one liquid passing over between 37 and 44°C and another between 170 and 180°C. The boiling points of these fractions were then

---

* A detailed account of the early work on the constitution of rubber may be found in the book by Dubosc and Luttringer (16).

† The elemental analysis by Faraday gave C = 87.20% and H = 12.80%, corresponding to $C_{5.0}H_{8.8}$ or $C_{4.54}H_{8.0}$.

lowered by further rectification over sodium. From the lower boiling fraction Williams thus obtained a liquid distilling between 37 and 38°C which he called *isoprene*. Although undoubtedly purer than Himly's *faradyine*, the *isoprene* probably also contained significant quantities of 2-methyl-2-butene as indicated by its boiling point and elemental composition, the latter corresponding approximately to $C_5H_{8.3}$.*

Williams observed that, when exposed to air for several months, his isoprene became viscous and absorbed about 10% oxygen. On distilling this oxidized or peroxidized (Williams called it "ozonized") isoprene, he found that unaltered isoprene came off first, whereupon the fluid thick-. ened and the boiling point rose rapidly until a sudden reaction occurred which Williams described as follows:

The ozone at this point instantly begins to act with energy, a cloudy vapor rises, accompanied by an intensely sharp odor, and the contents of the retort instantly solidify to a pure white spongy, elastic mass, having, when successfully prepared, but slight tendency to adhere to the fingers. When pure it is opaque, but if allowed to become exposed to the air, especially when warm, it becomes transparent first on the edges, and subsequently throughout the whole mass. When burnt it exhales a peculiar odor hitherto considered to be characteristic of caoutchouc itself. It is not easy to prepare this substance of definite composition.

The composition of the elastic mass was then shown to correspond to $C_{10}H_{16}O$.

On redistilling the higher boiling fraction Williams obtained a liquid boiling between 170 and 173°C which was identical with Himly's *caoutchine* and had a vapor density twice that of isoprene. From the compositional similarity between isoprene, *caoutchine*, and caoutchouc, Williams concluded that the action of heat on the latter compound resulted merely in "the disruption of a polymeric body into substances having a simple relation to the parent hydrocarbon." Williams also established that both *caoutchine* and "its isomer turpentine" would take up four equivalents of bromine and he demonstrated that both could be converted into cymene by the alternate action of bromine and sodium.

In 1875, at the suggestion of Berthelot, F. G. Bouchardat took up the study of the decomposition products of rubber and thus continued in the footsteps of his father, A. Bouchardat. From 5 kg rubber he obtained the

---

* Williams believed carbon to be divalent and to have an atomic weight of 6. He therefore gave the formula of isoprene as $C_{10}H_8$.

following major products (22):

1. Isoprene: 250 g, b.p. about 18–100°C
2. Caoutchine: 2000 g, b.p. about 100–200°C
3. Hévéène: 600 g, b.p. above 200°C

Since he found that the composition of *hévéène* essentially corresponded to $C_{15}H_{24}$ and since Williams had previously shown that the *caoutchine* consisted primarily of $C_{10}H_{16}$, Bouchardat, like Williams, concluded that a close relationship existed between the hydrocarbons obtained by distillation of rubber and that they consisted of either isoprene or its isomerides and polymerides. He then showed that, when heated in a sealed tube, isoprene is converted primarily into dimers but also into products having a considerably higher boiling point.

Four years later, Bouchardat demonstrated the compositional similarity between turpentine and isoprene, thereby pointing out a definite connection between the terpenes and isoprene (23). Shortly thereafter, when trying to prepare a hydrochloride of isoprene, he discovered (24) that a rubberlike material was also formed when this hydrocarbon was contacted with five times its own weight of concentrated hydrochloric acid in a sealed tube. He noted that a violent reaction with considerable liberation of heat took place when the ice-cooled tube was shaken shortly after having been sealed. However, he allowed the tube to stand for several weeks with occasional shaking before opening it and diluting its contents with water. Fractional distillation then yielded the mono- and dihydrochlorides of isoprene and a solid residue which was freed of the hydrochlorides by washing with boiling water. The solid product was then analyzed: C 87.1, H 11.7, and Cl 1.7%, or $C_5H_{8.00}Cl_{0.03}$.

Bouchardat found that it had the elasticity and other physical properties of rubber itself, that it would swell in ether and carbon disulfide, and that it could be dissolved in the distillation products of Pará rubber. Finally he showed that, on heating, it yielded the same volatile products as rubber, namely, isoprene, *caoutchine*, dipentene, *eupione*, and *hévéène*. Bouchardat therefore concluded that he had obtained a product which was analogous to natural rubber and that isoprene is the "mother substance" of the latter.

In 1882 Tilden became interested in isoprene as a starting material for making rubber. He wrote (25):

Isoprene presents two characters which distinguish it from the terpenes. One is the peculiar explosive property of the white syrupy substance which results from its oxidation by air. The other peculiarity—its conversion into true india-rubber or caoutchouc when brought into contact with certain chemical reagents, for example, strong aqueous hydrochloric acid as noticed by Bouchardat, or nitrosyl-chloride as observed by myself. It is this character of isoprene which gives it a somewhat practical interest, for, if it were possible to obtain this hydrocarbon from some other and more accessible source, the synthetical production of india-rubber could be accomplished.

Starting from Bouchardat's disclosures (23,24), Tilden then began looking at turpentine as a raw material for making isoprene. In a paper published in 1884 (26) he showed that isoprene could be obtained from turpentine in about 5% yield through pyrolysis. He further stated that the isoprene thus obtained could be polymerized to caoutchouc according to the methods disclosed by Bourchardat.

Tilden also demonstrated that isoprene was not an acetylenic compound and that as a consequence it had to contain two double bonds. After originally having suggested the correct structure (25) he proposed five different formulas for the compound, including the correct one, though at that time he seems to have favored an allenic structure (26). The correct structure of isoprene was fairly well established by Kondakov (27) and definitely proven by Ipatieff and Wittorf (18) and by Euler (28).

In 1892 Tilden (29) reported that he had found isoprene to undergo spontaneous polymerization during extended storage. When examining bottles containing isoprene prepared from turpentine several years earlier, Tilden found that, "in place of a limpid, colorless liquid, the bottle contained a dense syrup, in which was floating several large masses of solid of a yellowish color." This, Tilden, claimed, turned out to be india-rubber. He demonstrated that it combined with sulfur in the manner of ordinary rubber to give a tough elastic compound.

The first recorded observation of spontaneous polymerization of isoprene seems to have been made by Wallach (30), however, who reported in 1887 that isoprene will polymerize in sealed tubes when exposed to light for an extended period of time. He isolated the polymer by precipitating it with alcohol and recovered a "tough rubber-like substance."

As the nineteenth century was drawing toward its end it became almost generally recognized that natural rubber was built up from isoprene units, but the manner in which this occurred remained a mystery. Since terpenes, $C_{10}H_{16}$, but not isoprene, existed in nature, and since the distillation of rubber yielded much larger quantities of dipentene (*caoutchine*) than of isoprene, it was generally thought that the fundamental building block was a dimer of isoprene rather than isoprene itself. The demonstration by Wallach in 1885 (31) that the dimer (*terpilene*) prepared by

Bouchardat (23,24) and Tilden (26) through heat treatment of isoprene was identical with both the *caoutchine* obtained directly by distillation of rubber and the *cinene* obtained from wormseed oil gave further support for the view, especially since much more drastic conditions were required for obtaining isoprene from *terpilene* than for obtaining good yields of *caoutchine* from rubber. As a consequence, rubber was usually written as $(C_{10}H_{16})_n$ without anything being known about the structure of the $C_{10}H_{16}$ units and the manner in which these units were combined.

Wallach's demonstration (31) that the $C_{10}H_{16}$ hydrocarbon from rubber, which shortly thereafter became generally known as dipentene, gave a crystallizable tetrabromide indicated the presence of two double bonds in the molecule and thus a cyclic structure. The view that the rubber molecule was made up of a cyclic $C_{10}H_{16}$ units was not generally accepted, however, and in 1888 Gladstone and Hibbert (32) published an article in which they claimed to have found evidence indicating the $C_{10}H_{16}$ unit to be of the open acyclic type. Optical measurements (refraction and dispersion) on rubber solutions caused them to conclude that three double bonds per $C_{10}H_{16}$ unit were present in the molecule. Chlorination was then shown to yield a substance of the composition $(C_{10}H_{14}Cl_8)_n$ which was taken as further proof for the presence of three double bonds in the $C_{10}H_{16}$ unit.

Bromination of chloroform solutions of rubber gave less conclusive results. Slightly less than four atoms of bromine per $C_{10}H_{16}$ unit were taken up with normal reaction times, i.e., until the bromine uptake appeared to have ceased. Gladstone and Hibbert did not consider this determination significant, since they believed that the action of the bromine had either caused an isomeric change or that the "tetrabromide" $C_{10}H_{16}Br_4$ was still an unsaturated compound. They therefore subjected chloroform solutions of rubber to a prolonged (3 days) action of bromine and recovered a substance which had a composition corresponding fairly closely to $C_{10}H_{15}Br_5$. They concluded that this substance had been formed by HBr elimination from $C_{10}H_{16}Br_6$ and that the bromination experiments thus also supported the notion that rubber contains three double bonds per $C_{10}H_{16}$ unit.

The results of Gladstone and Hibbert were essentially confirmed in 1900 by Weber (33), who studied rubber primarily for the purpose of determining its molecular weight. He also concluded that the $C_{10}H_{16}$ unit in rubber contained three double bonds and stated:

Accordingly, india-rubber (polyprene) must be olefinic and cannot be a cyclic compound, like the ordinary terpenes; in other words, the constitution of polyprene would

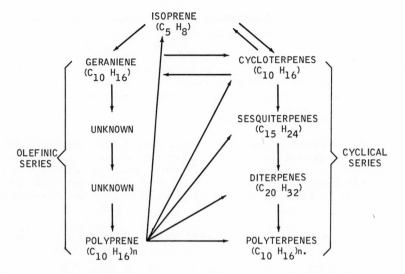

Fig. 2. Relationship between isoprene, terpenes, and natural rubber (polyprene).
(According to Weber (33).)

be that of an olefinic polyterpene, standing at the end of the olefinic terpene series just as the ordinary polyterpenes stand at the end of the cycloterpenes, the hemi-terpene C_5H_8 —isoprene— representing the connecting link between the two series in this manner.* (Fig. 2)

The relationship between isoprene and natural rubber (polyprene) shown in this diagram, together with Weber's comments about the $C_{10}H_{16}$ unit being of the acyclic olefinic type, should not be taken as an indication that he believed rubber to be a molecule with a large linear carbon chain as we know it today. On the contrary, Weber, like everyone in his day, believed that the primary units of a colloid like rubber, e.g., $C_{10}H_{16}$, were held together by certain ill-defined but rather weak forces. Thus it was

---

* In a footnote to his article in Berichte, Weber points out that the relationship between isoprene and the various terpenes as shown in the diagram should only be taken in a chemical sense. He further states, "That isoprene is first produced in the plant metabolism and that various terpenes are then formed from this compound through polymerization is highly unlikely." The correctness of Weber's assumption has, of course, now been proven. The biochemical reactions leading to the formation of natural rubber are extremely interesting, but they are outside the scope of this book. The reader interested in this subject is referred to the excellent article by Lynen and Henning (34).

believed that the same colloid could exhibit different molecular weights in different solvents. Weber states:

That the colloids as a class are characterized by enormous molecular weights is generally admitted. It should, however, not be forgotten that this admission involves the assumption that the colloids in the colloidal state, possess finite and definite molecules. The evidence for this assumption is none too convincing.

Then, in a footnote, Weber adds,

Many colloids are known to exist in both the colloidal and the crystalloidal state. In the latter they possess, of course, finite and definite molecules. The change of a substance from the crystalloidal to the colloidal state appears to be determined by an extensive aggregation of crystalloidal molecules, but in what this aggregation differs from that leading to the formation of crystalloid solids is only very imperfectly understood at present.

Aside from some disagreement that existed regarding the cyclic or acyclic character of the $C_{10}H_{16}$ unit, Weber's view is quite representative of that prevailing among rubber chemists at the turn of the century. However, the character of the $C_{10}H_{16}$ unit was soon to become the matter of a rather intense controversy, when Harries began his monumental investigation of the structure of rubber (35).

In the beginning, Harries employed methods which had been used by other investigators like Weber, namely, reaction with nitrogen oxides or nitric acid (36) and oxidation with permanganate (37). The results obtained in these experiments seemed to support Weber's conclusion that the $C_{10}H_{16}$ unit was of the open-chain type. For instance, by passing moist N_2O_3 into a benzene solution of rubber, Harries was able to prepare a compound of the composition $(C_{10}H_{15}N_3O_7)_2$ which he called *Nitrosite C* and believed to be a derivative of dimyrcene. Thus it was believed that the fundamental $C_{10}H_{16}$ unit consisted of myrcene

$$CH_3—C{=}CH—CH_2—CH—C—CH{=}CH_2$$
$$\quad\ \ |\qquad\qquad\qquad\quad \|$$
$$\quad\ \ CH_3\qquad\qquad\qquad\ CH_3$$

or a closely related acyclic terpene, but the manner in which the $C_{10}H_{16}$ units were held together remained a mystery. A striking change in Harries' opinion regarding the structure of these units occurred, however, when he started utilizing his previously developed ozonization technique (38) for elucidating the structure of rubber.

Harries (37, 39) found that ozonization of rubber produced a diozonide $C_{10}H_{16}O_6$ which upon hydrolysis yielded levulinic aldehyde and levulinic acid together with minor amounts of carbon dioxide, formic acid and succinic acid. He visualized the first step as both a depolymerization and an ozonization. From the character of the main products of the second step, he concluded that both the original $C_{10}H_{16}$ unit and its ozonide had to be cyclic. The overall reaction could then be written as in reaction 1.

$$(C_{10}H_{16})_x + 2xO_3 \rightarrow \left[\begin{array}{c} CH_3 \\ \diagdown \\ C\cdot CH_2\cdot CH_2\cdot CH \\ O \diagup \quad \diagdown \quad \diagdown O \\ \diagup \quad | \quad | \quad \diagdown \\ O \quad \quad \quad O \\ \diagdown \quad | \quad | \quad \diagup \\ O \diagdown \quad \diagup O \\ C\cdot CH_2\cdot CH_2\cdot C \\ \diagdown \\ CH_3 \end{array} \right]_x \rightarrow$$

Rubber

$$\left[\begin{array}{c} O{=}C(CH_3)\cdot CH_2\cdot CH_2\cdot CH{=}O \\ || \qquad \qquad \qquad || \\ O{=\!=\!=\!=\!=\!=\!=\!=\!=\!=\!=}O \end{array} \right]_x + xCH_2\cdot CO\cdot CH_2\cdot CH_2CHO \quad (1)$$

Levulinic aldehyde peroxide Levulinic aldehyde

$$\begin{array}{c} O{=}C(CH_3)\cdot CH_2\cdot CH_2\cdot CH{=}O \\ || \qquad \qquad \qquad || \\ O{=\!=\!=\!=\!=\!=\!=\!=\!=\!=\!=}O \end{array} + H_2O \rightarrow CH_3CO\cdot CH_2CH_2COOH + H_2O_2$$

Levulinic acid

As a consequence he believed the basic $C_{10}H_{16}$ unit to be 1,5-dimethyl cyclooctadiene,

$$\begin{array}{c} CH_3\cdot C\cdot CH_2\cdot CH_2\cdot CH^* \\ || \qquad \qquad \quad || \\ HC\cdot CH_2\cdot CH_2\cdot C\cdot CH_3 \end{array}$$

The appearance of dipentene as one of the major distillation products of rubber was explained as the result of an intermediate ring opening which could also lead to the formation of isoprene if a second C—C bond were broken.

Harries believed that the rubber molecules were formed by the $C_{10}H_{16}$ units linking up with each other by means of partial valences of the type suggested by Thiele. The rubber molecules could therefore be written

$$\begin{array}{cccc}
CH_3 & CH_3 & & CH_3 \\
| & | & & | \\
\cdots\cdot HC\cdot CH_2\cdot CH_2\cdot C\cdot & \cdots\cdots\cdot C\cdot CH_2\cdot CH_2\cdot CH\cdots & \cdots\cdot HC\cdot CH_2\cdot CH_2\cdot C\cdot & \cdots \\
|| & || & || & || \\
\cdots\cdot C\cdot CH_2\cdot CH_2\cdot CH\cdot & \cdots HC\cdot CH_2\cdot CH_2\cdot C\cdots & \cdots\cdots C\cdot CH_2\cdot CH_2\cdot CH\cdots \\
| & | & | \\
CH_3 & CH_3 & CH_3
\end{array}$$

* Harries' conclusion that 1,5-dimethyl cyclooctadiene would yield levulinic aldehyde and levulinic acid as the major ozonization products is basically correct although the reaction is somewhat more complicated than visualized by him. See, e.g., the excellent review of ozonolysis reactions by Bailey (40).

or

$$
\begin{array}{ccc}
\text{CH}_3 & \text{CH}_3 & \text{CH}_3 \\
| & | & | \\
\cdots\cdot\text{C}\cdot\text{CH}_2\cdot\text{CH}_2\cdot\text{CH}\cdots\cdots\cdot\text{C}\cdot\text{CH}_2\cdot\text{CH}_2\cdot\text{CH}\cdots\cdots\cdot\text{C}\cdot\text{CH}_2\cdot\text{CH}_2\cdot\text{CH}\cdots\cdots \\
\cdots\text{HC}\cdot\text{CH}_2\cdot\text{CH}_2\cdot\text{C}\cdots\cdots\text{HC}\cdot\text{CH}_2\cdot\text{CH}_2\cdot\text{C}\cdots\cdots\text{HC}\cdot\text{CH}_2\cdot\text{CH}_2\cdot\text{C}\cdots\cdots \\
| & | & | \\
\text{CH}_3 & \text{CH}_3 & \text{CH}_3
\end{array}
$$

The possibility of having at least two stereoisomers of the polymer molecule was taken as important support for Harries's theory, since his ozonolysis studies had shown that rubber and gutta-percha contained the same basic $C_{10}H_{16}$ unit and therefore had to be stereoisomers of the same basic molecule.

While Harries' theory became almost generally accepted in most of the world, and especially in Germany and Russia, it was rapidly challenged by English chemists, notably by Pickles (41) who pointed out that the cyclooctadiene formula proposed by Harries demanded "the employment of vague and unnecessary conceptions of polymerization." Pickles referred in particular to the results of bromine addition to rubber which yielded a polymeric tetrabromide of the composition $(C_{10}H_{16}Br_4)_n$, while on treatment with bromine a polymer of the structure suggested by Harries "should either (1) remain in the polymerized form and absorb only two atoms of bromine for each $C_{10}H_{16}$ group, or (2) become depolymerized and give simple $C_{10}H_{16}Br_4$ molecules." At the same time he agreed that Harries' ozonolysis of rubber had proven that it was built up from

$$
\begin{array}{c}
\text{CH}_3 \\
| \\
-\text{CH}_2\cdot\text{C}:\text{CH}\cdot\text{CH}_2-
\end{array}
$$

units. Pickles suggested therefore that these units were united by normal valence bonds to form long chains of the structure

$$
\begin{array}{c}
\text{CH}_3 \\
| \\
(-\text{CH}_2\cdot\text{C}:\text{CH}\cdot\text{CH}_2-)_n
\end{array}
$$

He also suggested that the chain length might vary from rubber to rubber and be responsible for the observed differences in physical properties.

As is readily recognized, this picture of the rubber molecule would have been rather accurate in the light of present knowledge, but unfortunately, Pickles did not stop his description at this point. The failure of Harries to isolate any ozonolysis products of the type expected from the endgroups of an open polyisoprene chain caused him to add:

The oxidation results require that the two ends of the chain should be linked together, which, of course, leads to the formation of a ring, but it is proposed that in each rubber molecule there is only one such ring. Rubber probably contains at least eight C_5H_8 complexes connected as above indicated.

One of the major problems facing Pickles and other contemporary investigators interested in the molecular structure of polymers was the absence of satisfactory methods of determining the molecular weight of such compounds. The measurements which had been made on rubber had given highly varying and poorly reproducible results in the range of about 1000–7000 (42–44), and nobody believed that this polymer could have such a high molecular weight that the end groups would hardly be detectable if the molecule were linear.

Harries did not wait long to attack Pickles' position (45), but he changed his own position somewhat in 1914 (46) after he had found that measurements carried out on benzene solutions indicated the ozonide to have a higher cryoscopic molecular weight than had been found previously with solutions in glacial acetic acid. Since the new determinations indicated the ozonide to have a molecular weight approaching that of $C_{25}H_{40}O_{15}$, Harries assumed that the basic unit in the rubber molecule could vary in size from an 8-membered to at least a 20-membered ring, but in contrast to Pickles he still believed that each rubber molecule contained several units of this type which were held together by partial valences according to Thiele.

The disagreement between the chemists in England and Germany continued essentially unchanged until Staudinger (47–49) in 1920 finally showed that both groups were wrong. However, by that time many years had already passed since the first major effort to synthesize rubber on an industrial scale was started.

II. THE FIRST SYNTHETIC RUBBER PERIOD

A. Early Developments in Germany

In 1906 a German chemist named Fritz Hofmann (50) approached his management at the Fabenfabriken vorm. F. Bayer & Co. in Elberfeld with the suggestion that they should start synthetic rubber research. The idea for this suggestion had come to Hofmann shortly before, when he had read an article by an English botanist, Professor Dunstan, in a magazine dealing with colonial affairs. The professor had pointed at the scarcity of rubber and besought his countrymen to counteract this situation by synthesizing the "colloid." At this time Hofmann did not know more about rubber than did the average chemist—which is to say very little—since, as should be clear from the preceding discussion, the knowledge of the leading rubber chemists of the time was very meager by modern standards.

Although it was fairly generally recognized that rubber was built up from isoprene units, the manner in which this occurred was poorly understood and was the subject of considerable controversy. To be sure, rubberlike polymers had been obtained from isoprene according to a number of methods, and some workers believed these polymers to be identical with the natural product. The possibility of preparing synthetic rubber from isoprene had occurred to Tilden as early as 1882 (25), but all the polymerization methods uncovered after that time were characterized either by exceedingly slow rates or poor yields, or both. This had caused Tilden himself to become rather pessimistic. In 1908 he wrote (51):

The conversion of isoprene into rubber occurs, so far as observed, under two conditions: (1) when brought into contact with strong aqueous hydrochloric acid or moist hydrogen chloride gas; (2) by spontaneous polymerization.

In the former case, the amount of rubber produced is small, as it is only a by-product attending the formation of isoprene hydrochlorides, which are both liquid. In the latter case, the process occupies several years.

Of course many attempts were made by me to hasten the process, but it was found that contact with any strong reagent, such as oil of vitriol, pentachloride of phosphorous, and others of milder character, led only to the production of a sticky "colophene" similar to the substance which results from the polymerization of the terpenes, and after a course of experiments which were carried on for about two years, I was reluctantly obliged to abandon the subject. It is, however, a question whether the process could be made commercially productive even if a suitable reagent could be found to effect the transformation, because the yield of isoprene from turpentine is very small, probably not exceeding 10 percent under favorable conditions. In my experiments it was less. If isoprene were obtainable at low cost from other sources it might be found possible to utilize the hydrochloric acid process, though I doubt it.

By the early years of this century it had also been observed that conjugated dienes other than isoprene could be polymerized to yield elastomeric or plastic polymers.

The first observation along this line was probably made by A. W. von Hofmann (52), who prepared piperylene (1,3-pentadiene) through repeated exhaustive methylation of piperidine. Although he did not immediately interpret the reaction properly,* he discovered that the final reaction products were trimethylamine and a hydrocarbon of the composition C_5H_8 which he called piperylene. When distilling this hydrocarbon, which was shown to contain two double bonds, Hofmann noticed that it had a tendency to polymerize during heating.

A year later, when trying to prepare piperylene from nitrosopiperidine by treatment with dehydrating agents (P_2O_5, $ZnCl_2$), Schotten (56) found that only traces of piperylene were liberated and that instead a resinous mass was formed in the reaction flask.

The observations by Hofmann and Schotten might have contributed greatly to the understanding of the polymerizability of unsaturated hydrocarbons had it not been for the contempt in which most organic chemists held resinous materials at that time. As it was, neither of these chemists did anything to follow up on his observation and the disclosures in their publications went unnoticed by others.

In 1892 Couturier (57) found that 2,3-dimethyl-1,3-butadiene could be made to polymerize either by heating it with dilute sulfuric acid or by treating it with $CaCl_2$ at room temperature. He did not examine the polymer closely but established that it was soluble in ether, chloroform and acetic acid.

More detailed observations on the polymerizability of this monomer were made about eight years later by Kondakov. He first discovered that dimethylbutadiene would partially polymerize to a leather-like, elastic, almost white polymer when heated with alcoholic KOH at 150°C for 5 hr (58). The polymer obtained was soluble in hydrocarbons, alcohols and ethers. Subsequently he also discovered that this monomer would polymerize quantitatively to a white spongy mass if exposed to diffuse daylight for an extended period of time (about a year) (59). The polymer was tasteless and odorless, was insoluble in hydrocarbons, carbon disulfide, chloroform, ether, alcohol, acetone and oil of turpentine, and would only swell in benzene.

In 1906, after Thiele (54) had established the correct structure of piperylene and had found that it could be polymerized to a rubber-like pol-

---

* The explanation of the reaction sequence leading to the formation of trimethylamine and piperylene from piperidine was furnished two years later by Ladenburg (53), who believed the structure of piperylene to be $CH_2{=}CH{-}CH_2{-}CH{=}CH_2$, a formula for which Hofmann is often erroneously credited. The correct composition of piperylene was established by Thiele (54) in 1901, although Armstrong and Miller (55) had isolated and correctly determined the structure of this compound in 1886 without recognizing it as identical with Hofmann's piperylene.

ymer, it was thus known that at least three simple conjugated dienes could be transformed into polymers having elastomeric properties. In addition, Harries (60) had found that dihydrotoluene, Klages (61) that phenylbutadiene, and Kronstein (62) that cyclopentadiene exhibited a tendency to polymerize when heated. The polymerizability in general of hydrocarbons having conjugated double bonds had therefore been fairly well established.*

After he had presented his written proposal for synthetic rubber research to the top management at the Farbenfabriken in Elberfeld, Fritz Hofmann was called in by Friedrich Bayer and Carl Duisberg to present more details about his suggestion. At the end of the presentation, these two well-known and influential chemists and industrialists gave Hofmann their decision in the following words (50):

"It will naturally be an expensive matter, and it will certainly not be accomplished rapidly, but the possibility exists that your plans will be successful. We will take the risk therefore. The project may cost a maximum of 100,000 marks per year—you must have it in ten years, or (may the devil take you!)."

It was only natural that Hofmann should start his rubber research by trying to find a practical and reproducible method for making a product closely resembling or, if possible, identical with natural rubber. As a consequence of Harries' great influence, Hofmann first tried to prepare rubber via dimethylcyclooctadiene. When this attempt was unsuccessful, he decided to start with isoprene. This, as Hofmann pointed out later on (50), was not at all as obvious an approach as it may seem today. All of his co-workers except one thought much more highly of Harries than of Bourchardt. This was rather natural since many of Hofmann's co-workers had been students of Harries, and their professor enjoyed a considerable international reputation as a rubber chemist. The fact that Hofmann was a neophyte as far as rubber chemistry is concerned may actually have been helpful in this connection. At any rate, he and his colleague Coutelle decided to start directly from isoprene. However,

---

* It has been claimed that Kondakov recognized the general polymerizability of conjugated dienes as early as 1900 (63,64) but the periodical (65) in which this disclosure is supposed to have been made is very difficult to find. Kondakov repeated his claim to the early disclosure in a book published in 1912 (66), but by that time the general polymerizability of simple conjugated dienes had been well established and the matter of priority in the synthetic rubber field had become the subject of an intense controversy with strong nationalistic overtones, as will be discussed later. In an article published in 1935, Lebedev (67) indicates that Kondakov's disclosure is concerned with the ability of conjugated dienes, including butadiene, to form dimers.

the poor availability and high cost of this monomer posed a problem even before polymerization experiments could be started.

Destructive distillation or cracking of terpenes was probably the cheapest method of preparing isoprene, but as pointed out by Tilden (51) the yields obtained in this process were so poor that the isoprene cost approached the price of the best natural rubber. In addition, the cracking process yielded a rather impure product. Purer isoprene had been obtained through organic synthesis, notably by Kondakov (68), Ipatieff (18), and Euler (28), but nobody had obtained more than 10 grams of this highly volatile hydrocarbon, and the synthesis methods used did not seem suitable for large scale production at a moderate cost. Consequently, the search for improved methods of preparing isoprene became a very important part of the synthetic rubber program.

As a matter of fact, the total effort spent on monomer synthesis, initially involving only isoprene but later on 2,3-dimethylbutadiene and butadiene also, was probably considerably greater than that spent on direct polymerization research during the first synthetic rubber programs—not only in Germany but in every other country that became involved in such research before World War I. Besides contributing to our knowledge of organic chemistry in general, the monomer synthesis efforts had a very stimulating effect on the development of both industrial organic chemistry and fermentation biochemistry. Since this particular part of the early rubber research does not involve polymer chemistry and since it has been treated in considerable detail elsewhere (69,70), it will not be discussed further here except where necessary for proper understanding of polymerization phenomena. It is important, however, that the reader be well aware of the great effort that was spent on monomer synthesis during the early attempts to prepare synthetic rubber.

Since hardly anything was known about the manner in which isoprene was converted into a rubberlike product, Hofmann and coworkers started by repeating the rubber-yielding experiments reported in the literature, in the hope of gaining information which could be used for improving the rubber formation. After three years, process conditions were found which allowed the preparation of larger quantities of a rubberlike polymer; samples of this were sent to Harries for ozonolysis and to the Continental Rubber Works in Hanover for practical evaluation. Harries concluded from the composition of the ozonolysis products that the synthetic material was composed of the same structural units as natural rubber, and the evaluation indicated it to be similar to natural rubber of African origin, which had less attractive properties than the Pará rubber.

The process used involved simply heating isoprene for an extended period of time (10–150 hr) in an autoclave at temperatures in the range of 90–200°C, with or without an inert diluent. An application for a patent on the new process was filed in Germany in September 1909, and a patent was granted three years later with a rather broad claim covering heating of isoprene at a temperature below 250°C and in the presence or absence of polymerization-promoting agents (71). Similar patent applications covering the preparation of polymers from butadiene (72) and 2,3-dimethylbutadiene (73) were filed a few months later, after experiments with these monomers had also been concluded.

Meanwhile, Harries had also got interested in finding an improved method of preparing rubber from isoprene. In a lecture presented at Vienna in March 1910, he disclosed that he had been able to prepare rubber by heating isoprene with glacial acetic acid at about 100°C for 8 days in a sealed tube (74). Although he was well aware of Hofmann's success in preparing a similar product by simply heating isoprene to a suitable temperature, he described his own method as the first one for making synthetic rubber, since he was not at liberty to discuss the results obtained at Elberfeld. These matters were discussed in more detail in a paper (75) published the following year. At that time Hofmann's polymerization procedure was already known on the outside through the French patent which had issued the year before (71). In his very comprehensive paper of 1911 (71 pages!) Harries also described the heat polymerization of butadiene and dimethylbutadiene to rubbery products according to the two methods developed by Hofmann and himself. His paper contained a vastly more important disclosure, however.

He revealed that he had discovered that conjugated dienes could be polymerized to rubberlike products by sodium metal at moderate temperature. He described the almost quantitative preparation of polymers from butadiene, isoprene, and 2,3-dimethyl-1,3-butadiene. Characterization of the polymers with the help of the ozonides, nitrosites and bromides revealed that the isoprene polymer differed from natural rubber and from the polymers prepared by Hofmann and himself by heat treatment of isoprene in the absence or presence of glacial acetic acid. The butadiene and 2,3-dimethylbutadiene polymers also seemed to differ more strongly from natural rubber than was expected from the differences between the monomers.

In contrast to the newly discovered heat polymerization methods, but like all preceding methods, the sodium polymerization was discovered entirely by chance. It was quite usual in those days to remove impurities, including moisture, of course, from liquid organic compounds by storing them over sodium wire. Harries, like many colleagues, had undoubtedly been using this method for purifying dienes long before 1910 without having recognized the polymerizing activity of sodium. This may seem strange in view of the high activity of this metal toward butadiene in particular, but it should be remembered that this monomer was not readily available and that the handling of such a low boiling hydrocarbon was then rather difficult. For these reasons much less work had been carried out with butadiene than with its three higher boiling homologs. As a consequence, it was not until 1910 that Lebedev (76) published the first paper disclosing that butadiene could also be polymerized to a rubberlike product by subjecting the monomer to the action of heat.*

Many of the workers who had used sodium for purifying other dienes had almost certainly noted on occasion that considerable reaction was taking place between the metal and the liquid, but since the monomers were usually rather impure any reaction was probably believed to have been caused by the impurities. In addition, the presence of impurities might have prevented extensive polymerization and an accompanying significant increase in the viscosity of the liquid. It is indeed interesting that Harries specifically states both in his paper (75, p. 213) and his book (35, p. 214) that the sodium-initiated polymerization was first observed with butadiene. He also points out that the purity requirements for the isoprene are much higher in this type of polymerization than in the older ones, and that the best results are obtained with pure isoprene.

Harries was not interested in filing a patent application on his discovery; therefore, according to his own statement (77), he disclosed and released the invention to representatives of the Elberfelder Farbenfabriken on October 28, 1910. A patent application was filed by the company in Germany on December 12, but no patent was granted there for reasons which will now be described. A U.S. patent did issue, however (78).

B. Early Developments in England

Shortly after the first effort to produce synthetic rubber had been started in Germany, it occurred to several chemists in other countries, notably in England, that such work might furnish a solution to the increasingly difficult world rubber supply situation. In 1908, E. H. Strange of Strange and Graham, Ltd, in London, decided to have his company look into the possibility of synthesizing rubber, and such work

* For discussion of a claim by Kondakov to have predicted the polymerizability of butadiene in 1900, see footnote on p. 44.

was started under the direction of F. E. Matthews. Later on, Strange sought assistance from Professor W. H. Perkin, Jr. in working out a method of preparing isoprene originally suggested by Matthews. Subsequently, as has been described by Perkin (79), several other chemists and bacteriologists joined the project, and eventually at least 14 professional people were involved in it.

As in Germany, most of the early synthetic rubber effort in England was devoted to finding improved methods of synthesizing isoprene. In the summer of 1910, Weizmann* suggested to Matthews that it may be possible to convert dimethylallene to isoprene by the action of sodium. It was found that some isomerization did take place, but to isopropylacetylene rather than to isoprene, a rearrangement that had been observed considerably earlier by Favorsky (80).

This gave Matthews the idea to investigate the action of sodium on isoprene, supposedly to see if any isomerization would occur also in this case. He therefore sealed a tube containing isoprene and sodium and set it aside in July, 1910. The following month he noted that the contents of the tube had become viscous, and on examining the tube again in September, he found that it contained a solid mass of amber-colored rubber. Further experiments demonstrated that sodium would also cause polymerization of other monomers belonging to the butadiene group, and a patent application covering this invention was filed for Matthews and Strange on October 25, 1910, about 1½ months before the corresponding German patent application by Farbenfabriken, Elberfeld, but only three days before Harries' supposed disclosure of his invention to representatives of that company. The British patent to Matthews and Strange (81) did not issue until after Harries' first paper (75) on the same subject had appeared.

While it may look like a remarkable coincidence that Matthews and Harries should independently have made the same accidental discovery at almost exactly the same time, the whole matter becomes even more astonishing if one considers that the simultaneous discovery was made as the result of two entirely different approaches, and not as the result of similar experiments being carried out at the same time. While Harries would probably not have made his discovery at that time had he not begun to work with the more reactive butadiene monomer, Matthews would probably not have made his unless the experiment carried out by him had

---

* Dr. Chaim Weizmann, who later on became better known as the inventor of the fermentation process for preparing acetone and n-butanol from starch with the help of *Clostridium acetobutylicum*, and then as the first president of Israel.

been of a kind that could have been expected to require an extended period of time.

C. The Great Controversy

At the time of the discovery of the sodium-initiated polymerization of dienes by Matthews and Harries, chemists in many countries, but especially in England, France, Germany, and Russia, had made important contributions in the field of rubber chemistry. Differences of opinion had of course existed from time to time, and there had been a few minor controversies, like the one between the English and German schools regarding the character of the rubber molecule, but the discussion had generally been kept on a high level and free of nationalistic overtones. In 1912, however, this situation changed drastically.

Suddenly much nationalism, often of a rather extreme character, appeared in the scientific literature. This manifested itself in two primary ways. First, there was a tendency of the chemists in each nation to take credit on behalf of their country for as much as possible of the discoveries in the rubber field, and second, there was a common tendency within the other nations to slight the German contributions in general. Since the controversies became very intense and tended to distort the facts, and since they caused considerable experimentation and led to many publications in reputable scientific journals, this matter is also an important part of the early history of synthetic rubber. While it was, of course, a very sensitive subject which could hardly have been treated in earlier years without somebody feeling offended, we can now look back dispassionately on those events, which are an important chapter in the history of rubber chemistry, and a commentary on the times.

To begin with, one must ask: Why did such a great controversy start so suddenly, and why did it become largely a matter of national prestige? The first obvious answer is that synthetic rubber had rapidly become something very important among chemists, something really worth fighting for, a far cry from the situation in the early 1880's when Hofmann and Schotten paid little attention to the polymer they apparently obtained from piperylene. Also, there was the generally increasing tension in the world with the accompanying strongly nationalistic feelings which were undoubtedly also felt among chemists. This was, after all, only about two years before the start of World War I.

As to the rather widespread attempts to slight the German contributions, one might suppose that this was a reaction to some provocation from the Germans. However, a study of the scientific literature gives little support to such a supposition. On the other hand, newspapers in Germany as well as in other countries may have helped to stir up the controversy. This is at least indicated by Harries' heated reply to various foreign publications in 1913, in which he severely criticizes the careless reporting by science and technology writers in the daily press (77).

A certain degree of nationalism is indicated in the introductory paragraph of some U.S. Patents issued to German citizens, but this is nothing that should have provoked a strong counterattack. The introduction to Harries' U.S. Patent (issued in 1913) covering the polymerization of dienes with alkalimetals is interesting in several respects, however. It reads (78):

To all whom it may concern:
Be it known that I, Carl Harries, professor of chemistry, doctor of philosophy, citizen of the German Empire, residing at Kiel, Germany, have invented new and useful Improvements in Caoutchouc Substances and Processes of Making Same, of which the following is a specification.

In the case of inventors who were not professors, "chemist" was substituted for "professor of chemistry," indicating the high esteem in which the title "chemist" was held in Germany at that time.

The fact that the Germans were the first to start a major effort to prepare synthetic rubber and to produce a useful material in quantities sufficient for technical evaluation certainly led to much publicity which, even when not distorted by careless reporting, tended to emphasize the German contributions. That rubber chemists in other nations were not particularly happy about this situation is understandable. The German pre-eminence in many areas of science and technology, and particularly in chemistry, which became more and more pronounced as time passed, was undoubtedly another important factor. In many respects the Germany of those days was undoubtedly looked upon by other nations in much the same way as the United States has been looked upon after World War II, that is with a mixture of envy and admiration.*

The lack of understanding of polymer structure and particularly of the chemical reactions involved in polymer formation helped to aggravate the situation. For instance, it was difficult if not impossible to establish if polymer formation was the result of a particular treatment of the monomer or the consequence of a side reaction caused by the presence of

* In an article describing among others an early synthetic rubber effort starting in 1911 in which he was involved at the Hood Rubber Co. in Watertown, Mass., Kyrides (82) makes a reference to the envy with which the German chemical industry of those days was regarded in the United States.

impurities. Reproducibility of the reaction may have been considered significant in this connection, but the differences in monomer treatment and purity in various laboratories made comparisons difficult. To this should be added that many chemists of that time, particularly those from academic institutions, had great difficulties in appreciating the difference between a reaction which led to the formation of a small amount of polymer and a reproducible reaction which could be used on a technical scale for producing a useful rubber-like material.

The matter of who had prepared the first synthetic rubber became the primary issue in the controversy that started in 1912. By that time the first successful preparation of practically useful synthetic polyisoprene had been claimed by the Germans in 1910–11 through Harries' disclosures (74,75), through patents in England and France (71) and apparently also through articles in the daily press. Perkin, Jr. (79) vigorously contested the German claims and made the following statement:

> There can now be no doubt that rubber may actually be obtained synthetically by the polymerization of isoprene and its homologues and that the synthetic product is really rubber and strictly comparable with natural rubbers, these facts having been confirmed by a large number of chemists working independently.
>
> As regards the history of this synthesis, it seems to me desirable, in view of statements which have been made abroad, to review the work of earlier investigators, and particularly that of W. A. Tilden in England, and G. Bouchardat in France, in order that I may emphasize the fact, as I particularly wish to do, that much of the credit of the pioneer work in this subject belongs to this country and to France.
>
> It is surprising that Prof. C. Harries (Gummizeit, 1907, 823 and 1910, 850) should doubt whether Tilden really had synthetic rubber in his hands in 1882 and 1884 and should advance the view that certain samples produced by Bayer und Co., of Elberfeld, were the first true samples of synthetic rubber. This criticism is quite beside the mark, and the specimens of crude synthetic rubber which Tilden undoubtedly prepared were made many years before Bayer und Co. produced rubber by their process.
>
> It would seem hardly necessary to point out the importance of a thorough study of the literature before attempting to decide claims for priority such as these. Especially in the case of a subject of such wide interest and importance as rubber, the literature of which is often contained in journals difficult to access, very careful search is imperative in order that credit for discovery may be given where it is due. For this reason, Mr. H. J. W. Bliss had been engaged for several months [sic.] in compiling from the literature what seems to me to be a very valuable and necessary sketch of the progress of rubber research from the commencement.
>
> It is of very great interest to notice that an Englishman, Greville Williams—who, in 1860, was the first to isolate isoprene from the products of the destructive destillation of rubber in a fairly pure state, was also the first to observe the transformation of this hydrocarbon into a rubber-like body.

Perkin then went on to praise the contributions of Tilden in particular, whereupon he pointed out that the process used by the Bayer Company for making polyisoprene was essentially anticipated by a British patent to Heinemann (83).

Also in 1912 Kondakov published his book (66) in which he claimed extensive priorities in the synthetic rubber field for Russian workers and for himself in particular.

The publications by Perkin and Kondakov drew an almost immediate response from Hofmann (84) and Harries (77). Hofmann's reply was short and factual and concerned itself exclusively with Perkin's remarks regarding the priority of the Heinemann patent, while Harries' reply was very long and temperamental.

After having expressed his general regret that priority questions in the synthetic rubber field had been turned into a matter of national prestige among English and Russian chemists,* Harries went into a detailed discussion of the claims made by Lebedev, Kondakov, Ostromisslensky and Perkin, Jr. His comments regarding the latter begin as follows:

W. H. Perkin, Jr. has given a lecture before the London Section of the Society of Chemical Industry, the contents of which require a rebuttal for many reasons. He is apparently pursuing the goal of representing all discoveries in the fields of caoutchouc and gutta-percha as purely English accomplishments.

One can really only shake one's head, when Perkin now tries to interpret the old meager disclosures by Williams regarding the autooxidation product of isoprene in such a manner as if this worker already had synthetic caoutchouc in his hands. I have previously pointed out that isoprene becomes viscous, indeed, when it is allowed to stand for a period of time together with air in a closed bottle, no caoutchouc is formed, however, but instead a highly explosive peroxide. If the latter is made to explode, an odor like that of burnt caoutchouc can be noticed to be sure, the residue contains no caoutchouc, however, contrary to the assumption by Perkin.

The major French contribution to *The Great Controversy* came in 1913 with the publication of the very comprehensive and, because of its many references, almost invaluable book by Dubosc and Luttringer, of which the English edition did not appear until 1918 (16). In their chapter on "The Preparation of Synthetic Rubber," these authors support Perkin's attack against the Germans while simultaneously expressing some resentment at the fact that Perkin seemed to slighten Bouchardat's work in favor of that of Williams and Tilden. They also indicate that the literature survey carried out by Bliss is "somewhat incomplete." In their attack on the Germans, Dubosc and Luttringer even go so far as to accuse

---

* It should be noted that neither Harries nor Hofmann claimed their contributions as specifically German.

the German Patent Office of refusing a justified patent to Heinemann for the sake of national interest (English edition, p. 346).

With the advent of World War I people had more important things to worry about than priority questions in the rubber field. The controversy did not stop with the war, however, and has flared up now and then during subsequent years, albeit with somewhat less intensity. Since the good possibility exists that persons reading only one side of the story may get a rather distorted picture of what various scientists contributed, as well as of their personal integrity, and since we are now in a much better position than only a few years ago to judge what was really accomplished, a short review of the facts behind the main issues of *The Great Controversy* will be given below. However, before starting the review we should emphasize that the nationality of the various contributors is of no consequence as such, but that on the other hand the reputation of no individual should suffer because some people chose to make the early contributions to rubber chemistry a matter of national prestige.

As already mentioned, the primary issue in the controversy was the question of who had prepared the first synthetic rubber. To resolve this question, one would actually first have to define "synthetic rubber". If we require that such a material be essentially equivalent to natural rubber, then nobody prepared synthetic rubber until about ten years ago, at best (cf. Part II, Chapters 6 and 7). If, as Perkin apparently was willing to do, we are prepared to concede priority on the basis of a general disclosure of a process in a patent or elsewhere which would lead to the production of "synthetic natural rubber," then a little-known German named Hans Labhardt (85) would be the man to get the credit for the first synthetic rubber, since he suggested the polymerization of isoprene with lithium alkyls in 1912. This particular matter will be discussed in more detail below. However, if we require only that the synthetic material be practically useful as a rubber, then the first synthetic rubber was undoubtedly prepared even earlier.

There is little doubt that some polymerization took place when Greville Williams tried to distil his oxidized ("ozonized") isoprene, but he certainly did not obtain any rubber, since the amorphous mass remaining after removal of the volatile components had a composition closely corresponding to $C_{10}H_{16}O$. Furthermore, Williams himself did not suggest that he had obtained any synthetic rubber. He did some very fine work, however, and he clearly suggested a close relationship between isoprene and rubber almost twenty years before Bouchardat. Williams was also a gentleman who gave ample credit to earlier workers. The scornful remarks made by Harries were not fair to Williams, therefore, even though they were justified on the basis of the misleading statements by Perkin.

There is also no reason whatsoever to doubt that Bouchardat (24) obtained some polymeric material in 1879 when he contacted his isoprene with five times its weight of concentrated hydrochloric acid. Likewise, there is little reason not to believe that Tilden (26) was able to obtain some rubbery material by the same method from the isoprene he had prepared by pyrolysis of turpentine. At the same time, Harries (77) and others were undoubtedly correct in challenging the correctness of the statement that strong aqueous hydrochloric acid will effect the polymerization of isoprene to a product closely resembling natural rubber. In 1918, after having failed to obtain any rubberlike material during extensive studies of the addition of hydrogen chloride to isoprene, the highly regarded Finnish chemist O. Aschan (86) made the following statement:

> During the above experiments I have never observed the condensation to a substance completely like caoutchouc analogous to the manner in which it is supposed to take place according to Bouchardat by the action of hydrochloric acid on isoprene. The same experience has presumably been had by everyone among one's professional colleagues who have been occupied with the problem of making caoutchouc from isoprene during recent years. It is suggested that this false statement be deleted from the literature from now on.

Thus it seems quite clear that the polymer formation observed by Bouchardat and Tilden resulted from the presence of impurities in the isoprene, such as peroxides, which on decomposition may have initiated some free radical polymerization. The hydrogen chloride treatment of isoprene cannot be considered a method of preparing synthetic rubber, therefore. It is also clear from his statement of 1908 that Tilden (51) did not think very much of this method. Furthermore, it should be noted that he considered the rubber-like material to be formed most readily if the acid were allowed to act on the oily polymeride (primarily dimers) resulting from the action of heat on isoprene. This method, which also crops up in a later patent to Lilley (87), will, of course, not yield a useful synthetic rubber as we now know.* In addition to this, it should be noted that natural rubber itself reacts vigorously with hydrogen chloride

---

* It is worth noting that Himly (17) as early as 1838 observed the formation of a "brown, resinous material" when *caoutchine* (= impure dipentene) obtained by destructive distillation of rubber was treated with excess "dry hydrochloric acid."

yielding a chlorinated product, the vulcanizate of which is brittle and therefore useless as a rubber (cf. Part II, Chapter 10.A).

The polymer obtained by Wallach in 1887 from isoprene which had been exposed to light for a long time at room temperature may have been of a useful type, even though it could certainly not have had the steric regularity and the attractive physical properties of natural rubber. Nevertheless, if it had been properly evaluated, it might have qualified as the first synthetic rubber. The rubbery product isolated by Tilden five years later from isoprene that had been kept at room temperature for several years was probably of a similar nature, since both presumably had resulted from free radical-initiated polymerization. However, both processes were exceedingly slow and could not qualify as useful methods for making synthetic rubber.

For these reasons, it appears correct, indeed, to call Hofmann's heat polymerization of isoprene the first useful method of preparing synthetic rubber. The inventive aspects of it must be considered marginal, however, in view of all the prior art disclosing the polymerizing or rather oligomerizing effect of heat on various dienes. Hofmann (50,71) quite correctly points out that previous workers had primarily obtained dimers upon heat treatment of the monomers and that the heating conditions are critical. This notwithstanding, the German Patent Office allowed a claim with a broad temperature range, something that certainly does not appear justified, at least by modern standards.

On the other hand, Perkin's (79) statement that Heinemann's British Patent 21,772/07 (83) and his German Patent Application H44823 anticipated the Bayer process is incorrect. As stated by Hofmann (84) in his rebuttal, the passage in Heinemann's German patent application which disclosed polymerization "by heating in a sealed tube with or without the addition of acidic, alkaline or neutral catalysts" was added on December 14, 1910, four months after the corresponding French patent to Bayer (71) had been issued. Heinemann's original documents disclosed only the polymerization of isoprene with concentrated hydrochloric acid. These facts also indicate how wide of the mark Dubosc and Luttringer (16) were in their accusation of the German Patent Office of unfair practices against foreign applicants.

In this connection if should be mentioned that a controversy also developed between Harries and the Bayer Company as to who invented the heat polymerization. Harries claimed that he was the inventor and referred to an article (88) in which he described the results of heating pure isoprene to 300°C. The results obtained by Harries (mostly dimer formation) were quite similar to those obtained previously by Bouchardat and no rubberlike material was indicated. Nevertheless, Harries repeated his claims to the invention as late as 1919 in his book "*Untersuchungen*" (35, p. 133).

It should be emphasized that Harries, while not trying to claim anything as specifically German, tried to claim as much as possible for himself. He was apparently a master at "hedging," i.e., he tried to present and interpret his results in such a manner that he could claim to be right no matter what turned out to be right in the final analysis. A typical example was his behavior in connection with his discussion of the structure of the rubber molecule.

On the basis of the composition of the ozonolysis products of natural rubber, Harries (89) mentioned several structural possibilities for the rubber molecule. On the basis of further evidence he then suggested his dimethylcyclooctadiene formula. When Pickles (41) then remarked that the evidence much better supported a much larger ring-structure, Harries (45) first attacked Pickles' suggestions, but then he added that even if the latter were right, the credit for the correct structure should be given to Harries who had been the first to suggest such a structure, too. The same reasoning was employed by Harries (46) in 1914 when he abandoned his cyclooctadiene structure in favor of a larger ring.

In the case of the alkali metal initiated polymerization of dienes, which also became a subject in the international controversy, Perkin (79) actually treated Harries rather kindly and acknowledged his discovery of the method independently of Matthews. This was not sufficient for Harries, however, especially since he felt that Perkin's emphasis of the fact that the patent had been granted to Matthews tended to indicate that the latter was the first discoverer. Harries (35,77) considered the sodium-initiated polymerization the most important matter in his rebuttal, therefore.

Even in 1912 it would probably have been very difficult to establish with absolute certainty who had been the first to realize the polymerizing activity of alkali metals. Matthews made the invention some time during September–October 1910 and definitely before October 25. Harries seems to have recognized the invention before October 28, 1910, when he disclosed it to representatives of the Bayer Company. Since one of Harries' most typical characteristics was great thoroughness, it may seem reasonable to assume that he had been working on this invention for a considerable time before disclosing it to anybody else. However, in his first paper on the subject, Harries (75) stated that the discovery was made toward the end of 1910. It appears quite unlikely, therefore, that Harries made the invention before Matthews. At any rate, Matthews got the patent he deserved, and the world recognized Harries as an independent and essentially simultaneous inventor of the sodium polymerization.

Another controversy that Harries became involved in dealt with the first polymerization of butadiene to an elastomeric product. Lebedev (76) published the first paper on this subject in 1910 when he had successfully employed the heat polymerization technique. Harries' first formal disclosure of the same preparation as well as of the polymerization of butadiene with sodium came in his comprehensive paper of 1911 (75). However, he later claimed that he had disclosed the heat polymerization in his lecture in Vienna (35,77) some months before the appearance of the paper by Lebedev. The report of his lecture (74) discloses only that he was working at that time on the polymerization of butadiene but it has of course been fairly well established that Harries discovered the sodium polymerization not later than October 1910 and it is not unlikely that he had been successful in polymerizing butadiene according to the heating technique before that time. In the absence of better evidence favoring Harries, most non-German authors have been considering Lebedev the inventor (69,90–92). However, they have all forgotten that the Bayer Company applied for a patent (72) on the heat polymerization of this monomer on September 20, 1909. Thus it is strongly indicated that neither Lebedev nor Harries but Hofmann and Coutelle at the Farbenfabriken in Elberfeld prepared the first rubberlike polybutadiene.

The question of who had prepared the first useful rubber from 2,3-dimethylbutadiene also became a matter of controversy between German and Russian chemists. The Germans claimed that the two polymers prepared by Kondakov in 1900 and 1901 did not qualify as rubbers, since the first of these was soluble in alcohols and ether while the second was insoluble also in hydrocarbons.

In agreement with the ideas regarding the molecular structure of rubber that prevailed in those days, the Bayer Company (73) claimed in its patent application that the degree of polymerization of Kondakov's first polymer was too low, and that that of his second was too high; only the material prepared by Bayer through heat polymerization was of the right molecular weight and could qualify as a rubber, since it was soluble in hydrocarbons but insoluble in alcohols. Today we know, of course, that this reasoning was wrong. Kondakov's first product must have contained some polar groups, probably hydroxyl groups, and may have been disqualified on that basis. His second product, on the other hand, may originally have qualified as a rubber according to the German definition, but it had probably become crosslinked upon standing, since it has a great tendency to react with oxygen and form peroxides.

The controversy between Harries and Kondakov became a particularly ugly one. Kondakov had undoubtedly provoked it by claiming much too much for himself in his book of 1912 (66). However, this cannot be considered sufficient reason for Harries not to have behaved like the gentleman he expected everybody else to be. Kondakov made many great contributions working as he did in the little Estonian university town of Tartu (Dorpat). In his later years, Fritz Hofmann (50), whose writings were always characterized by dignity, gave considerable credit to Kondakov.

Many other issues were brought up in connection with *The Great Controversy*, which reached its peak just before World War I, but the above examples should suffice to give the reader a reasonably good idea of one of the most controversial chapters in the history of modern chemistry.

D. The Discovery of Vulcanization Accelerators and the Start of a Second Controversy

When the Great Controversy reached its peak in 1912–13, the rubber price had dropped drastically from its peak level in 1910 (Fig. 1), primarily because of the appearance of plantation rubber in significant quantities on the world market. The incentive for synthetic rubber research had therefore decreased strongly, and as a consequence, such research was severely curtailed in most countries. Only in Germany where, among others, the Bayer Company had committed itself to a long-term project, did the work continue at a significant level. However, the stark realities of economics made themselves felt there, too, and it became clear that the research effort had to be thoroughly reevaluated and limited to those approaches that seemed to offer the most economical solutions.

The heat polymerization of isoprene yielded a useful product but only in poor yields. The sodium polymerization promised to give much higher yields, if properly developed, but the product thus obtained was different from and inferior to the polymer obtained by heat as demonstrated by Harries (75). In addition, when prepared by the most economical method that could be visualized, isoprene remained so expensive that an economical rubber synthesis seemed impossible, even if the conversion to useful polymer could be made close to quantitative.

As a consequence, considerable attention was given to polymers from other dienes, primarily butadiene and 2,3-dimethylbutadiene, even though such polymers had not exhibited properties as attractive as the polyisoprenes. Of the two monomers, butadiene seemed to yield the better polymers, but it too was very difficult and expensive to prepare. Dimethylbutadiene, on the other hand, could be prepared much more easily and at a considerably lower cost from acetone, according to a method developed by Meisenburg and Delbrück (50). This monomer could also be polymerized in good yield according to the heating technique, albeit at a low rate. However, the polymers thus obtained turned out to be very difficult to evaluate because of their great tendency to react with atmospheric oxygen, particularly during compounding and milling. The latter operation caused the polymer to change into a completely useless sticky, acid-smelling mass. The same problem was encountered with the polymers from isoprene and butadiene, though to a considerably lesser extent. Since the tendency of natural rubber to become oxidized under the same conditions was slighter yet, it was indicated that some of the impurities in the natural product acted as oxidation inhibitors or *antioxidants*. However, it had long been noticed that articles made from natural rubber, whether vulcanized or not, deteriorated with time. That oxidation played an important role in this aging process had been indicated long before synthetic rubber research was started, and the use of antioxidants of the phenolic type had been suggested as early as 1870 (93).* The smoke treatment of rubber articles by the Indians of Central and South America before the arrival of the first Europeans may be considered an even earlier utilization of the preserving effect of phenolic substances. The particulary attractive properties of rubber obtained from latex dried over a smoking fire, "smoked sheet," can also be taken as early indication of the oxidation inhibiting properties of phenols.

In 1908, Wolfgang and Walter Ostwald (95) discovered that neutral or basic nitrogen-containing aromatic or heterocyclic compounds were useful for retarding or preventing the aging of rubber articles, and they applied for patents on their discoveries. Hofmann and coworkers at the Bayer Company based their search for suitable antioxidants on the disclosures in the Ostwald patents, the first of which had issued in 1910. A systematic investigation soon uncovered several compounds which were effective oxidation inhibitors for *methyl rubber*, a name given to the polymer from dimethylbutadiene because of the presence of one methyl group more per monomer unit than in natural rubber. Piperidine and derivates of this compound turned out to be particularly effective in this respect.

---

* The historic background of the use of materials which improve aging has been treated in considerable detail by Semon (94).

A surprising and much more important discovery was made when vulcanization tests were carried out with the stabilized polymers. Until that time it had been virtually impossible to get satisfactory vulcanizates from methyl rubber. However, when polymer samples containing certain antioxidants, piperidine and its derivatives in particular, were tested in otherwise normal recipes, excellent and rapid vulcanization was accomplished. Tests carried out with other synthetic diene polymers as well as with natural rubber soon revealed that many of the antioxidants acted as very potent and generally useful vulcanization *accelerators*. This important discovery led to the filing of several patent applications, and patents were granted in Germany and many other countries (96,97).

For a time it was believed that the discovery of the organic vulcanization accelerators was the one outstanding result of the first synthetic rubber effort in Germany; however, in 1917, some three years after the issuance of the first patent on this matter and after several articles had appeared (98–101), Spence (102) published a note in which he claimed that everything the Germans had discovered had been discovered considerably earlier in America. The beginning of Spence's note reads as follows:

In view of the world-wide publicity given in recent times to the above subject it would seem to be no longer amiss, but, indeed, desirable to put on record a few facts in refutation of some of the very erroneous statements which have appeared in the literature of this subject. The various writers reviewing this subject, and in particular Ditmar, Gottlob, King, Peachey, have all, without exception, assumed that that which is not proclaimed from the house top is not known, and from these faulty premises have drawn certain equally erroneous yet sweeping conclusions which it has been my intention sooner or later to nullify and correct. I shall take the various references in the order in which they appear.

Ditmar in that part of his article which deals with the subject of organic accelerators refers to the first patent application on this subject as follows:

The 16th of November, 1912, marks an important turning point in the development of the rubber industry. For on the publication of this patent there began a systematic hunt in German, Russian, Austrian, and in American rubber factories for organic vulcanizing accelerators. The ice was broken; the practical value was recognized.

This patent deals with the vulcanization of natural or of synthetic rubber using piperidine or its homologues as accelerators.

After having discussed the contents of the other articles and after having stated that he discovered the accelerating effect of piperidine on the vulcanization of natural rubber considerably before the workers at Bayer, Spence continues:

Now in regard to the discovery of vulcanization of synthetic rubber by means of organic accelerators, also accredited to the Elberfeld Farbenfabrik. The vulcanization of these products by this means was actually carried out and successfully accomplished by me both for isoprene-caoutchouc and for the dimethylcaoutchouc from pinacone at a time when the large industrial concerns in Germany engaged in the study of the synthesis of india-rubber were in the dark as to their lack of success in the vulcanization of their products. The proof of this assertion and the complete vindication of my position with reference to the subject in general I am content to leave over until a later date. At that time I will also bring evidence to show that even the reagents claimed by Peachey (English Patent 4263 of 1914) were known to me and had been used by me both scientifically and industrially several years before his application was applied for.

Spence's note is, of course, interesting not only because of the claims to priority regarding vulcanization acclerators but also because of its disclosure of another early but until that time unpublished synthetic rubber effort in the United States. Details of this effort, which took place at the Diamond Rubber Company in Akron, Ohio, now part of the B. F. Goodrich Company, were published at a considerably later date by Semon (103).

Spence did not have to wait long for a reply from Peachey (104). In a note on "Organic Vulcanization Accelerators" published six weeks later, the latter stated:

In reference to the above subject a communication from Dr. Spence has recently appeared in this Journal. The statements made therein appear to me to be of such an intemperate character that I feel constrained to challenge them.

After having discussed Spence's various claims to priority, which according to Peachey "certainly cannot be said to err on the side of modesty," the latter continued:

I may add that the nitroso-compounds, the use of which was patented by me, constitutes a class of accelerators quite distinct from those discovered by Bayer and Co. in that the mechanism of the acceleration differs in the two cases; it would therefore be a remarkable coincidence if both types of catalyst should have been anticipated, and one is tempted to ask why Dr. Spence refrained from either publishing or patenting his discoveries.

I suggest that having put forward his claim in a scientific journal he should without delay submit the evidence which he has promised to reveal at a future date.

Another controversy was thus started in the rubber field, but it did not assume the same proportions as the first one, and the nationalistic aspects became less intense.

With few exceptions, the Germans quite simply ignored the American claims. This may not seem particularly nice in view of the evidence that was eventually furnished by the American workers, but one should perhaps ask oneself what the reaction would have been in America, if some Germans or Russians had suddenly claimed priority for a very important and apparently totally novel discovery published more than three years earlier in America.

Although many articles were published in American journals during the years that immediately followed (105–108), truly convincing evidence supporting the American claims was slow in coming forth. However, the validity of U.S. Pat. 1,149,580 (corresponding to Ger. Pat. 280,198 (97)) issued to Bayer in 1915 was challenged in court and the patent was declared invalid in 1926. The decision was probably based more on the incorrectness of a claim covering any organic base having a dissociation constant greater than 1×10^{-8} than on the fact that vulcanization accelerators had been in widespread use in the American rubber industry prior to the discovery of their use by Delbrück and Gottlob. Nevertheless, the court action did bring out more facts about prior American activities.

As things now stand it appears established that Oenslager, working at the Diamond Rubber Company, discovered the first organic vulcanization accelerators in 1906, namely aniline and its carbon disulfide derivative, thiocarbanilide. Much work was subsequently carried out along these lines at the same company, among others by Spence, but the results were kept as a manufacturing secret. This notwithstanding, information about the improved vulcanization process began leaking out, and after a few years several other rubber companies in the Akron area had begun using organic accelerators. The early work on such compounds in the United States has been described in considerable detail by Oenslager (109) who eventually received both the Perkin Medal and the Goodyear Medal for his discoveries.

However great these American discoveries may have been, it should be emphasized that it was the German disclosures that informed the world about the importance of organic vulcanization accelerators. Little if anything was added to our knowledge by the subsequent American disclosures, which could not have been expected to create anything but ill feelings among those who, as far as could be judged, had been the first to make the discovery. The intemperate character of Spence's (102) first communication may be understandable, but it was entirely unjustified. It shows, however, what can happen if a scientist is not given the opportunity or otherwise fails to publish an important discovery within a reasonable period of time.

E. First Production of Methyl Rubber and Continued Search for Improved Polymerization Methods

Production of methyl rubber in developmental quantities was started by the Bayer Company in 1911, and various application studies were begun as soon as its poor processing properties had been improved through the use of antioxidants and vulcanization accelerators. The new material was tested both for soft and hard rubber applications and found satisfactory for certain purposes. Thus it was demonstrated that fully acceptable automobile tires could be made from a mixture of this rubber with natural rubber. Carl Duisberg traveled to several technical meetings in a car equipped with such tires, and he also succeeded in getting the Kaiser to try them on one of his automobiles. However, after some time it was noticed that tires containing methyl rubber were aging considerably faster than those made from natural rubber, especially during storage.

The results obtained in hard rubber applications were more encouraging, but even so, the incentive for large scale production disappeared when the price of natural rubber dropped to about 7 marks per kilogram in 1913. The production of methyl rubber was therefore terminated well in advance of World War I after only a few tons had been produced.

The shift in emphasis from polymers of isoprene to those of dimethyl butadiene (methyl rubber) during the years just preceding World War I did not cause the search for improved polymerization methods in general to be completely discontinued in Germany.

In 1910 Heinemann (110), who was living in England, had applied for a patent on the production of India rubber from isoprene by the polymerizing action of ozone. The application was extended within a few months to include the polymerizing action of oxygen. From the way Heinemann's patent application reads (heating of the product after exhaustive treatment with oxygen or ozone), there can be little doubt that it was entirely inspired by the disclosures of Williams in 1860, a point made also by Perkin (79) in his article of 1912. The amounts of oxygen or ozone employed according to the specification of the Heinemann patent are such that a useful rubbery material can hardly have been obtained, let alone "India-rubber," and no examples are furnished in the patent. However, it is interesting in that it suggests peroxides as the source of oxygen for the oxidation of the isoprene.

That peroxides are formed when isoprene is treated with air had been established earlier by Harries, and since it was indicated or at least believed that Williams had obtained some polymerization when his air-oxidized ("ozonized") isoprene was heated for the purpose of removing unreacted monomer, the polymerizing activity of peroxides had undoubtedly been suggested by this time. It may therefore seem surprising that nobody had applied for a patent on the use of peroxides as polymerization initiators until such a patent application was submitted by Holt and Steimmig in 1911 (111). However, it should be remembered that nothing was known at that time about free radical-initiated polymerization of olefins.

In their patent, Holt and Steimmig claim as polymerization initiators compounds which are able to "give up" or "transport" oxygen. As useful compounds of this type are mentioned: old oxidized rubber of natural or synthetic origin, isoprene which has undergone incipient oxidation, peroxides such as benzoyl peroxide, as well as analogous compounds of organic or inorganic nature such as perborates and ozonides of terpenes and caoutchouc. Ceric sulfate and the blue salts of vanadium are also mentioned as oxygen-transporting compounds.

At about the same time Hofmann and coworkers at the Bayer Company conceived that certain advantages could be realized from carrying out the polymerizations in emulsion rather than in bulk or in solution. It was apparently believed that the presence of a protective colloid in natural rubber caused this material to be superior to the synthetic product and that the properties of the latter could be improved by the proper addition of such a colloid. Although now largely proven erroneous, this belief was not unreasonable at a time when chemists were essentially ignorant of the importance of naturally occurring antioxidants to the aging properties of the rubber and the influence of molecular structure on the physical properties of the polymer. Another advantage which Hofmann and coworkers apparently also visualized for emulsion polymerization was correct; namely, that the polymerization charges are easier to handle than are the bulk or solution systems. The protective colloids mentioned in the examples of the patent applications submitted early in 1912 (112) were all proteinaceous materials of natural origin, such as blood serum, egg albumen and gelatin.

Efforts to improve the alkali–metal-initiated polymerization of dienes were also made. Probably the most important improvement was that of Holt (113), which involved polymerization in the presence of carbon dioxide. This modification was claimed to yield an isoprene polymer more closely resembling natural rubber than that obtained in sodium polymerization according to Matthews and Harries.

A somewhat related and much more important discovery was made by Labhardt early in 1912, but it apparently passed rather unnoticed at that time. This chemist discovered that butadiene and its homologs can be polymerized with alkali metal alkyls, and patents were applied for on behalf of his company, BASF (85). The German patent is reproduced in Fig. 3. It is an outstanding example of simplicity and clarity, quite different from most modern patents which frequently contain lengthy specifications necessary for distinction from prior art. Yet, Labhardt's patent broadly covers a process that would lead to the formation of high *cis*-1,4 polyisoprene, i.e., "synthetic natural rubber." The corresponding British patent is actually a little more informative in that it definitely states that in the sodium trialkyl zinc compound used in the example, "the actual polymerizing agent appears to be the alkali metal alkyl." The British specification also suggests a relationship between the alkali metal and the alkali–metal-alkyl initiated polymerizations of dienes.

It is readily recognized that the patents just mentioned describe polymerization procedures which are closely related to some methods currently used for large scale manufacture of high quality general purpose

KAISERLICHES PATENTAMT.

AUSGEGEBEN DEN 20. JANUAR 1913.

PATENTSCHRIFT

— № 255786 —

KLASSE **39 b.** GRUPPE 1.

BADISCHE ANILIN- & SODA-FABRIK in LUDWIGSHAFEN a. Rh.

Verfahren zur Darstellung von kautschukähnlichen Produkten.

Patentiert im Deutschen Reiche vom 27. Januar 1912 ab.

Es wurde gefunden, daß sich Butadien, seine Homologen und Analogen und Gemische solcher Verbindungen durch die Einwirkung von metallorganischen Verbindungen, insbe-
5 sondere von Alkylverbindungen der Alkali- und Erdalkalimetalle und des Magnesiums oder deren Zinkalkyl-Doppelverbindungen, in ausgezeichneter Weise zu kautschukähnlichen Produkten polymerisieren lassen. Schon in
10 der Kälte tritt die Polymerisation in verhältnismäßig kurzer Zeit ein; durch Erwärmen wird der Vorgang noch weiter beschleunigt.

Beispiel.

15 20 Teile Isopren werden mit 4 bis 5 Teilen der Doppelverbindung aus Zinkäthyl und Na-

triumäthyl im verschlossenen Gefäß bei Luftabschluß geschüttelt oder gerührt. Schon nach wenigen Stunden ist eine erhebliche Volumenabnahme und Verdickung der Masse einge- 20 treten. Sobald diese ihr Maximum erreicht hat, wird in üblicher Weise aufgearbeitet.

PATENT-ANSPRUCH:

Verfahren zur Darstellung von kaut- 25 schukähnlichen Produkten, darin bestehend, daß man auf Butadien, seine Homologen und Analogen oder Gemische solcher Körper metallorganische Verbindungen, insbesondere die Alkylverbindungen der Alkali- 30 und Erdalkalimetalle oder des Magnesiums, einwirken läßt.

Fig. 3. German Patent 255,786 disclosing first polymerization with alkylmetals. Inventor: Hans Labhardt. (Original furnished by courtesy of BASF Colors and Chemicals, Inc.)

rubbers such as SBR and *cis*-1,4 polyisoprene; yet, no truly effective use
was made of any of these improvements until more than 20 years later.
The almost complete lack of understanding of the basic structure of rubber molecules was probably the most important reason for this delay,
since it completely precluded any understanding of the polymerization
reactions. However, the poor availability of isoprene and butadiene also
played a role in this connection, since it is primarily these monomers, and
not dimethyl butadiene, that provide superior polymers by the newer
polymerization methods.

F. Production of Methyl Rubber in Germany during World War I

General recognition by the major nations prior to 1914 that a war could
be precipitated at any time caused them to make more or less extensive
war preparations. However, none of the belligerents to be had prepared
for a long war. Previous experiences, like the Franco-Prussian War of
1870–71, and an acute awareness of the great technological advances that
had been made during the preceding decades, had caused statesmen and
military strategists to believe that an extended war was no longer
possible. Even if they did not think in terms of a modern *Blitzkrieg*,
for which the means of transportation were then still too slow, they
nevertheless believed quite generally that the destructive power of
modern arms had become so great that no nation could stand a prolonged
war, and a definite decision would have to be gained rapidly. This
line of thinking was particularly prevalent in France and Germany and
was to be disastrous to the latter country.

The supplies of natural rubber in Germany were rather small when that
country entered the war in August 1914. The situation was apparently
still not considered serious after Germany had failed to gain a decisive
victory during the first month of the war. A Rubber Commision was
established in Berlin with Emil Fischer as chairman, and Fritz Hofmann
(50) was called to testify before this commission at the end of October, but
the rubber industry persisted in its opinion that existing rubber supplies
and proper utilization of reclaimed rubber would make synthetic rubber
production unnecessary (114). The attitude of the industry changed
drastically over the next year and a half as the rubber supplies grew steadily smaller, and Hofmann, Gottlob and coworkers at the Bayer Company
were again asked to start the manufacture of methyl rubber. The
severity of the situation at that time is indicated by the fact that Germany
also decided to build large merchant submarines for the purpose of trans-
porting strategically important goods, primarily natural rubber, through

the British naval blockade. The resulting two submarines of this type, the *U-Deutschland* and the *U-Bremen*, made a few trips to America, where they created a sensation, before the United States broke off diplomatic relations with Germany in February 1917. The need to produce synthetic rubber in Germany would undoubtedly have been minor if this traffic had been allowed to continue, for the cargo capacity of the *U-Deutschland* was apparently sufficiently large to satisfy the most urgent rubber needs of the Central Powers for about half a year (13).

When the decision to restart the production of synthetic rubber was finally made, there was no time for developing new polymers, and it was decided to begin production of methyl rubber of the types previously developed. Even so, as has been described in great detail by Gottlob (115), many serious problems faced the chemists and engineers at the Bayer Company. Not only had the old facilities for production of both monomer and polymer been converted for other purposes even before the war, but many of the professional people involved in the early production had been called into military service and some of them, like Delbrück, had lost their lives. In addition, it was no easy task to procure suitable equipment for proper handling of the highly volatile and inflammable chemicals involved in the monomer and polymer syntheses. Nevertheless, the problems were rapidly solved and methyl rubber production was started again in 1917. The two types of polymer made were designated as W (for *weich* = soft) and H (for *hart*). The over-all synthesis scheme for monomer and polymers is shown in Fig. 4.

Methyl rubber W was prepared by charging about 3000 kg monomer to a cast iron vessel, preferably coated with water-glass, and then blowing air over the monomer for some days until it had become oxidized sufficiently for good initiation of the polymerization. The monomer was then heated to its atmospheric boiling point, i.e., 69–70°C, and kept there for about 5 months or until about 97% of it had polymerized. Later on it was found that the polymerization period could be shortened by about one month if 0.6% benzoic acid was added to the monomer. The polymer, which consisted of a tough, transparent, rubberlike mass, was removed from the vessel by means of a special rotating cutting device. It was used almost exclusively in soft rubber applications, where it proved to be only a poor substitute for natural rubber.

Methyl rubber H was prepared in thin-walled sheet metal vessels at 30–35°C. The vessels were about 1 m tall and 0.7 m in diameter, and they were charged to about a third with 40 kg monomer and about 4% of previously prepared polymer as a "seed." The vessels were then soldered

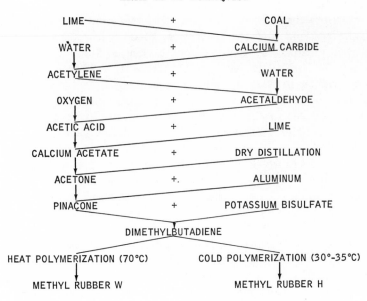

Fig. 4. Schematic representation of process for making methyl rubber in Germany during World War I.

air-tight like giant tin cans and brought to a heating chamber, where they were allowed to stand for 3–4 months. A quantitative yield of a cauliflowerlike polymer which filled almost the whole space in the vessel was formed during this period. The polymer was removed from the reactor under nitrogen protection and was then rapidly mixed with an effective antioxidant. This procedure was necessary because of the great air sensitivity of the polymer. Methyl rubber H was used primarily in hard rubber applications, such as storage battery boxes for submarines, where it turned out to be a very good substitute for natural rubber.

A small amount of methyl rubber, designated BK, was also made by BASF in Ludwigshafen according to the sodium-carbon dioxide method invented by Holt (113). It was used primarily for insulation of electric cable and wire.

Methyl rubber production reached almost 150 tons per month at the end of the war, at which time a total of 2350 tons had been produced. This is not a large amount by present standards, but it should be kept in mind that the polymerization periods were so long that the production of even such modest quantities required the simultaneous handling of very

Fig. 5. Plant for making methyl rubber during World War I. (Courtesy of Farben-fabriken Bayer AG.)

large quantities of monomer. The picture of the polymerization plant shown in Fig. 5 gives a better idea of the size of the operation than do the quoted production figures.

When natural rubber became available again at a moderate price (cf. Fig. 1 and Table II), the incentive for both methyl rubber production and synthetic rubber research vanished and both were terminated in Germany, not to be resumed until 1926. Thus ends the First Synthetic Rubber Period, but a second more successful period was soon to follow.

III. THE SECOND SYNTHETIC RUBBER PERIOD

The end of World War I definitely signaled the end of the first era of rubber synthesis, but it did not permanently terminate the synthetic rubber effort even though methyl rubber production in Germany ceased. The war provided the incentive for the production of this rubber, but the important effect of the end of hostilities, as far as the history of synthetic elastomers is concerned, was that the subsequent economic conditions made it impossible to continue synthetic rubber research in Germany for several years. Then, during this period of little or no research activity, something happened which caused the situation to be entirely different when synthetic elastomer research was started again early in 1926.

Some minor research efforts in this field were apparently being made in some countries, notably in the United States (116) and the Soviet Union, between the end of the war and 1926, but little or nothing was published about them, and they had no significant effect on subsequent developments.

The occurrence that caused the background for synthetic elastomer research to change completely between World War I and 1926 was the appearance of Staudinger's first papers on the structure of polymers (47,49,117). These papers, which furnished very strong evidence for the view that most naturally occurring polymers are very large linear molecules in which the monomer units are attached to each other by normal valence bonds, completely revolutionized rubber chemistry and polymer chemistry in general. Staudinger's view strongly resembled the one brought forth by Pickles (41) in his controversy with Harries, but Staudinger further demonstrated that the apparent absence of endgroup reactivity need not be explained on the basis of a cyclic structure, since a linear polymer would have the same property if its molecular weight were sufficiently high. He also disclosed recent measurements which indicated that the molecular weight of some polymers was adequately high in this respect, e.g., 50–100 thousand.

By hydrogenating natural rubber, Staudinger and Fritschi (49) furnished convincing evidence in support of a linear structure. Hydrogenation yielded a completely saturated high molecular weight polymer of the composition $(C_5H_{10})_x$. This polymer was found to be much more stable than the original natural rubber toward thermal breakdown. Since the double bonds had been removed, poorly understood "partial valences" could no longer be held responsible for the polymeric character of the hydrogenated product. This eliminated Harries' dimethylcyclooctadiene formula from further consideration and established fairly well that natural rubber consists of linear macromolecules in which the monomer units are linked together by normal carbon-to-carbon bonds in a 1,4 fashion, i.e.,

$$+ CH_2 - \overset{\overset{\displaystyle CH_3}{|}}{C} = CH - CH_2 +_x$$

A. Second Synthetic Rubber Effort in Germany

It was decided to restart synthetic rubber research in Germany early in 1926, as a consequence of the rapidly rising price of natural rubber caused by the Stevenson plan. At that time the major chemical companies of that nation were newly united in a cartel or trust called the *I. G. Farbenindustrie AG.* This combine, which was often referred to as "I. G. Farben" or "I. G." for short, was a very resourceful organization even by modern standards, with a capability of attacking any research problem on a very broad front. Thus it was decided that the Leverkusen Works (Farbenfabriken Bayer) would carry out the polymerization studies, since they had considerable previous experience in that area, and that the other major units at Ludwigshafen (BASF) and at Hoechst would concentrate on monomer synthesis (118). Many other decisions had to be made before the work could be started. For instance, although hardly anything was known about structure-property relationships in polymers, it was concluded on the basis of past experience that dimethylbutadiene was not capable of yielding a good elastomer. The other two monomers under consideration, butadiene and isoprene, had yielded polymers of almost equivalent properties, but since the development of new synthesis methods had made it possible to prepare butadiene at a much lower cost and in a higher purity than isoprene, it was decided to concentrate on polymers of this diolefin.

The choice of suitable methods of polymerization also presented a problem. Although dimethylbutadiene had been chosen as the monomer for the synthetic rubber production during the war, considerable laboratory experimentation had been carried out also on the polymerization of isoprene and butadiene according to the methods used in the methyl rubber production. It had been found quite generally that these monomers required reaction periods of about the same length as dimethylbutadiene for maximum yields, but their conversion to elastomeric products was lower. This was particularly the case for butadiene which as an example, gave an elastomer yield of less than 50% when polymerized for 4 months, that is, for 2 months at 60°C and 2 more months at 80°C. With respect to polymer properties, considerably improved results had been obtained for isoprene and butadiene by polymerizing these monomers in emulsion, particularly with edestin from hemp seeds as the protective colloid. It had also been observed that the addition of 1% benzoylperoxide sometimes had a beneficial effect on butadiene polymerization (119). Nevertheless, it was apparent that even with these improvements the methods used for making methyl rubber during the war would not be satisfactory for economical production of rubber from butadiene.

The sodium-initiated polymerization, on the other hand, had been found to proceed much more rapidly and with considerably better yield, especially for butadiene, e.g., 93% polymer in 14 days at 30°C. Although relatively little work had been spent on this polymerization process in Germany during World War I, there were indications that it could produce a polybutadiene of reasonably good properties. It was decided, therefore, to try to develop sodium polymerization into a commercially useful process. The lasting effect of this decision was to furnish a generic name for all synthetic elastomers developed by I. G. Farben, namely *Buna*, from *Bu*tadien and *Na*trium, the New Latin name for sodium used in many countries including Germany.

Toward the end of the 1920's, sodium polybutadienes had been prepared in sufficient quantities for testing in passenger car tires. The results were fairly encouraging, since the new tires exhibited wear properties almost equivalent to those of natural rubber tires. However, other characteristics of the new polymers were not equally good, notably their processability. Great efforts were made to improve the sodium technique to yield a truly satisfactory general purpose rubber, but when these turned out to be unsuccessful the emphasis was channelled into other directions, particularly emulsion polymerization, a method that had been greatly improved during the intervening period.

Nevertheless, two commercial polybutadienes designated as Buna 32 and Buna 115 were eventually prepared in Germany by the sodium technique. Also, a third polymer called Buna 85 was prepared by initiation with potassium. The sodium polymers were produced in a batch operation, while the potassium polymer was manufactured in a continuous process, which was possible because of the higher rates obtained with potassium. The numbers used for identifying the polymers were closely related to their Staudinger "molecular weights" and were thus an indication of their degree of plasticity.

Buna 32 was so soft that it was used as a softener for other elastomers, while Buna 115 displayed borderline processability. Buna 85 exhibited particularly good properties in hard rubber applications, e.g., for lining equipment used in the chemical industry. The production of Buna 115 was terminated before the end of World War II, but the other two grades continued to be produced throughout the war.

The great improvements in the area of emulsion polymerization that were being made simultaneously with the development of the sodium processes began in 1927, when Luther, Heuck, Tschunkur and Bock (120,121) found that butadiene could be polymerized to true latices with the help of suitable *emulsifiers*, especially sodium salts of alkylnaphthalene sulfonic acids, known under the trade name *Nekal*. At about this time it was also established with certainty that peroxides such as potassium peroxybisulfate had a beneficial effect on the polymer formation (122) and such compounds were henceforth used in all emulsion polymerization studies.

In view of all the previous indications that peroxides could initiate the polymerization of olefins, dating back as far as the days of Greville Williams, it may seem strange that this effect was not clearly established until the late 1920's. However, it should be kept in mind that the important papers by Taylor and coworkers (123,124) on the free-radical initiated polymerization of ethylene began to appear at about this time and that those by Kharasch and coworkers (125,126) on the "peroxide effect" had not yet been published; when they were published in 1933 they marked only the beginning of our understanding of peroxide-initiated free radical reactions.

The polybutadienes obtained through peroxide-initiated emulsion polymerization in the presence of emulsifiers had rather disappointing properties and were almost impossible to process even on a hot mill. When after thousands of experiments it appeared that no significant improvements in the polymer properties could be realized from changes in the polymerization conditions, the idea occurred that improved properties might result from the copolymerization of butadiene with another monomer. Thus it was conceived in 1929 by Tschunkur and Bock (127) that styrene might be used as the second monomer, and the following year by Konrad and Tschunkur (128) that acrylonitrile might be used for the same purpose.

The simultaneous polymerization of two monomers was not a new idea in itself, since it had been employed as early as 1912 both by Kondakov (129) and by Gottlob (112) for the simultaneous polymerization of two or more dienes. However, the idea of using a comonomer which would not yield an elastomer when polymerized alone was novel.

Copolymers of the new type, which were named Buna S and Buna N, from styrene and nitrile, respectively, turned out to have interesting and important properties; Buna S was a general purpose rubber and Buna N was an oil-resistant rubber. When the drastic drop in the price of natural rubber resulting from the international economic crisis (cf. Table II) made it impossible to commercialize a general purpose rubber, emphasis was put on the development of Buna N since this elastomer, which later was named *Perbunan* and today is generally known as *nitrile rubber*, had valuable vulcanizate properties not obtainable with natural rubber (130) (cf. Chapter 4.B). By 1935 the development reached a stage where industrial production of Buna N could be started on a small scale. Since the National Socialist regime, which had come into power shortly before, also desired to make Germany self-sufficient with respect to rubber in general, more effort was put on the development of Buna S also. This resulted in the construction of a plant at Schkopau which produced 200 tons per month early in 1937 and eventually reached a capacity of 70,000 tons per year. Thus began the industrial production of the first truly important general purpose rubber, which is known today as *styrene-butadiene rubber*, or SBR for short (cf. Chapter 4.A). Much effort was then devoted to improving these elastomers and the processes of manufacture, and considerable progress was made until the end of World War II. This more recent history has been described in considerable detail elsewhere, notably by Bebb and Wakefield (131), and by Hofmann (118).

B. Russian Developments

Relatively little is known about the work on synthetic rubber in the Soviet Union subsequent to World War I. Such work was probably started on a small scale shortly after the war, but it did not reach significant proportions until 1926. Large scale production was not started until 1932 (132). The production was then rapidly expanded, reportedly reaching a level of 53,000 tons per year in 1938. However, all the Russian rubber, which is also known as *SK rubber* or *Sovpol*, was supposedly made by sodium polymerization of butadiene. As a consequence, nothing fundamentally novel was developed, even though many technical innovations were undoubtedly made, as described in some detail by Taft and Tiger (133).

C. Developments in the United States

Very little work was spent on the synthesis of general purpose rubber in the United States before the middle 1930's, when such work was started on a modest scale based on German disclosures. In spite of this, some very important and novel contributions to the field of synthetic elastomers were made by American workers between the two World Wars.

1. Thiokols

The first major contribution was made by Patrick and Mnookin, who reacted ethylene and propylene dichloride with sodium polysulfides and obtained a rubberlike material to which they gave the name *Thiokol*. Their discovery came as an outgrowth of an attempt to utilize refinery olefin streams for making glycols by chlorinating the olefins and then reacting the chlorination products with alkali. When this particular approach turned out to be unsuccessful, Patrick and Mnookin went on to treat the chlorinated olefins with sodium di- and polysulfides and thus made their discovery, supposedly in 1924 (116),* although 1920 (134) has also been given as the date of the discovery. No patent application was filed until 1927 (135,136) however, at which time Baer (137) had already filed a patent application for a similar invention in Switzerland. The reaction between ethylene dihalides and alkali sulfides had supposely been studied even earlier by others (116,134), but Baer and Patrick were apparently the first to appreciate the value of the rubbery material obtained in the reaction.

Commercial production of polysulfide elastomers began in 1929 in the United States, hence they may be considered the oldest synthetic rubberlike materials still in production, even though they have changed much in character during subsequent years and their use is restricted to a few specialized applications. The polysulfide elastomers are, of course, also of historical significance in that they furnished the first example of a rubberlike material made from a saturated monomer, i.e., through a polycondensation.

2. Neoprene

The second major American contribution to the synthetic elastomer field was made by Carothers and Collins (138,139) at the du Pont Company when they discovered polychloroprene, or *neoprene* as it is now generally known. This discovery came as an outgrowth of work on acetylene chemistry started by Nieuwland in 1906 which, some twenty years later, led to the development of a process of making vinylacetylene from acetylene (140). By adding hydrogen chloride to vinylacetylene, Carothers and Collins obtained 2-chloro-1,3-butadiene, chloroprene, $CH_2:CCl\cdot CH:CH_2$. This compound was found to undergo spontaneous polymerization with formation of a rubbery material.

Since neoprene was commercialized by du Pont in 1932 under the trade name Duprene, well ahead of Buna N, and since it is much more generally useful than the Thiokols in applications previously belonging to natural rubber, there is considerable justification for stating that it was the first commercially successful synthetic elastomer (cf. Chapter 4.C). However, because of its relatively high cost and its particular usefulness in certain applications, neoprene must still be considered a specialty rubber, albeit a very versatile one.

* See also Part II, Chapter 8.C of this book.

3. Butyl Rubber

The third major American contribution to the field of synthetic elastomers between the two World Wars came as an outgrowth of work on the polymerization of isobutene started by I. G. Farben in Germany, yet it was an entirely independent discovery. Although the tendency of isobutene to polymerize to low molecular weight products in the presence of strong acids had been observed by Butlerov and Goryainov in 1873 (141) no interesting polymerization products were obtained until Otto and Mueller-Cunradi at the Oppau Works of I. G. Farben discovered that the molecular weight of the polyisobutene increases with decreasing temperature of polymerization. By using a strong Lewis acid such as BF_3 as catalyst and by carrying out the polymerization at about $-75°C$, the German workers were able to prepare a high molecular weight elastomeric polyisobutene. However, they were not able to find a method of vulcanizing the new material and could hardly find any use for it.

As has been described in considerable detail by Howard (142), the German discoveries were disclosed to the management of the Standard Oil Company of New Jersey early in 1932 under terms of a research agreement entered into by the two companies in 1930. According to this agreement the two companies were to assist each other in developing new products and processes based upon raw materials from petroleum. For this purpose, a separate company, the Joint American Study Company, usually known as Jasco, was organized in September 1930.

The managements of the Standard Oil Company and its research subsidiary, the Standard Oil Development Company (now Esso Research and Engineering Co.) were particularly interested in the new polymeric material as a fuel additive (143). However, two of the Standard Oil chemists working on improved methods of making polyisobutene, R. M. Thomas and W. J. Sparks, became particularly intrigued by the interesting elastomeric properties of the high molecular weight polymer. They recognized that the material was useless as a rubber, so long as it could not be vulcanized, but that this deficiency might be overcome if some unsaturation could be introduced into the polymer chain, for instance by copolymerization with a diene. The problem facing the Standard Oil chemists was essentially the opposite of that encountered earlier by the workers at I. G. Farben who were trying to make a good elastomer by emulsion polymerization of butadiene. The Germans obtained a polymer containing much unsaturation or *functionality* for crosslinking but having poor elastomeric properties, while the Americans obtained a polymer of good elastomeric properties but completely deficient in functionality.

The idea of copolymerizing isobutene with a diene was not novel as such (144); however, nobody had succeeded in incorporating a diene into a high molecular weight elastomeric polyisobutene until Thomas and coworkers accomplished this by employing new polymerization methods. The first copolymer was made with butadiene and with ethyl chloride solvent for the $AlCl_3$ catalyst (145,146) but it was soon found that isoprene was a better comonomer and methyl chloride a better diluent for making a vulcanizable elastomer (147). The new elastomer was announced under the name of *butyl rubber* in 1940 (148), but commercial production was not started until late in 1942.

Although butyl rubber has a broad spectrum of application which overlaps that of natural rubber to a considerable extent and although it is superior to the natural product in many applications, it may also be considered a specialty elastomer (cf. Chapter 5.A).

The above resumé covers the discovery and development of essentially all synthetic elastomers that were of practical importance not only before and during World War II but also during the first decade after the war. Nothing was mentioned about the construction of the gigantic facilities for producing general purpose rubber in the United States during the war. This omission is justified by the fact that the SBR produced under the designation GR-S (Government Rubber-Styrene) was a direct descendant of Buna S. Unquestionably, many important improvements were made on the original German product, particularly through the use of various modifiers and other additives, just as many improvements were simultaneously being made in Germany, but the product remained principally the same. The design and construction of the synthetic rubber plants during the war was an enormous chemical engineering achievement, but this is outside the scope of this book; besides, this subject has been covered in considerable detail by Dunbrook (116), and by Livingston and Cox (149). Thus, as things stood only about ten years ago there would hardly have been anything else to add to this chapter.

However, in 1955, rubber chemists throughout the world were startled by the announcement that workers in various laboratories had been successful in preparing cis-1,4 polyisoprene or "synthetic natural rubber," thus accomplishing the original and ultimate goal of chemists attempting to prepare synthetic rubber. A variety of catalysts had been found useful for this purpose, but all of them had one thing in common: they contained an organometal which had either been added as such or had been formed *in situ*. While the new disclosures certainly caused much surprise, chemists familiar with the early history of synthetic rubber could not but recognize a similarity to some very early work dating back to 1910. Yet, only a few of these chemists were probably aware that the polymerizing activity of organometals, notably the alkali metal alkyls, had been the subject of considerable study during the intervening period. Such unawareness would be expected from the fact that much of the work between the two wars had been directed toward elucidating the nature of the elementary reactions involved in the alkali metal alkyl initiated polymerization of olefins. This work was largely carried out by Karl Ziegler and coworkers, and it is particularly interesting since it also led to extremely important results which must be treated to some extent in any chapter dealing with the history of synthetic elastomers.

IV. THE DEVELOPMENT OF ORGANOMETALLIC CATALYSTS AND THE FIRST PREPARATION OF "SYNTHETIC NATURAL RUBBER"

The independent discovery of the alkali metal initiated polymerization of conjugated dienes by Matthews (81) and Harries (75,77,78) in 1910, which was discussed in a previous section of this chapter, seemed to offer the first efficient method of selectively converting conjugated dienes into useful rubberlike polymers and created wide interest among rubber chemists. Considerable effort was spent during the next few years on improving both the polymerization process and the physical properties of the polymers obtained (113,150,151), but the results were largely disappointing. After ten years of little or no activity in this area, work on sodium-initiated polymerization of butadiene was taken up again in both Germany and the Soviet Union in the mid-1920's. As has already been described, this work led to some practical success, even though no truly satisfactory general purpose rubber was ever obtained.

By modern standards (or perhaps one should consider it a matter of hindsight), the industrial research on the sodium process appears to have been very empirical. There are hardly any indications that any serious efforts were made in the industrial laboratories to establish the manner in which sodium and other alkali metals caused conjugated dienes to polymerize; yet, some clues to the reaction mechanism involved had been furnished even before World War I.

The German patent of 1912 assigned by H. Labhardt to BASF (85), in which the alkali-metal-initiated polymerization of conjugated dienes is broadly disclosed (Fig. 3), was discussed previously in this chapter. It was also mentioned that the corresponding English patent suggested a relationship between the alkali metal and the alkali metal alkyl initiated polymerizations of dienes. However, the Labhardt patent apparently never received the attention it deserved. Therefore, many years passed before a definite statement about such a relationship appeared in the literature, even though some work published only a little over a year later made further suggestions in the same direction.

In a paper appearing early in 1914, Schlenk and coworkers (152) reported that sodium is capable of adding to certain carbon to carbon double bonds, and that aryl substituents on the carbon atoms of a double bond greatly enhance its reactivity. They prepared the sodium addition products of stilbene and tetraphenylethylene,

$$\underset{\underset{\text{Na}}{|}}{\overset{\overset{\text{H}}{|}}{C_6H_5-C}}-\underset{\underset{\text{Na}}{|}}{\overset{\overset{\text{H}}{|}}{C}}-C_6H_5 \quad \text{and} \quad (C_6H_5)_2CNa\cdot CNa(C_6H_5)_2$$

and proved their compositions by hydrolysis, which yielded the corresponding saturated hydrocarbons.

Addition of sodium to 1,1-diphenylethylene yielded a dimeric adduct,

$$\underset{\underset{\text{Na}}{|}}{(C_6H_5)_2C}\cdot CH_2CH_2\cdot \underset{\underset{\text{Na}}{|}}{C}(C_6H_5)_2$$

which Schlenk explained as being formed from an intermediate radical, or radical anion according to modern terminology, $(C_6H_5)_2CNa\cdot CH_2\cdot$. The formation of an intermediate radical (anion) was shown even more clearly for the sodium addition to anthracene in ethyl ether. A deep blue color appeared almost immediately upon addition of the sodium. However, this color very soon became impure and turned into violet after a while. From the absorption spectra, it was shown that the two colors were caused by different compounds, and since it was demonstrated that the final product was the expected disodium adduct, Schlenk concluded that the intermediate reaction product was a monosodium adduct of the composition

In the case of styrene, the phenomena initially observed were the same as with stilbene in that the liquid turned reddish-yellow after a short time indicating that a small amount of sodium adduct was being formed. However, this reaction did not continue. Instead, upon examination several weeks later, it was found that the styrene had polymerized into glassy metastyrene "apparently through the catalytic action of the sodium."

Several years later, in 1927, Ziegler and Bähr (153), who were familiar with Labhardt's patent and with the work of Schlenk et al., accidentally discovered that the C—K bond in 2-phenyl isopropylpotassium was capable of adding to the double bond of stilbene as shown in reaction 2. Subsequently, they demonstrated that the potassium com-

$$C_6H_5(CH_3)_2 \cdot C \cdot \cdot \cdot K^*) + C_6H_5 \cdot CH : CH \cdot C_6H_5 \rightarrow C_6H_5(CH_3)_2C \quad \overset{\displaystyle C_6H_5CH \cdot CH \cdot C_6H_5}{\underset{\displaystyle K}{\cdot \ \cdot}} \qquad (2)$$

pound would add also to the double bonds of styrene, 1,1-diphenylethylene, anthracene and 2,3-dimethylbutadiene and concluded that the reaction was general for compounds having double bonds which were either of the conjugated type or adjacent to a benzene ring. Since such addition resulted in the formation of a new metal to carbon bond which would again be able to add to a suitable double bond, they concluded, quite correctly, that this reaction was responsible for the polymerizing activity of both the alkali metal alkyls and the alkali metals, since the latter would first form the corresponding alkyls according to the reaction discovered by Schlenk.

Ziegler and Bähr also illustrated how a high molecular weight polybutadiene could be formed by repeated addition of the diene to the metal to carbon bond of the growing molecule† and how frequent or predominant 1,2 in place of 1,4 addition could explain the structural differences between, for instance, sodium-polymerized isoprene and natural rubber. They also pointed out that the polymerization would continue as long as free monomer was present in the system, provided the metal to carbon bonds were not destroyed by impurities such as oxygen or moisture. However, they also recognized that, in the absence of impurities, the

* The three dots were used by Ziegler to denote a heteropolar bond.
† In polymer chemistry, it is always convenient to consider the monomer as the added species even when the opposite is the case from a mechanistic point of view.

molecular weight of the polymer might be reduced by metal–hydrogen exchange (metallation) reactions of the type discovered by Schorigin (154).

In a somewhat later article, Ziegler and Kleiner (155) expanded on this subject and demonstrated that fairly high molecular weight polymers of butadiene could be formed by phenyl isopropylpotassium initiated polymerization of butadiene. They also discussed several of the phenomena involved in the polymerization and introduced the concepts of *initiation* (*Primärreaktion*), and *propagation* (*Folgereaktionen*). They pointed out that the rate of initiation depended upon the character of the metal–carbon bond in the original alkali metal compound, while the rate of propagation depended upon the type of monomer used, since the properties of the metal–carbon bond in the growing polymer molecule was determined only by the character of the monomer as soon as the first unit had been added.

The use of lithium alkyls for initiating the polymerization of butadiene was mentioned in this connection, but it was pointed out that these compounds were unsuitable because of their low reactivity, i.e., rate of initiation. Ziegler and Kleiner also discussed the structure of the polymer molecules obtained and pointed out that no simple method was available for a reasonably accurate structure determination, i.e., with respect to 1,2 and 1,4 addition.

At about the same time, Friedrich and Marvel (156), who were studying reactions involving lithium alkyls, accidentally discovered that ethylene could also be polymerized by lithium alkyls. The reactions did not proceed as expected; instead reactions of the following type took place:

$$LiC_2H_5 + (C_2H_5)_3 C_4H_9AsBr \rightarrow LiBr + (C_2H_5)_2C_4H_9As + C_2H_4 + C_2H_6$$

The observation that the gaseous products contained less than 5% unsaturated and more than 95% saturated hydrocarbons caused considerable concern until they found that the ethylene produced in the reactions was polymerized rather rapidly by the lithium alkyl, which had always been added in excess.

While Marvel and coworkers did not follow up on their discovery, Ziegler and his coworkers (157–161) devoted considerable time during the following ten years to the study of the mechanism involved in the alkali metal alkyl initiated polymerization of conjugated dienes. In these studies, Ziegler et al. began extensively using lithium alkyls, *n*-butyl lithium in particular, after they had discovered that these compounds were rather unreactive only in the initiation, and that they exhibited a fairly good propagation rate once the first monomer addition had taken place. Although the polymerization of isoprene was mentioned briefly, almost all the work involved butadiene and was primarily directed toward proving that the metal alkyls (as well as the alkali metals through their alkyls) were polymerizing the diene through stepwise and repeated addition of the monomer between the metal and the alpha-carbon of the alkyl.

Other work was devoted to the determination of the structure of the polymer formed. It was shown that butadiene is incorporated into the polymer both by 1,4 and 1,2 addition, with the first type dominating. It was also demonstrated that the polymerization temperature has a pronounced influence on the mode of addition, the 1,4 type increasing with increasing temperature. This caused Ziegler and coworkers (161) to suggest that a 100% 1,4 polymerization might be possible even though they themselves had certainly not yet reached that goal. They also mentioned that certain experiments had indicated the presence of *cis-trans* isomerism in the polymers and promised to discuss this matter in a later article. However, the Second World War apparently interfered with these plans. Their extensive and certainly very time consuming structure work did not yield much information by present-day standards, primarily because modern indispensable tools for determining molecular structure, such as IR and NMR instruments, were not available.

In the mid-1930's Scott and coworkers (162) expanded on the work started by Schlenk before World War I. They demonstrated that sodium and other alkali metals were capable of forming intensively colored 1:1 complexes with certain aromatic compounds which had previously been considered unreactive, such a naphthalene, methyl napthalene, biphenyl, acenaphthene and phenanthrene, if dimethyl ether or a fully alkylated glycol or polyhydric alcohol, dimethoxy ethane in particular, was used as the solvent. Subsequently, Scott (163) discovered that cyclic ethers such a dioxane were also suitable solvents for the complex-forming reaction and that the complexes were capable of initiating the polymerization of essentially the same monomers as had been polymerized previously with alkylmetals by Ziegler and coworkers.

After Ziegler's studies of the alkyllithium-initiated polymerization of butadiene had been terminated just before the start of World War II, little work was done in this field for almost 10 years, although two interesting patents appeared. The first of these, by Ellis (164), claimed the polymerization of ethylene to high molecular weight polymers with a catalysts consisting of lithium and nickel powder. The second one by Hanford, Roland, and Young (165) described the polymerization of ethylene to a hard waxy polymer with alkyl or aryl lithium compounds at high pressures. The authors clearly and correctly described the effect of monomer pressure and reaction temperature on polymerization rate and polymer molecular weight. They also described the chain terminating effect of oxygen and emphasized the importance of using ethylene containing very little oxygen. Because of the exceptional conditions in the world at the time when the patent issued (June 5, 1945), it may not have become generally known in Europe until at a much later date. Otherwise it would certainly have amounted to a clear and general disclosure of the ability of organolithium compounds to polymerize ethylene.

A few years after World War II, Ziegler resumed his studies of lithium alkyls. During an investigation of the thermal stability of these compounds, he and Gellert (166) rediscovered, again almost accidentally, that these compounds were able to polymerize ethylene. They made an observation which indicated that butene-1 formed in the decomposition of butyllithium added to the latter. This observation turned out to be wrong, but it did cause Ziegler and Gellert to start investigating the addition of ethylene to lithium alkyls and thus to rediscover this reaction which Ziegler gave the name *growth reaction* (*Wachstrumreaktion*).

In contrast to Friedrich and Marvel, Ziegler immediately recognized the importance of his findings and started extensive experimentation to find a low pressure method of making high molecular weight polyethylene plastic. Ironically, at about the same time, or even a little earlier, Goubeau and Rodewald (167) found that butene-1 and higher aliphatic hydrocarbons were formed during the thermal decomposition of diethyl beryllium. In view of present knowledge, these compounds had apparently been synthesized through a beryllium alkyl catalyzed growth reaction involving ethylene formed in the decomposition of diethyl beryllium. However, this (correct) observation did not cause these workers to study the addition of ethylene to beryllium alkyls.

The alkyllithium based polymerization of ethylene fell far short of Ziegler's expectations because of the appearance of true termination reactions. The most important of these seemed to be monomolecular elimination of olefin according to the following equation:

$$\text{LiEt} + n\text{C}_2\text{H}_4 \rightarrow \text{Li}(\text{C}_2\text{H}_4)_n\text{Et} \rightarrow \text{LiH} + \text{H}_2\text{C}{=}\text{CH}(\text{C}_2\text{H}_4)_{n-1}\text{Et}$$

Because of its very poor solubility in all diluents used, the lithium hydride precipitated out and was removed from further reaction. This made the process look uninteresting also for making higher, C_8-C_{20}, linear alphaolefins, which might otherwise have been valuable intermediates for making detergents, etc. It was noticed, however, that ether promoted the addition of ethylene to alkyllithium to a remarkable extent. This turned out to be an extremely important observation, since it suggested the use of the recently discovered, easily prepared and ether-soluble $LiAlH_4$ as the starting material for making lithium alkyls (168). It was found that this compound would first react with ethylene to form $LiAlEt_4$ and that a growth reaction would then take place. At first Ziegler thought that the growth took place only on the LiEt portion and that the $AlEt_3$ was just an inactive *"Statist"* ("Extra") whose role consisted primarily in preventing the insoluble lithium hydride from precipitating out and being withdrawn from further reaction. However, he soon got reason to suspect that the $AlEt_3$ played a major role also in the growth reaction. Experiments with pure $AlEt_3$ confirmed not only that this was indeed the case but also that the growth on the trialkylaluminum was actually more efficient than on the alkyllithium. The emphasis was therefore shifted to the potentially cheaper aluminum compounds for polymerizing ethylene.

During this work, which still had as its ultimate goal the preparation of a higher molecular weight product, i.e., true plastic grade polyethylene, it was noticed one day that the ethylene charged to an autoclave containing $AlEt_3$ had been converted almost quantitatively to 1-butene instead of undergoing the normal growth reaction. A thorough search for the cause of this unexpected result revealed that traces of metallic nickel present in the autoclave from a preceding hydrogenation experiment were responsible for the high rate of displacement or monomolecular elimination leading to the almost exclusive formation of 1-butene (169,170). This phenomenon was therefore named the *nickel effect*.

Another investigation was started in order to find out if other transition metals or transition metal compounds would also promote the formation of 1-butene. This revealed that cobalt and platinum gave the same effect as nickel (170). However, something quite different and entirely unexpected happened when compounds of metals of Group IV were tested together with aluminum alkyls.

Zirconium acetylacetonate was the first member of this group to be used. When the autoclave to which this compound had been added was opened, it was found to be completely filled with a solid white mass consisting of polyethylene (170). Further experiments soon revealed that several other compounds of transition metals belonging to Groups IV to

VI, when combined with aluminum alkyls, formed catalysts for polymerizing ethylene to a high molecular weight solid polymer. It was also established that the transition metal was present in the active catalyst not as the free metal but in some partially reduced form, i.e., as a compound in which its valence state was at least one unit below its normal maximum. Early experiments also revealed that compounds of titanium, especially $TiCl_4$, gave particularly active ethylene polymerization catalysts which were capable of initiating a rapid reaction even at atmospheric pressure. Thus the Ziegler-type catalysts were discovered. Naturally, they created a sensation when they were revealed to the scientific community; yet, observations had already been published which, to the alert and *informed* mind, might have suggested the existence of such catalysts.

In 1937, Hall and Nash (171) discovered that a mixture of alkylaluminum chlorides was formed when Al powder, $AlCl_3$ and ethylene were reacted at 50–100 atm and 100–200°C. They demonstrated that sodium chloride formed an insoluble double salt with ethylaluminum dichloride and utilized this complex formation for separating diethylaluminum chloride formed in the original reaction mixture. They also studied the polymerization of ethylene with alkylaluminum chlorides at elevated temperatures and found that diethylaluminum chloride tended to produce large amounts of linear α-olefins containing even numbers of carbon atoms. Since there was some indication that the diethylaluminum chloride contained traces of triethylaluminum, Hall and Nash may actually have carried out the first growth reaction with the latter compound, even though the major portion of their products was probably produced by other reactions.

Although it yielded much new and highly valuable information, including the first report on the existence of ethylaluminum chlorides, the investigation of Hall and Nash apparently received little attention for many years. However, it did not pass entirely unnoticed as evidenced by a patent issued in 1942 to Ruthruff (172) claiming an improved process of making alkylaluminum halides.

In 1943, while attempting to prepare lubricating oils from ethylene through polymerization with the help of Friedel-Crafts type catalysts, Max Fischer (173) of the BASF unit of I.G. Farben discovered that the course of the reaction could be altered drastically to yield substantial amounts of solid polyethylene, if some aluminum powder was added (as an HCl scavenger) to a catalyst consisting of $AlCl_3$ and $TiCl_4$. The discovery was considered startling enough to justify a patent application, but otherwise little seems to have been done at BASF to follow up on this discovery, which is perhaps not so surprising considering what was happening in Germany at that time.

The patent application, which was filed in December 1943, apparently, escaped the attention of the Allied investigating teams searching Germany for scientific and technical information after World War II and was thus saved from becoming public property. The patent was issued on April 20, 1953, after having been published for opposition (as *Auslegeschrift*) on June 19, 1952.

At the time the patent application was published, it was well known that aluminum powder alone as well as $AlCl_3$ and $TiCl_4$ either alone or in combination, would not catalyze the formation of solid polyethylene. From the work of Hall and Nash it was also known that aluminum powder, $AlCl_3$ and ethylene would react with formation of ethylaluminum chlorides under the conditions of temperature and pressure disclosed in the Fischer patent. Thus it could have occurred to people familiar with the works from both sources that a combination of at least $TiCl_4$ and aluminum powder was necessary for the production of solid polymer, and as is now well known, some solid polymer would have been obtained if such a combination had been tried.

On the other hand, it might have suggested to those who believed all three components of the Fischer catalyst were necessary for solid polymer formation that this catalyst depended on the intermediate (*in situ*) formation of ethylaluminum chloride for its activity. Today, it is of course well known that this is indeed the case (174,175).

It is rather interesting that, almost simultaneously with the Fischer patent, a patent assigned to du Pont (176) appeared which disclosed and claimed the use of a catalyst comprising a metal alkyl in combination with a transition metal (compound). Apparently this was at best a "Freudian Slip," since the catalyst was supposed to initiate the polymerization of ethylene through a free radical mechanism.

In 1952, shortly before the complex catalysts containing transition metals were discovered by Ziegler and coworkers, the Firestone Tire and Rubber Company took up the study of diene polymerization with lithium metal or alkyllithium, which Ziegler had more or less discontinued at the beginning of World War II. This time emphasis was put on the preparation of high molecular weight polymers, especially polyisoprene, rather than on mechanism studies. The effort was crowned with success within a year or two when high *cis*-1,4 polyisoprene was obtained (177) (see Part II, Chapter 6). This may have represented the first time a polymer closely resembling natural rubber had been prepared.

At this point it seems worth mentioning that another highly interesting but more complex alkali metal alkyl containing catalyst system, the so-called *alfin* catalysts, had been discovered by Morton and coworkers (178–180) about five years earlier. These catalysts were made from an alkyl chloride, sodium, a secondary alcohol and an olefin containing the structural unit —CH:CH·CH₂—, e.g., propylene, and derived their name from *al*cohol and ole*fin*. The alfin catalysts exhibit high activity but tend to give polymers of a very high molecular weight which are difficult to process. Since polymerization catalysts which were capable of producing polymers of much more attractive properties appeared shortly after the alfin catalysts, the latter never received the attention they might otherwise have got.

Very early in 1954, Giulio Natta, as a consultant of the Montecatini Company, was informed by Ziegler (181) about the discovery of the new transition metal containing catalysts capable of producing solid polyethylene. The information was disclosed under terms of an agreement between Ziegler and Montecatini. At that time Natta (182) had already become so interested in the ethylene addition to pure aluminum alkyls that he had begun studying the kinetics of this reaction. He was in an excellent position, therefore, to take up the study of the new catalysts disclosed by Ziegler, which he apparently did without delay. Within a few months it was discovered that crystalline polymers of α-olefins such as propylene, 1-butene and styrene could be obtained in good yield with the help of properly modified Ziegler catalysts (183–185). From x-ray diffraction measurements it was concluded that the crystallinity of these polymers was a consequence of their steric regularity, whereupon Natta (183) coined the words *isotactic, syndiotactic,* and *atactic* for describing the basic structures obtainable from linear head-to-tail polymerization of alpha-olefins.

Like the polyethylene prepared previously by Ziegler, the crystalline poly-α-olefins prepared by Natta and coworkers were of course plastics rather than elastomers; however, polymerization experiments involving monomers which had long been known to yield elastomers, e.g. butadiene, isoprene and piperylene, were started shortly after the stereoregular character of the crystalline poly-α-olefins had been recognized. Experiments on the preparation of elastomers through the copolymerization of two or more olefins were also started. This led very rapidly to the synthesis of a large number of new elastomers as described in Part II, Chapter 7.

The first preparation of a stereoregular polydiene with a Ziegler catalyst appears to have been made at the Goodrich-Gulf Chemicals Company, however.* Like Montecatini, this company was informed by Ziegler (181) about his catalyst discoveries shortly after they had been made and was, therefore, in a position to start work in this area before most chemical companies. By using a catalyst based on TiCl$_4$ and trialkylaluminum, i.e. the two components involved in Ziegler's most active polyethylene catalyst, Horne (186) succeeded in obtaining high cis-1,4 polyisoprene. Thus two different methods of preparing "synthetic natural rubber" were discovered almost simultaneously at two different laboratories in the United States, and the original and ultimate goal of the early chemists interested in synthetic rubber had finally been reached.

While this development would undoubtedly have caused nationalistic claims had it come at the time of *The Great Controversy* or even several years later, by 1955 much too much had been accomplished in the field of synthetic elastomers for any such claims to be made. It was also well recognized that this climactic development was only the final link in a whole chain of important developments to which chemists in many nations had contributed.

References†

1. A. de Herrera y Tordesilla, *Historia general de los hechos de los Castellanos en las islas y tierra firme del mar oceano 1492–1554*, Madrid, 1601–1615, Antwerp, 1728.
2. J. de Torquemada, *De la monarquia indiana*, Vol. II, Madrid, 1615, p. 663; 3rd ed., Nicholas Rodrigues Franco, Madrid, 1723, p. 621; facsimile ed., Salvador Chavez Hayhoe, Mexico, 1943.
3. P. M. d'Anghiera, *De orbe nuovo decades*, 1530.

* It is, of course, not the purpose of the writer to pass judgment on who was first to prepare each of the many new, often commercially valuable, polymers that appeared subsequent to Ziegler's discovery of the complex transition metal containing catalysts. Several new polymers were discovered independently and almost simultaneously in two or more laboratories. Consequently, many complicated and costly legal proceedings are currently in progress to establish who is entitled to patent coverage on various aspects of these polymers. If and when all patent questions have been resolved or settled, it may still be difficult to establish true scientific priority in connection with some of the new polymers. What is much more important from a historical point of view is to establish how the latest developments are related to previous developments, and this we can presently do with reasonable accuracy in the field of stereospecific polymerization.

† If other than the inventor(s), the owner of a patent is cited within parentheses, independent of whether the patent rights have been obtained through direct grant or through assignment.

The following abbreviations have been used for patent references: f. = date of filing; conv. = convention date; i. = date of issuance.

4. G. F. de Oviedo y Valdez, *Historia general y natural de las Indas*, Vol. 5, II, Seville, 1535, Madrid, 1851, p. 165.

5. C. M. de la Condamine, *Mem. Acad. Roy. Sci.*, **1751**, pp. 17, 319.

6. J. Priestley, *Familiar Introduction to the Theory and Practice of Perspective*, London, 1770, p. XV.

7. M. Speter, *Gummi-Ztg.*, **43**, 2270 (1929).

8. L. A. P. Hérissant and P. J. Macquer, *Mem. Acad. Roy. Sci.*, **1763**, p. 49.

9. F. Lüdersdorff, *J. Tech. Ökonom. Chem.*, **15**, 353 (1832).

10. C. N. Goodyear, *Gum Elastic and Its Varieties*, New Haven, Conn., 1855.

11. T. Hancock, *Personal Narrative of the Origin and Progress of the Caoutchouc or India Rubber Manufacture in England*, London, 1857.

12. H. Wolf and R. Wolf, *Rubber: A Story of Glory and Greed*, Covici, Friede, New York, 1936.

13. W. Jünger, *Kampf um Kautschuk*, Wilhelm Goldmann, Leipzig, 1940.

14. C. M. Wilson, *Trees and Test Tubes, The Story of Rubber*, H. Holt and Co., New York, 1943.

15. R. Ditmar, *Gummi-Ztg.*, **18**, 1013 (1904).

16. A. Dubosc and A. Luttringer, *Rubber: Its Production, Chemistry and Synthesis*, English ed. by E. W. Lewis, Charles Griffin & Co., Ltd., London, 1918, Chapter X.

17. F. C. Himly, *De Caoutchouc ejusque destillationis siccae productis et ex his de Caoutchino, novo corpore ex hydrogenio et carboneo composito*, Dissertation, Göttingen, 1835; *Ann.*, **27**, 40 (1838).

18. V. N. Ipatieff and N. Wittorf, *J. Prakt. Chem.*, **55**, 1 (1897).

19. M. Faraday, *Quart. J. Sci. Arts*, **21**, 19 (1826).

20 A. Bouchardat, *J. Pharm.*, **1837**, 454; *Ann.*, **27**, 20 (1838).

21. C. G. Williams, *Proc. Roy. Soc. (London)*, **10**, 516 (1860); *Phil. Trans.*, **1860**, 241; *J. Chem. Soc.*, **1862**, 111.

22. F. G. Bouchardat, *Bull. Soc. Chim.*, **24**, 108 (1875); *Compt. Rend.*, **80**, 1446 (1875).

23. F. G. Bouchardat, *Compt. Rend.*, **87**, 654 (1878); *ibid.*, **89**, 361 (1879).

24. F. G. Bouchardat, *Compt. Rend.*, **89**, 1117 (1879).

25. W. A. Tilden, *Chem. News*, **46**, 120 (1882).

26. W. A. Tilden, *J. Chem. Soc.*, **45**, 410 (1884).

27. I. Kondakov, *J. Russ. Soc.*, **21**, 39 (1889).

28. W. Euler, *Ber.*, **30**, 1989 (1897).

29. W. A. Tilden, *Chem. News*, **65**, 265 (1892).

30. O. Wallach, *Ann.*, **238**, 78 (1887).

31. O. Wallach, *Ann.*, **227**, 277 (1885).

32. J. H. Gladstone and W. Hibbert, *Trans. Chem. Soc.*, **53**, 679 (1888).

33. C. O. Weber, *J. Soc. Chem. Ind.*, **19**, 215 (1900); *Ber.*, **33**, 779 (1900).

34. F. Lynen and U. Henning, *Angew. Chem.*, **72**, 820 (1960).

35. C. D. Harries, *Untersuchungen über die natürlichen und künstlichen Kautschukarten*, Julius Springer, Berlin, 1919.

36. C. D. Harries, *Ber.*, **34**, 2991 (1901); *ibid.*, **35**, 3256, 4429 (1902); *ibid.*, **36**, 1937 (1903); *ibid.*, **38**, 87 (1905).

37. C. D. Harries, *Ber.*, **37**, 2708 (1904).

38. C. D. Harries, *Ber.*, **36**, 1933 (1903); *ibid.*, **37**, 839 (1904).

39. C. D. Harries, *Ber.*, **38**, 1195 (1905).

40. P. S. Bailey, *Chem. Rev.*, **58**, 925 (1958).

41. S. S. Pickles, *Trans. Chem. Soc.*, **97**, 1085 (1910).
42. F. W. Hinrichsen and E. Kindscher, *Ber.*, **42**, 4329 (1909).
43. P. Bary, *Compt. Rend.*, **154**, 1159 (1912).
44. J. H. Gladstone and W. Hibbert, *Phil. Mag.*, **28**, 38 (1914).
45. C. D. Harries, *Ann.*, **383**, 222 (1911).
46. C. D. Harries, *Ann.*, **406**, 173 (1914).
47. H. Staudinger, *Ber.*, **53**, 1073 (1920).
48. H. Staudinger, *Die hochmolekularen organischen Verbindungen Kautschuk und Cellulose*, Julius Springer, Berlin, 1932.
49. H. Staudinger and J. Fritschi, *Helv. Chim. Acta*, **5**, 785 (1922).
50. F. Hofmann, *Chem.-Ztg.*, **60**, 693 (1936).
51. W. A. Tilden, *India-Rubber J.*, **36**, 321 (1908).
52. A. W. von Hofmann, *Ber.*, **14**, 659 (1881).
53. A. Ladenburg, *Ber.*, **16**, 2057 (1883).
54. J. Thiele, *Ann.*, **319**, 226 (1901).
55. H. E. Armstrong and A. K. Miller, *J. Chem. Soc.*, **49**, 74 (1886).
56. C. Schotten, *Ber.*, **15**, 421 (1882).
57. F. Couturier, *Ann. Chim. Phys.*, **26**, (6), 485 (1892).
58. I. Kondakov, *J. Prakt. Chem.*, **62**, 166 (1900).
59. I. Kondakov, *J. Prakt. Chem.*, **64**, 109 (1901).
60. C. Harries, *Ber.*, **34**, 300 (1901).
61. A. Klages, *Ber.*, **35**, 2649 (1902).
62. A. Kronstein, *Ber.*, **35**, 4150 (1902).
63. E. Grandmougin, *Chem.-Ztg.*, **90**, 862 (1912).
64. Reference 16, pp. 208, 331.
65. I. Kondakov, *Uchenyia Zapisky Yurjevskogo Universiteta* (*Academic Records of the Tartu* (*Dorpat*) *University*), **1901**, No. 1, p. 7.
66. I. Kondakov, *Sinteticheski Kautchuk: ego gomologi i analogi*, Tartu (Dorpat), 1912.
67. C. B. Lebedev and C. R. Sergienko, *Zh. Org. Khim.*, **5**, 1839 (1935), see also: A. J. Favorsky, Ed., *C. B. Lebedev*, Onti Khimteoret, Leningrad, 1938, p. 147.
68. I. Kondakov, *J. Russ. Soc.*, **21**, 57 (1889).
69. Reference 16, Chapters XII–XIV.
70. H. L. Fisher, "Preparation and Production of Dienes by Other Methods," in *Synthetic Rubber*, G. S. Whitby, Ed., Wiley, New York, 1954.
71. Ger. Pat. 250,691, f. Sept. 9, 1909, i. Sept. 6, 1912; Fr. Pat. 419,316, Aug. 12, 1910; Brit. Pat. 17,734/10, June 8, 1911 (Farbenfabr. vorm. F. Bayer & Co.), Inventor: F. Hofmann: See also: F. Hofmann, *J. Soc. Chem. Ind.*, **30**, 226 (1911).
72. Ger. Pat. 235,423, f. Sept. 30, 1909, i. June 10, 1911 (Farbenfabr. vorm. F. Bayer & Co.), Inventors: F. Hofmann and C. Coutelle.
73. Ger. Pat. 250,335, f. Dec. 27, 1909, i. Sept. 6, 1912 (Farbenfabr. vorm. F. Bayer & Co.), Inventors: F. Hofmann and C. Coutelle.
74. C. Harries, *Gummi-Ztg.*, **24**, 850 (1910).
75. C. Harries, *Ann.*, **383**, 184 (1911).
76. C. B. Lebedev, *J. Russ. Phys. Chem. Soc.*, **42**, 949 (1910).
77. C. Harries, *Ann.*, **395**, 211 (1913).
78. C. Harries, U.S. Pat. 1,058,056, f. Nov. 4, 1911, i. April 8, 1913 (to Farbenfabr. vorm. F. Bayer & Co.).

79. W. H. Perkin, Jr., *J. Soc. Chem. Ind.*, **31**, 616 (1912).
80. A. J. Favorsky, *J. Russ. Soc.*, **19**, 558 (1887).
81. F. E. Matthews and E. H. Strange, Brit. Pat. 24,790, f. Oct. 25, 1910, i. Oct. 25, 1911; Ger. Pat. 249,868, July 30, 1912.
82. L. P. Kyrides (Kyriakides), *Chem. Eng. News*, **23**, 531 (1945).
83. A. Heinemann, Brit. Pat. 21,772, f. Oct. 2, 1907, i. Oct. 1, 1908.
84. F. Hofmann, *Gummi-Ztg.*, **26**, 1794 (1912).
85. Ger. Pat. 255,786, f. Jan. 27, 1912, i. Jan. 20, 1913, Br. Pat. 5667/12, Mar. 6, 1913 (Badische Anilin- & Soda-Fabrik), Inventor: H. Labhardt.
86. O. Aschan, *Ber.*, **51**, 1303 (1918).
87. G. Lilley, Brit. Pat. 29,277, f. Dec. 14, 1909, i. Mar. 14, 1911.
88. C. Harries, *Ber.*, **35**, 3256 (1902).
89. C. Harries, Lecture at Danzig, 1907; *Z. Angew. Chem.*, **20**, 1265 (1907).
90. F. Marchionna, *Butalastic Polymers*, Reinhold, New York, 1946, p. 16.
91. H. Barron, *Modern Synthetic Rubbers*, Chapman & Hall Ltd., London, 1949, p. 191.
92. H. L. Fisher, *Chemistry of Natural and Synthetic Rubbers*, Reinhold, New York, 1957, p. 4.
93. J. Murphy, U.S. Pat. 99,935, Feb. 15, 1870.
94. W. L. Semon, "History and Use of Materials Which Improve Aging," in *The Chemistry and Technology of Rubber*, C. C. Davis and J. T. Blake, Ed., Reinhold, New York, 1937
95. W. Ostwald and W. Ostwald, Ger. Pats. 221,310, f. Nov. 1, 1908, i. Apr. 29, 1910, and 243,346, f. Dec. 1, 1909, i. Feb. 7, 1912.
96. Ger. Pats. 265,221, f. Sept. 16, 1912, i. Oct. 3, 1913, and 266,619, f. Dec. 25, 1912, i. Oct. 27, 1913 (Farbenfabr. vorm. F. Bayer & Co.), Inventors: K. Delbrück and K. Gottlob.
97. Ger. Pats. 269,512, f. Feb. 26, 1913, i. Jan. 23, 1914, and 280,198, f. Jan. 1, 1914, i. Nov. 3, 1914 (Farbenfabr. vorm. F. Bayer & Co.), Inventors: K. Gottlob and M. Bögemann.
98. R. Ditmar, *Gummi-Ztg.*, **29**, 425 (1915).
99. K. Gottlob, *Gummi-Ztg.*, **30**, 303, 326 (1916).
100. A. H. King, *India-Rubber J.*, **52**, 440 (1916).
101. S. J. Peachey, *India-Rubber J.*, **52**, 603 (1916).
102. D. Spence, *J. Soc. Chem. Ind.*, **36**, 118 (1917).
103. W. L. Semon, *Chem. Eng. News*, **21**, 1613 (1943).
104. S. J. Peachey, *J. Soc. Chem. Ind.*, **36**, 321 (1917).
105. D. Spence, *Ind. Eng. Chem.*, **10**, 115 (1918).
106. G. D. Kratz, A. H. Flower, and C. Coolidge, *Ind. Eng. Chem.*, **12**, 317 (1920).
107. G. D. Kratz, A. H. Flower, and B. J. Shapiro, *Ind. Eng. Chem.*, **13**, 67 (1921).
108. C. W. Bedford and L. B. Sebrell, *Ind. Eng. Chem.*, **13**, 1034 (1921).
109. G. Oenslager, *Ind. Eng. Chem.*, **25**, 232 (1933).
110. A. Heinemann, Brit. Pat. 14,041, f. June 10, 1910, i. June 12, 1911.
111. Fr. Pat. 440,173, conv. Aug. 28, 1911, i. July 3, 1912 (Badische Anilin- & Soda-Fabrik), Inventors: A. Holt and G. Steimmig; U.S. Pat. 1,189,110, June 27, 1916.
112. Ger. Pats. 254,672, f. Jan. 26, 1912, i. Dec. 11, 1912, and 255,129, f. Mar. 13, 1912, i. Dec. 20, 1912 (Farbenfabr. vorm. F. Bayer & Co.) Inventor: K. Gottlob.

113. Ger. Pat. 287,787, f. Sept. 4, 1912, i. Oct. 5, 1915 (Badische Anilin- & Soda-Fabrik), Inventor: A. Holt.

114. F. Hofmann, *Mitt. Schles. Kohlenforschungsinst. Kaiser-Wilhelm Ges.*, **2**, 235 (1925).

115. K. Gottlob, *India-Rubber J.*, **58**, 305, 348, 391, 433 (1919); *Gummi-Ztg.*, **33**, 508, 534, 551, 576, 599 (1919).

116. R. F. Dunbrook, "Historical Review," in *Synthetic Rubber*, G. S. Whitby, Ed., Wiley, New York, 1954, p. 32.

117. H. Staudinger, *Ber.*, **57**, 1203 (1924).

118. W. Hofmann, *Beiträge zur hundertjährigen Firmengeschichte 1863–1963*, Farbenfabriken Bayer, Ludwigshafen, 1963, p. 208.

119. W. Zieser, *Synthetischer Kautschuk (Eine Zusammenstellung bis Ende 1926)*, Report from the Rubber Testing Laboratory, Elberfeld, 1927.

120. M. Luther and C. Heuck, U.S. Pat. 1,864,078, conv. Jan. 8, 1927, i. June 21, 1932; Ger. Pat. 558,890 (to I. G. Farben); U.S. Pat. 1,896,491, conv. Feb. 8, 1928, i. Feb. 7, 1933 (to I. G. Farben).

121. W. Bock and E. Tschunkur, U.S. Pat. 1,924,227, conv. Sept. 21, 1928, i. Aug. 29, 1933; Ger. Pat. 555,585 (to I. G. Farben); U.S. Pat. 1,898,522, conv. Jan. 15, 1930, i. Feb. 21, 1933 (to I. G. Farben).

122. Brit. Pat. 292,103, conv. June 13, 1927, i. Aug. 29, 1929 (I. G. Farben).

123. J. R. Bates and H. S. Taylor, *J. Am. Chem. Soc.*, **49**, 2438 (1927).

124. H. S. Taylor and W. H. Jones, *J. Am. Chem. Soc.*, **52**, 1111 (1930).

125. M. S. Kharash and F. R. Mayo, *J. Am. Chem. Soc.*, **55**, 2468 (1933).

126. M. S. Kharash and M. C. McNab, and F. R. Mayo, *J. Am. Chem. Soc.*, **55**, 2521, 2531 (1933).

127. Ger. Pat. 570,980, f. July 20, 1929, i. Feb. 27, 1933 (I. G. Farben), Inventors: E. Tschunkur and W. Bock; U.S. Pat. 1,938,731, Dec. 12, 1933.

128. Ger. Pat. 658,172, f. Apr. 20, 1931, i. Mar. 25, 1938 (I. G. Farben), Inventors: E. Konrad and E. Tschunkur; U.S. Pat. 1,973,000, Sept. 11, 1934.

129. I. Kondakov, *Rev. Gen. Chim.*, **15**, 408 (1912).

130. P. Stöcklin, *Proc. 1st Intern. Rubber Technol. Conf., London, 1938*, p. 434.

131. R. L. Bebb and L. B. Wakefield, "German Synthetic Rubber Developments," in *Synthetic Rubber*, G. S. Whitby, Ed., Wiley, New York, 1954, p. 937.

132. *India Rubber World*, **102**, 63 (1940).

133. W. K. Taft and G. J. Tiger, "Diene Polymers and Copolymers Other Than GR-S and the Specialty Rubbers," in *Synthetic Rubber*, G. S. Whitby, Ed., Wiley, New York, 1954, p. 682.

134. C. H. Fisher, G. S. Whitby, and E. M. Beavers, "Miscellaneous Synthetic Elastomers," in *Synthetic Rubber*, G. S. Whitby, Ed., Wiley, New York, 1954, p. 892.

135. J. C. Patrick, U. S. Pat. 1,890,191, conv. Dec. 13, 1927, i. Dec. 6, 1932.

136. J. C. Patrick and N. M. Mnookin, Brit. Pat. 302,270, conv. Dec. 13, 1927, i. Aug. 15, 1929.

137. J. Baer, Swiss Pat. 127,540, f. Oct. 20, 1926, i. Sept. 1, 1928; Brit. Pat. 279,406, June 28, 1928; Ger. Pat. 526,121, f. June 30, 1928, i. June 2, 1931.

138. W. H. Carothers and A. M. Collins, U.S. Pat. 1,950,431, f. Oct. 22, 1930, i. Mar. 13, 1934 (to DuPont).

139. W. H. Carothers, I. Williams, A. M. Collins, and J. E. Kirby, *J. Am. Chem. Soc.*, **53**, 4203 (1931).

140. J. A. Nieuwland, W. S. Calcott, F. B. Downing, and A. S. Carter, *J. Am. Chem. Soc.*, **53**, 4197 (1931).
141. A. Butlerov and V. Goryainov, *Ber.*, **6**, 561 (1873).
142. F. A. Howard, *Buna Rubber, The Birth of an Industry*, D. Van Nostrand, New York, 1947.
143. F. A. Howard, U.S. Pat. 2,049,062, f. Aug. 3, 1935, i. July 28, 1936 (to Standard Oil Development Co.).
144. H. Güterbock, *Polyisobutylen und Isobutylen-Mischpolymerisate*, Springer, Berlin, 1959, p. 122.
145. R. M. Thomas and W. J. Sparks, U.S. Pat. 2,356,127, f. Dec. 29, 1937, i. Aug. 22, 1944 (to Jasco).
146. R. M. Thomas and O. C. Slotterbeck, U.S. Pat. 2,243,658, f. Oct. 2, 1937, i. May 27, 1941 (to Standard Oil Development Co.).
147. W. J. Sparks and R. M. Thomas, U.S. Pat. 2,356,129, f. Aug. 30, 1941, i. Aug. 22, 1944 (to Jasco).
148. R. M. Thomas, I. E. Lightbown, W. J. Sparks, P. K. Frolich, and E. V. Murphree, *Ind. Eng. Chem.*, **32**, 1283 (1940).
149. J. W. Livingston and J. T. Cox, Jr., "The Manufacture of GR-S" in *Synthetic Rubber*, G. S. Whitby, Ed., Wiley, New York, 1954, p. 175.
150. Ger. Pat. 280,959, f. March 21, 1912, i. Dec. 1, 1914 (Farbenfabr. vorm. F. Bayer & Co.), Inventors: F. Hofmann, K. Delbrück, K. Meisenburg, and Wesenberg.
151. C. Coutelle, U.S. Patent 1,073,845, f. Apr. 10, 1913, i. Sept. 23, 1913 (to Farbenfabr. vorm. F. Bayer & Co.).
152. W. Schlenk, J. Appenrodt, A. Michael, and A. Thal, *Ber.*, **47**, 473 (1914).
153. K. Ziegler and K. Bähr, *Ber.*, **61**, 253 (1928).
154. P. Schorigin, *Ber.*, **41**, 2723 (1908).
155. K. Ziegler and H. Kleiner, *Ann.*, **473**, 57 (1929).
156. M. E. P. Friedrich and C. S. Marvel, *J. Am. Chem. Soc.*, **52**, 376 (1930).
157. K. Ziegler, F. Dersch and H. Wollthan, *Ann.*, **511**, 13 (1934).
158. K. Ziegler and L. Jakob, *Ann.*, **511**, 45 (1934).
159. K. Ziegler, L. Jakob, H. Wollthan and A. Wenz, *Ann.*, **511**, 64 (1934).
160. K. Ziegler, *Angew. Chem.*, **49**, 499 (1936).
161. K. Ziegler, H. Grimm and R. Willer, *Ann.*, **542**, 90 (1940).
162. N. D. Scott, J. F. Walker and V. L. Hansley, *J. Am. Chem. Soc.*, **58**, 2442 (1936).
163. N. D. Scott, U.S. Pat. 2,181,771, f. Dec. 10, 1935, i. Nov. 28, 1939 (to Dupont).
164. L. M. Ellis, U.S. Pat. 2,212,155, f. Nov. 1, 1938, i. Aug. 20, 1940 (to DuPont).
165. W. E. Hanford, J. R. Roland, and H. S. Young, U.S. Pat. 2,377,779, f. Feb. 18, 1942, i. June 5, 1945 (to DuPont).
166. K. Ziegler and H. G. Gellert, *Ann.*, **567**, 195 (1950).
167. J. Goubeau and B. Rodewald, *Z. Anorg. Chem.*, **258**, 166 (1949).
168. K. Ziegler, *Brennst.-Chem.*, **33**, 193 (1952).
169. K. Ziegler, *Brennst.-Chem.*, **35**, 321 (1954).
170. K. Ziegler, E. Holzkamp, H. Breil, and H. Martin, *Angew. Chem.*, **67**, 541 (1955).
171. F. C. Hall and A. W. Nash, *J. Inst. Petrol. Technol.*, **23**, 679 (1937); **24**, 471 (1938).
172. R. F. Ruthruff, U.S. Pat. 2,271,956, f. Sept. 27, 1939, i. Feb. 3, 1942.
173. Ger. Pat. 874,215, f. Dec. 18, 1943, published June 19, 1952, i. Mar. 12, 1953 (Badische Anilin- & Soda-Fabrik), Inventor: M. Fischer.

174. Brit. Pat. 779,540, conv. Nov. 27, 1954, i. July 24, 1957 (Badische Anilin-& Soda-Fabrik).

175. N. G. Gaylord and H. F. Mark, *Linear and Stereoregular Addition Polymers*, Interscience, New York, 1959, pp. 103 and 162.

176. Brit. Pat. 682,420, conv. June 10, 1949, i. Nov. 12, 1952 du Pont).

177. F. W. Stavely and Coworkers, *Ind. Eng. Chem.*, **48**, 778 1956).

178. A. A. Morton, E. E. Magat, and R. L. Letsinger, *J. Am. Chem. Soc.*, **69**, 950 (1947).

179. A. A. Morton, *Ind. Eng. Chem.*, **42**, 1488 (1950).

180. A. A. Morton, F. H. Bolton, F. W. Collins, and E. F. Cluff, *Ind. Eng. Chem.*, **44**, 2876 (1952).

181. K. Ziegler, *Intern. Symp. Macromol. Chem., Prague, 1957;* Main and Section Lectures, *Coll. Czech. Chem. Comm.*, **22**, 295 (1957).

182. G. Natta, P. Pino and M. Farina, *Proc. Intern. Symp. Macromol. Chem.*, Milan, 1954; *Ric. Sci., Suppl. A*, **1955**, pp. 120–133.

183. G. Natta, *Atti Acc. Naz. Lincei, Mem.*, **4**, 61 (1955); *J. Polymer Sci.*, **16**, 143 (1955).

184. G. Natta, P. Pino, P. Corradini, F. Danusso, E. Mantica, G. Mazzanti, and G. Moraglio, *J. Am. Chem. Soc.*, **77**, 1708 (1955).

185. G. Natta, *Angew. Chem.*, **68**, 393 (1956).

186. S. E. Horne, U.S. Pat. 3,144,743, f. Dec. 2, 1954, i. Dec. 17, 1963 (to Goodrich-Gulf Chemicals).

CHAPTER 3

THE STATUS AND FUTURE OF ELASTOMER TECHNOLOGY, 1966

GLEN ALLIGER AND F. C. WEISSERT

Central Research Laboratories, The Firestone Tire and Rubber Company, Akron, Ohio

Contents

I. INTRODUCTION

A. The Structure of Elastomers

Low shear and high compression modulus coupled with long range elasticity are the material properties which define a rubber. These features are observed in composites of very long chain molecules which are tied together at a few stable crosslinks, yet flexible enough so that the small chain segments are free to exhibit Brownian motion. Depending largely on the chemical makeup of these long chains, each macromolecule has a precise temperature at which, for a given rate of deformation, it is poised between glassy and rubbery behavior. This is the glass temperature (T_g) of the polymer. If the molecule is spatially regular so that it can order itself into a crystalline structure, then a crystalline melting point (T_m) is needed to further define the elastomer. The gross macromolecular features such as molecular weight distribution (MWD) and branching profoundly influence both the processing of the rubber compound and the quality of the vulcanizate. The chemical nature of the polymer backbone dictates the response of the elastomer to the effects of such environmental factors as oxygen, ozone, solvent, and temperature. These four measurable parameters go far toward characterizing the processing, flexing, and failure properties of a rubber, including those of the raw compound and the vulcanizate. The cost of the compound is, of course, an important consideration in the final selection of the most suitable elastomer for a given application.

The rubber industry is beginning to consider its materials and products in terms of the above T_g, T_m, MWD, and chemical parameters. To do so, it is equally necessary to specify—for each industrial application—the desired performance range of temperature, rate of deformation, and chemical environment. The rubber engineer still must depend to a large extent on knowledge of the state-of-the art, followed by the in-service testing of new plausible compounds. Nevertheless, current intense research on the relationships between molecular structure and mechanical behavior is paying off in improved performance in numerous specific areas of application, including, of course, the automotive tire.

The molecular structure of elastomers can now be controlled with an increasing degree of specificity. The scientific importance of this work was spotlighted by the awarding of the Nobel Prize in Chemistry for 1963 to Ziegler and Natta. This new-found ability of the polymer chemist to put together new polymers and copolymers in infinite variety must now be matched by laboratory evaluation broad enough to recognize the potential areas of utility for each new elastomer.

This review closely follows that of the present authors' *Annual Review of Elastomer Technology* (*1963–1966*) (1–4). Among the many fine reviews papers on elastomers, special mention is made here of a recent article by Cooper and Grace which provides an excellent summation of the development of the understanding of the nature of elastomers from 1920 to the present (5).

Along the way, sophisticated instrumentation has provided the necessary insight into both the micro- and macrostructure of elastomers. A partial list of these techniques includes infrared spectroscopy, nuclear magnetic resonance, differential thermal analysis, light scattering, osmometry, solution and bulk viscometry, molecular weight fractionation, birefringence, and x-ray analysis (6).

It is possible to relate the physical properties of the elastomer to its structure. It is now known that rubbery behavior is a direct consequence of the disorderly thermal motion of the flexible segments of the rubber molecule. The equilibrium energy of retraction of a stretched piece of rubber is quantitatively related to the entropy changes inherent in moving toward a most random configuration. However, at high rates of deformation, additional energy is required to overcome the viscous drag. Thus rubber is a viscoelastic material. As indicated previously, the T_g is the temperature above which the polymer begins to exhibit viscoelastic behavior. Borders and Juve in 1946 (7), showed that the tensile strengths of amorphous polymers were smoothly proportional to the difference between the test temperature and the T_g of the elastomer. This concept of comparing the properties of polymers at equal $T - T_g$, rather than at a given temperature, provides a powerful generalization of the mechanical behavior of elastomers (cf. Chapter 1).

It is now becoming increasingly clear that both branching and molecular weight distribution (MWD) vary widely in commercial elastomers. These parameters strongly affect the processibility of elastomers, in that they both increase the resistance to the "cold flow" at low shear rates and also decrease the resistance to high rate extrusion at high shear rates. Alkyllithium catalyzed polybutadiene, a highly linear and narrow

MWD elastomer, is Newtonian in that its bulk viscosity is relatively independent of shear rate. This elastomer can be modified to reduce the degree of Newtonian behavior for improved processibility and yet avoiding relatively low molecular weight polymer which causes high heat build-up in the vulcanizate. The processibility of Alfin polybutadienes has been greatly improved by modified polymerization to a lower average molecular weight (8). Gel permeation chromatography is an excellent technique for the measurement of MWD with a minimum of effort and time (9). Van der Hoff (10) has proposed that of two polymers with equivalent intrinsic viscosities, the one having the higher Mooney viscosity has the higher degree of branching. Kraus and co-workers (11) have shown that polybutadienes with either branched chains or broad MWD are non-Newtonian with respect to their bulk viscosities. A prime object of current research is the development of techniques to distinguish between the relatively parallel effects of MWD and chain branching in elastomers.

It is apparent that with each of the many homopolymers varying in T_g, T_m, MWD, branching, and chemical reactivity, a much wider spectrum of elastomers can be achieved by copolymerization. In addition it is now possible to achieve a great degree of specificity in comonomer sequence distribution from irregular to block formations as desired. Thus several new elastoplastic polymers are examples of unique behavior obtained by "tailor-making" an elastomer; flexible by virtue of long block sequences of a polymer with a low T_g, like polybutadiene ($-90°C$), and yet reinforced by suitable attendant blocks with high T_g, like polystyrene ($+100°C$). These elastoplastic materials exhibit rubbery behavior at room temperature without either filler or vulcanizing ingredients, yet they are thermoplastic and can be molded at elevated temperature.

B. The Tire and Other Uses for Elastomers

The automotive tire remains the most important rubber product. It accounts for over 60% of U.S. rubber consumption. It demands quality performance, but many of its requirements are so contradictory in nature—i.e., high strength vs low hysteresis—that most careful optimization of compounding ingredients is necessary. In addition, it is now clear that different tread compounds may be required for different service conditions such as low or high ambient temperatures, low or high rates of wear, or need for increased traction. The same care to fit the elastomer to the end use is required in all other applications, such as motor mounts and bushings, weather stripping, hose, adhesives, and flexible foams, to name just a few of the nontire rubber products. The accelerated development of new rubbers with special mechanical and chemical characteristics is resulting in new applications in areas where the product is borderline in function between that of a rubber and a plastic.

Today, 60% of the world demand for raw rubber is supplied by synthetic elastomers, only partly for economic reasons. The spectrum of available synthetic rubbers permits the adjustment of the basic T_g, T_m, MWD, and chemical parameters of the product either through selection of a single best elastomer, blending of several rubbers, or adjustment of monomer sequences in copolymers. The addition of carbon black, oil, and curatives to the rubber compound also affects the viscoelastic response of the composite and hence is of prime importance in the development of commerical rubber compounds. In fact, before there were so many varieties of elastomers, the compounders' art consisted almost solely of ways by which he could modify the few available rubbers by the addition of nonrubber ingredients.

In looking for the most common monomer units which make up the elastomer backbone, either in homopolymers or in copolymers, one finds that about 60% of all the weight of natural and synthetic rubber consumed today is derived from isoprene or butadiene. Polymers or copolymers based on high proportions of either isoprene or butadiene have suitably low glass temperatures and they exhibit rubbery behavior in flexural modes at temperatures from -40 to $150°C$ and at high rates of deformation. Copolymerization of isoprene or butadiene with various additional monomers serves to build elastomers specifically suited to a given application. This review article will emphasize the glass temperature, or more generally the modulus over a range of temperature and rate of deformation, as the single most important parameter to classify an elastomer with respect to its commercial utility.

Thus, to summarize: 60% of U.S. rubber consumption is for the manufacture of tires; 60% of the world usage consists of synthetic rubber; 60% of all of the world rubber is based on isoprene and butadiene.

II. ELASTOMERS

A. Polyisoprenes

Natural rubber and synthetic polyisoprenes prepared by either Ziegler or alkyllithium catalysts are high cis-1,4 polyisoprenes with T_g's near

$-72°C$ and T_m's near 30°C. Bruzzone (12) has compared synthetic polyisoprene of slightly varying cis-1,4 content with natural rubber. The higher cis-1,4 polymers held their properties over a wide range of cure levels, which he attributed to relative crystallization effects. At low states of cure, he found that better properties were obtained with the synthetic polyisoprenes containing the smallest amount of low molecular weight materials.

Synthetic polyisoprene made with a Ziegler catalyst is claimed to be an economic replacement for all but the lowest grades of natural rubber. Recent advances in the development of this synthetic polyisoprene centered around the minimization of both low molecular weight and crosslinked gel material. An easily processible synthetic polyisoprene with good physical properties was produced (13). Kerscher (14) predicts that the consumption of synthetic polyisoprenes will increase from 35,000 long tons in 1964 to 200,000 by 1975 in the United States, with increasing amounts employed in larger tires (cf. Chapter 7.A in Part II).

The alkyllithium catalyzed polyisoprenes are characterized by slightly lower cis-1,4 structure and especially by a much narrower molecular weight distribution than either natural rubber or the aluminum and titanium complex catalyzed polyisoprenes. Krol (15) has compared synthetic polybutadienes (PBD) and polyisoprenes (PI) in *truck* tread compounds. The PBD had the greater wear resistance at equal black levels. However, the wear resistance of the PI was improved by higher black loadings without excessive heat build-up. The synthetic polyisoprene prepared by alkyllithium catalysts is also finding commercial acceptance because of its good mold flow, color, low heat build-up, and flex resistance (cf. Chapter 6.A in Part II).

Research efforts in the development of natural rubber technology have been directed toward securing high yields per acre, uniform quality raw material and the modification of the natural rubber for special uses (16). ICR (initial concentration rubber) prepared by controlled coagulation of the latex without prior dilution so that the maximum amount of protective agents is retained, is more uniform with respect to accelerator and antioxidant activity than is normal smoked sheet (17). Prepeptized natural rubber, rubber stabilized against excessive "hardening" (attributed to small amounts of aldehyde groups), and an anticrystallizing natural rubber prepared by partial isomerization of the cis-1,4 structure are some of the new forms of natural rubber prepared for special applications.

The high oil levels used in the compounding of SBR are now being recommended for natural rubber. Mixing methods for improved black dispersion and proper black and curative levels for oil-extended natural rubber compounds have been presented by Moore (18). It is reported that oil-extended natural rubber gives better wear resistance than low oil natural rubber compounds under conditions of above average abrasion severity, and that natural rubber performs better than SBR in passenger tire treads at low speeds.

An important general feature of polyisoprenes as contrasted with polymers or copolymers of butadiene is that they break down more easily on milling and also tend toward chain scission rather than hardening during the oxidative aging of the vulcanizates.

A synthetic gutta-percha has been prepared by polymerizing isoprene to a very high *trans*-1,4 structure (19). This stereoregular polymer has a glass temperature of less than $-60°C$, but because of its high degree of crystallization in the unstretched state, it is solid at room temperature and softens at 65°C.

B. Polybutadienes

Polybutadienes are finding increasing use in applications requiring high abrasion resistance, especially under severe service conditions, tire crack growth resistance, and low temperature performance. These properties are directly related to the low glass temperatures of the 1,4 polybutadienes, with the *cis*-1,4 structure having a T_g near $-108°C$ and the *trans*-1,4 having a T_g near $-83°C$. The 1,2 polybutadiene, on the other hand, has a higher T_g near $-5°C$. The four polybutadiene types commercially available, and some of their distinguishing structural features are given

TABLE I
Polybutadienes

Catalyst	Ziegler	Alkyllithium	Free radical	Sodium alkenyl
Medium	Solution	Solution	Emulsion	Solution
Microstructure, %				
cis-1,4	90–95	36	10	10
trans-1,4	—	55	70	70
1,2	—	9	20	20
Approximate T_g	$-103°C$	$-93°C$	$-78°C$	—
Molecular weight distribution	Broad	Rel. narrow	V. broad	Broad
Branching	Some	Low	Medium	—

in Table I.   Because of their different microstructures, the T_g's of these polybutadienes vary from -78 to $-103°C$.

It is now generally recognized that opportunity for free chain segment rotation is reduced by either lowering the temperature or shorter relaxation times.   If very high rates of deformation arise in the tire abrasion process, then polymers with lower T_g values are also likely to exhibit higher abrasion resistance.   With each rate of deformation, there is likely to be an optimum T_g for maximum abrasion resistance as well as such other important performance values as skid resistance and heat build-up (20).   Thus Sarbach (21) has demonstrated that under conditions of low test severity (low speed or low cornering rates), polybutadiene tires are marginally poorer for wear resistance than SBR, while they wear very much better under conditions of high severity.   There has been, in the past, some concern over the chipping resistance of polybutadiene tread compounds.   Sarbach and Hallman (22), by suitable compounding, obtained excellent resistance to wear, cracking, and chipping in road tests of truck tires in Yugoslavia and the U.S.A. with tread compounds containing high levels of polybutadiene.   In general, most of the commercial use of polybutadienes has been in the form of blends with either SBR or natural rubber (23).

The skid resistance of polybutadiene tread compounds is measurably increased by the use of higher oil and black loadings, with aromatic oils being preferred to naphthenic oils (24).   Vohwinkel (25) has reviewed the present state of tire technology involving polybutadiene developments in the area of coefficient of friction.   cis-1,4 Polybutadiene has been reported (26) to resist changes in physical properties under high vulcanization or tire operation temperatures much better than natural rubber.   At the same time, tires with higher ratios of polybutadiene to natural rubber were characterized by lower operating temperatures.

Engel (27) has suggested that the highly linear 1,4 polybutadienes (with no side groups) can accept very large amounts of oils to give inexpensive but still highly serviceable compounds.   He has found it easier from a manufacturing point of view to produce high molecular weight polybutadienes in solution by a so-called "molecular weight jump reaction."   In this case polymerization is carried out in a normal manner, and then after completion, the molecular weight is "jumped" in a short time in a separate step by the addition of further catalysts.   In the Ziegler system, consisting of cobalt chloride and alkyl aluminum halide, a cocatalyst such as tertiary butylchloride is added to induce the "jumping reaction."   Butyllithium is not effective in this jumping reaction since it is not a Friedel-Crafts catalyst.   Hence for polymers prepared by butyllithium catalysis it is necessary to add both a Friedel-Crafts catalyst and such cocatalysts as ethylaluminum sesquichloride and water.   The alkyllithium catalyzed polybutadienes are particularly suited for extension with large amounts of black and oil because of the absence of low molecular weight fractions.

Polybutadiene made in an alkyllithium polymerization system has been found useful in sealants, surface coatings, textiles, thermoplastics, thermosets, chemical modifications to form new polymers, and cold curing liquid rubber compounds (28).   It has been suggested that these applications are possible because of high purity and good color, narrow molecular weight distribution, highly linear structure with the absence of gel, mixed microstructure (cis and trans) which precludes crystallization, low glass temperature, and the solubility and compatibility of these polymers (cf. Chapter 6.A in Part II).

Hansley and Greenberg (8) have reported that the wear and heat buildup performance for Alfin butadiene/styrene copolymers is intermediate between SBR and solution polybutadiene.   An important development in the Alfin system is the control of the high molecular weight fractions by the use of 1,4 dihydro derivatives of benzene and naphthalene so that these new Alfin polymers process readily on standard rubber machinery.

Refinements in the polymerization control of emulsion polybutadiene have resulted in more uniform rheological properties with improved processibility.   This new emulsion polybutadiene (EBR) has been found to be superior for wear to SBR under conditions of high severity at ambient temperatures.   Under conditions of low severity, EBR is superior to SBR at low temperatures, while the reverse is true at high temperatures.   Thus de Decker (29) has stated that the complex behavior of abrasion resistance is difficult to predict from basic physical data since the wear resistance is now known to be dependent on the exact tire test conditions.   Tire compounds prepared from blends of EBR with Hevea show less heat reversion than 100% Hevea compounds.

The present commercial polybutadienes are finding increasing use as general purpose rubbers because of their abrasion resistance and low temperature properties, their ability to tolerate large amounts of extender oils, and finally because their introduction into SBR and natural rubber applications require no major change in technology.   Blumel (30) has compared the polymer, compound, and vulcanizate properties of fourteen types of polybutadienes.

C. Copolymers Based on Butadiene

The glass temperatures of the polybutadienes range from about $-78°C$ for the emulsion polymer to about $-103°C$ for the high cis-1,4 polybutadiene.   In this review we are emphasizing the paramount importance of the T_g in characterizing a polymer.   In general it appears that in tire performance, the butadiene copolymer with the lowest T_g has better wear resistance but poorer wet traction.   Thus to meet the above and certain other requirements, it might be desirable to alter the T_g of the elastomer.   This can be done via copolymerization.   Thus emulsion polymerized 75/25 butadiene/styrene SBR with a glass temperature of about $-55°C$ is the number one elastomer with respect to usage in world markets.   In the last twenty years this rubber has been extensively developed to meet the increasing needs of the rubber industry for general purpose rubbers.   The quality of SBR has steadily improved; being aided in increasing degree by direct computer control of polymerization and other critical operations.   Furture expansion of SBR will likely be challenged by the increased use of newer synthetic rubbers such as solution polymerized polybutadienes and so-called solution SBR.   By polymerization in organic solution with organometallic catalysts, the chemist has the opportunity to control the amount of cis-1,4, $trans$-1,4, and 1,2 microstructure, the sequence distribution of butadiene and styrene in a copolymer, and the molecular weight distribution and branching. These accomplishments should result in new rubbers much superior to those which are currently available for each general area of application (cf. Chapter 4.A).

Two commercial copolymers of butadiene/styrene are now being prepared by alkyllithium solution polymerization.   Because of the difference in reactivity ratios between butadiene and styrene in this system, ordinarily the butadiene and styrene are not uniformly distributed along the polymer chain.   Special polymerization systems have been developed to eliminate the formation of long sequences of either monomer.

The microstructure of these two solution type SBRs differs in that the alkyllithium–ether modified catalyst (31) produces an elastomer with 27% 1,2 (vinyl) structure which is slightly higher than that of SBR, whereas the other copolymer contains 9% of the 1,2 structure.

Barlow (32) has reported that solution SBR is faster curing, higher in resilience, has less heat buildup, and greater abrasion resistance than emulsion SBR.   He suggested that the interaction of the black and rubber is more intense in the case of solution SBR.   Haws (33) has reviewed the role of both block and random butadiene/styrene polymers, the Phillips' Solprenes, in tire, plastic, sponge, and footwear applications. With the alkyllithium system it is relatively easy to prepare block copolymers by adding one monomer at a time to a "living polymer," or by depending on substantial differences in reactivity ratios of two monomers. For such applications as in shoe soles these butadiene/styrene block copolymers are stiffer at room temperature and hence require less filler reinforcement. Of course these block copolymers show two separate glass transitions, so that at room temperature, the polybutadiene blocks are in the rubbery state while the polystyrene blocks are in the plastic state.

Firestone's solution SBR (Duradene) with its somewhat lower 1,2 structure and styrene content than emulsion SBR, has a glass temperature intermediate between that of emulsion SBR and alkyllithium catalyzed polybutadiene. The degree of branching and broadness of molecular weight distribution of this polymer has been increased for improved processing characteristics. But this commercial alkyllithium elastomer still has a much narrower molecular weight distribution than emulsion SBR, and has less low molecular weight fractions. Weissert and Johnson (34) have suggested that improved wear resistance of this solution styrene/butadiene copolymer over that of emulsion SBR is related both to the difference in glass temperature and to the amount of low molecular weight polymer.

The Shell Chemical Company introduced Thermolastic, a butadiene/styrene block copolymer, which, without vulcanization or fillers, can be molded and processed like a thermoplastic and yet upon being cooled possesses the resilience, stiffness, and dimensional stability of a filled vulcanized rubber. Initially, Thermolastic is being used in adhesives and mechanical goods. The development of this polymer is yet another demonstration of the polymer scientists' ability to design polymers for special applications. It would appear that the basic elastomer design features are first, the construction of a long chain molecule which is free to undergo thermal segmental motion in the time interval of stress application and second, the provision of gross dimensional stability by judicious application of fillers, bulky groups, high T_g blocks, and crosslinking agents.

Carboxylated SBR latex, prepared to adhere to synthetic fibers can be crosslinked by heat alone without additional vulcanizing agents to give an elastomer with good physical properties (35).

Emulsion acrylonitrile/butadiene elastomers similar to SBR can be prepared with T_g's ranging from $-55°C$ for the 20% acrylonitrile elastomer to $-17°C$ for the 52% acrylonitrile copolymer. However, there is a significant difference in solubility parameters between the acrylonitrile and styrene copolymers with butadiene even though they may have similar T_g's at about the 20% butadiene level. The more polar (higher solubility parameter) characteristics of the acrylonitrile copolymers are disadvantageous if it is desired to prepare elastomer blends with general purpose hydrocarbon elastomers. However, this same property is desirable in that it renders these copolymers excellent for low swelling resistance to hydrocarbon oils and gasolines. These polar copolymers are more compatible in blends with other polar polymers such as neoprene, phenol formaldehyde resins, and polyvinylchloride (36) (cf. Chapter 4.B).

D. Butyl and Ethylene Propylene Elastomers

The glass temperature of butyl rubber is about −70°C, about the same as that of natural rubber, yet it is somewhat different in its viscoelastic response in that it has a broader range of temperatures and/or rates of vibration in which it possesses high damping characteristics. Butyl rubber has the unique property of high hysteresis combined with low dynamic modulus which, along with its good heat and age resistance, makes it highly desirable for antivibration and shock absorption applications. Booth (37) has discussed the effects of changing the level and type of both black and oil on the dynamic properties of butyl compounds over a broad range of frequencies and temperatures. The gas impermeability, flexibility, and weather resistance of butyl rubber makes it especially suitable in the form of sheets, extrusions, moldings, mastics, and paints in the building industry (38).

Zapp (39) has written an excellent review of the viscoelastic properties of butyl rubber relative to tire performance. The high skid resistant properties of butyl tires have been described by Umland (40). The good heat resistance of butyl inner tubes and tire curing bags has been further improved by the use of chlorobutyl compounds cured with zinc oxide (41). Chlorobutyl polymers are also more suitable than butyl for blending with the highly unsaturated elastomers because of their faster rate of cure. Resin cured butyl aircraft tires (42) have been found to be promising for use up to 450°F.

Butyl rubber is still one of the major general purpose elastomers. However as with the other two major general purpose rubbers, natural rubber and emulsion SBR, many of the newer synthetic rubbers reviewed here are often preferred for a specific application (cf. Chapter 5.A).

Ethylene propylene copolymers possess the properties of amorphous elastomers if the sequence distribution is sufficiently irregular to preclude the formation of crystalline blocks. Polyethylene probably has a glass temperature near −125°C, but readily crystallizes in a form which melts at 137°C so that experimentally one almost never observes an amorphous melt (an elastomer) of polyethylene below 137°C. Although it is easy to observe that the ethylene-propylene monomer (EPM) or ethylene propylene-diene monomer (EPDM) co- and terpolymers have nearly random sequence distributions, there is strong evidence for the presence of stereoblock and random block, as well as random distribution patterns (43). A minimum glass temperature of about −58°C is observed over a broad range of ethylene propylene compositions which suggest the presence of a degree of nonrandom structures (44). There is still a need for more precise techniques to characterize in detail the sequence distribution of copolymers. In a study of ethylene, propylene, and dicyclopentadiene (DCP) terpolymerization, it was found that the relative reactivities are influenced by the concentration of DCP in the liquid phase. The rate of vulcanization of ethylene-propylene-dicyclopentadiene terpolymer is slower than that of ethylene-propylene-dicylooctadiene-1,5 terpolymer (45).

The commercial EPM and EPDM elastomers have outstanding resistance to ozone, heat, and abrasion. Automotive engineers used almost 3 lb of EPDM in each 1965 car in the form of weather strips, wheel cylinder boots, and seals (46). EPDM sponge has been produced over a broad range of hardness values for application in automotive, footwear, and appliance parts which require ozone, heat, and low temperature resistance.

Crespi (47) related the potential uses of these essentially saturated elastomers to the excellent mechanical and dynamic properties of their vulcanizates plus the ability of these polymers to accept large amounts of oil and filler. Saturated hydrocarbons were preferred extending oils in EPDM compounds (48). Natta (49) reviewed the synthesis, structure evaluation, and uses of ethylene-propylene copolymers.

There is as yet no general agreement as to the future share of EPDM in the tire market (50). McCabe (51) projected a growth of EPDM in tires from less than 1% in 1965 to 33% in 1970. EPDM tires have been evaluated and are reported to have the ability to diffuse heat well at high speeds, to have improved sidewall cracking resistance, to show high resilience at low temperatures, and to possess tread wear and traction properties about equal to the current SBR/polybutadiene blend tread compounds (46). Current areas in the development of EPDM tires center around the solution of the problem of low building tack, the acceleration of the relatively slow cure rates, and the concern over possible in plant contamination of EPM and EPDM with the normally used unsaturated rubbers. Continued research in both the design of ethylene-propylene elastomers and the selection of compounding ingredients will likely lead to further improvements (cf. Chapter 7.B in Part II).

E. Polyurethanes

The molecular structure of polyurethanes can be readily altered so as to provide a variety of elastoplastics which span the range of rubbery to plastic behavior. In addition to the urethane groups (—CONH—), these polymers also contain hydrocarbon, ether, ester, aromatic, and amide groups in varying proportions. Thus studies of relations between polymer structure and properties is especially rewarding in polyurethanes, because of such a diversity of structural possibilities as reviewed by Saunders in Chapter 8.A in Part II and in Ref. 52. Thus the T_g's of toluene diisocyante (TDI)-poly (oxypropylene) glycol elastomers may vary from -24 to $-65°C$ depending upon the urethane and aromatic content. Saunders has also reviewed the effect of changing polyester type in urethane elastomers on torsion modulus–temperature curves which show T_g's varying from $-50°C$ to about $30°C$.

Testroet (53) has investigated the degradation of polyester foams at hydrolysis at $158°F$ and by attack of microorganisms at room temperature. Polyether foams were not adversely affected by the same conditions of high humidity. Flame-retarded polyurethane foams have been prepared from both reactive and additive type materials containing phosphorous or phosphorous-chlorine units (54).

The good tear, abrasion, and oil resistance of the individual polyurethanes account for their use in both flexible and rigid foams, elastomeric and elastoplastic products, propellants, mechanical goods, fibers, and sealants. Of course, there is no single polyurethane which can serve as the "general purpose polyurethane" for all of the above applications.

Whitman (55) has related polyurethane foam properties to polymer structure as reflected in T_g, tensile strength, modulus, and mechanical loss values. Harding (56) reported that the highest strength is obtained in rigid foams with the use of highly functional aromatic polyols of low molecular weight. Polyurethane foams can be produced and used with densities of 1.5–2.5 lb/ft^3 vs the 7–10 lb/ft^3 densities of the rubber latex and vinyl foams. The flexible foam is used in automotive parts, furniture, and foam backs as insulation for cold weather clothing. Insulation is the largest outlet for rigid foams, which may also be foamed-in-place.

Damusis (57) has described the swelling techniques, stress relaxation, and torsion modulus tests which have been used to arrive at a "mechanical spectral analysis" of the effects of formulations directed toward sealant applications. The T_g's varied from -50 to $-15°C$. The good wear, weathering, moisture, and chemical resistance of these polyurethanes were required for a sealant applications.

By the use of a flexible high molecular weight poly-ϵ-caprolactonediol, polyurethane elastomers were prepared with better low temperature flexibility than that of SBR 1500 and Neoprene GN and in the same range as natural rubber (58). High specific impulse solid propellant compositions, with safe and controllable burning rates, have been based on polyurethanes (59).

In addition to high tensile strength, good resistance to oxygen, ozone, and oil, good low temperature properties, and low impermeability to gases, solid polyurethanes are available with 2 to 5 times the load-bearing capacity of hydrocarbon rubbers (60,61). Urethane seals for ball joint suspension and linkages have extended car mileage between lubrications from 1,000 to 30,000.

Polyurethane elastic thread is stronger than rubber thread (up to 0.75 g/denier vs 0.25 g/denier) and is also more oil resistant. The effect of variations of polyurethane structure on the properties of these elastomeric spandex fibers has been described (62).

It is thus not surprising that, because of the urethane, hydrocarbon, ether, ester, aromatic, and amide segments which result in widely differing T_g's, crystallizability, and solubility characteristics, the polyurethanes are engineering materials which bridge the range from rubbery to plastic properties (cf. Chapter 8.A in Part II).

F. Oxygen-Containing Elastomers

As with the oxygen-containing urethane elastomers discussed above, other elastomers containing ether-O-chains are more flexible than chains containing O in ester linkages. Thus polyethers exhibit lower T_g's, while polyester linkages containing polymers generally have superior oil resistance.

1. Ethylene-Vinyl Acetate and Ethylene-Ethyl Acrylate Copolymers

This new family of tough polymers has the processing characteristics of thermoplastics but exhibits the resilience and flex resistance of an elastomer down to $-100°F$ but are not suitable for performance above $200°F$. The flexibility of these polymers is inherent in their structure and is not dependent on the addition of plasticizers (63,64). These polymers are expected to be used in applications served by plasticized poly(vinyl chloride).

2. Acrylic Elastomers

Improved temperature performance in the range of -40 to $300°F$ plus outstanding oil resistance characterize new acrylic-type elastomers. A total of four or more monomers, ethyl and butyl acrylates for low temperature and oil resistant properties, another monomer to supply cure sites, and still one or more additional monomers to improve low temperature performance are believed to have been required to achieve the desired balance of properties (65) (cf. Chapter 3.D).

Polyacrylate elastomers are attracting new interest, especially in response to the growing demand for heat resistant seals in transmission gears, because of the ability of the polyacrylates to resist sulfur-modified oils at temperatures over $350°F$ (57). The polyacrylates can serve as low cost replacements for fluorocarbons in many applications.

3. Polypropylene Oxide

Copolymers of propylene oxide with an unsaturated epoxide can be vulcanized by sulfur to give elastomers with good low temperature flexibility, ozone, and moderate oil resistance. They show good resistance to fatigue where high temperatures are developed and an outstanding constancy of properties over a range of temperatures, from the brittle point of $-80°F$ to $250°F$ (66,67) (cf. Chapter 5.C).

4. Polycarbonates

Perry (68) has described the development of new elastomeric polycarbonates where bulky three-dimensional norborane-type groups serve as "tie down" points having the same general effect as the classical crosslinks, hydrogen bonds, polymer crystals, or "hard fillers" often considered necessary to impart dimensional stability or "reinforce" an elastomer. A polycarbonate composition which compares favorably with commercial spandex fibers contains 65% poly(tetramethylene ether) glycol. It has an elongation of 500% and a tenacity of 0.6 g/denier. The T_g of these elastomeric polycarbonates is about $-70°C$, well within the range of T_g's of commercial rubbers, and is slightly higher than that of spandex type elastomers. Thus, any combination of high molecular weight polymers having a low T_g (below $-20°C$) and also held in shape by a relatively few "hard spots" has elastomeric properties.   The addition of polar groups such as oxygen to an elastomer will, as expected, reduce the degree of swelling in hydrocarbon oils.

G. Chlorine-Containing Elastomers

The addition of halogen to a hydrocarbon backbone will markedly affect the solubility properties of the rubber.   The replacement of a hydrogen by a chlorine atom also raises the T_g of the polymer far more than the smaller fluorine atom.   The substitution of halogen for hydrogen imparts flame resistance to the polymer.

1. Neoprene

The $-43°C$ T_g of the polychloroprenes is a direct consequence both of the presence of chlorine and the *trans*-1,4 microstructure.   Neoprene GP is a new polymer which performs all the functions of Neoprene GN, GNA, and GRT.   It combines the vulcanizate properties of the G neoprenes, the nonstaining characteristics of GN, and the raw polymer storage stability of GNA (69).   New neoprene polymers developed for adhesive applications include Neoprene AF cured at room temperature by metal oxide with rapid development of bond strength, Neoprene ILA, a copolymer of chloroprene and acrylonitrile having good oil resistance, and Neoprene HC which resembles polyisoprene in rate and degree of crystallization (70).   A study of the dependence of both raw and vulcanizate properties of polychloroprene upon temperature in the range of $-40°$ to $100°C$ detected two stiffening temperatures which correspond to the glassy (T_g) and crystalline (T_m) states.   Both of these effects are dependent upon the polymer structure and are a function of the deformation rate of testing (71) (cf. Chapter 4.C).

2. Chlorinated and Chlorosulfonated Polyethylenes

Oswald (72) has shown that chlorinated polyethylenes can be considered as terpolymers of ethylene, vinyl chloride, and 1,2 dichloroethylene with the expected correlations between structure, T_g, and mechanical behavior. Maynard (73) has also demonstrated how the properties of the elastomeric chlorosulfonated polyethylenes may be controlled by changes in the base polyethylene, chlorine distribution, and the choice of curing agent for the sulfonyl chloride crosslinking sites (cf. Chapter 10.A in Part II).

3. Chlorohydrin Elastomers

Two new chlorohydrin rubbers have been developed, CHR a polyepichlorohydrin and CHC a 1:1 copolymer of epichlorohydrin and ethylene oxide. The CHR has a brittle point of $-15°F$ and the CHC one of $-50°F$. These rubbers are candidates for applications requiring oil and flame resistance along with low temperature flexibility (74) (cf. Chapter 5.C).

H. Other Elastomers

1. Viton

A copolymer of hexafluoropropylene and vinylidene fluoride is suited to applications requiring exceptional resistance to heat, fluids, and compression set (75). Stress relaxation techniques have been used to clarify the process of thermal degradation in Viton vulcanizates. A new crosslinking system based on p-phenylenediamine has resulted in superior high temperature performance (76). Montermoso has reviewed the development from 1953 to 1961 of chemical resistant fluorine elastomers. During that time the low temperature serviceability was improved from $32°F$ to $-46°F$ with fluorinated nitroso and fluorinated silicone elastomers, while the upper limit for high temperature service was raised from $400°F$ to $600°F$ with fluoropolyamidine. Montermoso also pointed out that the fluorine polymers have an advantage in thermal resistance in that the bond energy of the C—F bond is about 17 kcal higher than that of the C—H bond (77) (cf. Chapter 4.E).

2. Nitroso Elastomers

Military and space requirements placed on rubber components include resistance to low temperature ($-70°F$), short impulses of high intensity heat, concentrated acids, propellants, and strong oxidizers. Nitroso rubbers, highly fluorinated structures containing repeated $[N(R)O(CF_2)_2]$ linkages where R is equal to perfluorinated alkyl or aryl groups, meet many of these requirements (62). The unvulcanized nitroso rubber gum is completely nonflammable, resistant to red fuming nitric acid, swells only 3% after 24 hr immersion in a hydrocarbon mixture, has a T_g of $-51°C$, and has negligible weight loss after 20 hr at 200°C (77).

3. Silicone Elastomers

Polydimethyl siloxane ($—O—Si(CH_3)_2—$) with a very low T_g of about $-123°C$ possesses excellent low temperature properties. Since the Si—O bond energy is 106 kcal/mole vs 85 kcal/mole for a C—C bond, silicone elastomers are also stable to very high temperatures. As with all of the other elastomers reviewed in this chapter, those noncrystallizable amorphous rubbers with a low T_g require filler reinforcement to obtain a high tensile strength at low rates of elongation.   Lewis (78) reviewed the range of physical properties that can be obtained by modification of chain and side groups, random copolymers, block copolymers which do have high tensile strength but suffer in flexibility, and by interaction with silica filler.   New silicone rubbers which require no postcuring have been prepared by the incorporation of a small amount of vinyl structure. Fibrous talc has been used to improve the high temperature tensile strength of silicone rubbers (79) (cf. Chapter 8.B in Part II).

4. Polysulfide Elastomers

Effective elastomeric polysulfide sealants for building and aircraft construction have been evaluated in terms of their chemical structure and crosslinking reactions (80).   Tobolsky (81) has related crosslink variation with viscoelastic properties of polyethylene tetrasulfide polymers.   Polysulfide elastomers have been modified by the incorporation of pendant unsaturated groups so that they may be cured with sulfur (82). These materials serve as solvent, heat, and weather resistant sealants (cf. Chapter 8.C in Part II).

5. Inorganic Polymers

With the limited ability of organic rubbers to stand extremely high temperatures, a widespread research effort is being directed toward the macromolecular chemistry of inorganic polymers prepared from P, N, Si, B, and Se containing monomers.   Eilar and Wagner (83) in reviewing inorganic polymers point out the general tendency of "inorganic monomeric" units to form low rather than high molecular weight polymer. Boron polymers may be formed through linkage of boron with oxygen, nitrogen, phosphorous, or another boron atom.   Phosphonitrile elastomers of 80,000 mol wt have been prepared with good physical properties and thermal stability.   However, they are readily susceptible to hydrolysis. (See Ref. 83.)

6. An Example of Molecular Architecture

Atoms with high polarity which provide oil resistance in most cases lead to reduced flexibility at low temperatures.   In searching for elastomers which are both low temperature and oil resistant, Ossefort and Veroeven have used diisocyanates to link up polyether glycols and diamines to form urethane-urea elastomers (84). Ossefort and Veroeven have listed eight key points in this development.

1. Higher proportions of ether oxygens, to provide greater polarity for high oil resistance, and yet a low T_g.

2. Macroglycol blends to interrupt chain regularity to prevent crystallization.

3. Pendant unsaturation to provide cure sites, yet to retain good ozone resistance in the polymer backbone.

4. Urea as well as urethane groups for greater chain interaction.

5, 6, 7. Proper selection of components for best balance of strength, oil resistance and low temperature flexibility.

8. Reinforcing fillers to yield best strength and elastic recovery.

This is an excellent example of molecular architecture where the chemical structure was well designed to fit specific requirements. In this case, for instance, it was desirable to avoid crystallization. However, in designing a rubber for a gumband, the crystallization upon stretching of natural rubber gum is the prime factor in reinforcement.

I. Elastomer Blends

It is often desirable to mix two elastomers as in the case of the solution polybutadienes with either SBR or natural rubber, to achieve a good balance of both good processibility and abrasion resistance. It is likely that there are no two high molecular weight elastomers which are completely compatible. However, if they have a similar solubility parameter, the blend may be sufficiently homogeneous to possess one rather than two T_g's and have the desired properties intermediate between those of the components. Corish presented a fundamental study of such rubber blends at the September, 1966 meeting of the Division of Rubber Chemistry of ACS (85).

J. Elastomers in High Impact Resins

The impact resistance of such rigid plastics as polystyrene is increased by a factor of 10 by the addition of about 8% of a low T_g elastomer. Angier (86) has described the preparation of these heterogeneous blends by which the elastomer is first dissolved in styrene. As the polymerization proceeds a phase inversion occurs where the low molecular weight polystyrene at first and then the elastomer-polystyrene graft polymer are present as the particulate phase. Better impact resistance is observed with increased particle size of the discrete elastomer. The amount of elastomer required in these blends is reduced if the elastomer is grafted to polystyrene. A general principle involved in this application is that for a broad range of usefulness, it is often desirable to use a composite with multiple T_g's and multiple relaxation mechanisms in order to induce "toughness" in a material.

III. COMPOUNDING

It can be seen in the previous section that the molecular architects are beginning to design elastomers which need neither curing agent nor filler to provide gross dimensional stability. However, it is necessary to keep the molecular weight low for processibility of most elastomers, and then to thermoset by vulcanization reactions after forming. In addition, pure amorphous elastomers with a low T_g and free of reinforcing crystalline or high T_g blocks also require the addition of reinforcing fillers.

A. Fillers and Extenders

The technological importance of adding fine fillers for the reinforcement of rubber products is well known, but there is no complete agreement as to the fundamental chemical and physical principles underlying reinforcement. It is known that small diameter (large surface area) particles reinforce rubber better than large particles and that the tendency of carbon black (or other pigment) particles to flocculate or to remain flocculated in the form of a "brushlike structure" affects both such processing characteristics as die swell and vulcanizate properties. Many of the experimental results appear to be consistent either with theories that postulate a strong chemical bond between rubber and filler or those that postulate weaker adhesion forces. Kraus has recently edited an excellent book, *Reinforcement of Elastomers*, which reviews current developments in the technology and understanding of this most important area of rubber technology (87).

Wake (88), in reviewing the nine papers on reinforcement presented at the DKG Munich conference, suggested that the problem of understanding rubber reinforcement is even more complex and the evidence more contradictory than was previously thought to be the case. Andrews (89) has postulated that the high rupture resistance of reinforced rubber is the result of the reduction of stress at the tip of the propagating crack. Brennan and Jermyn (90) have related the modulus of the rubber at low extension to the "structure" of the black, but have found the modulus at high extensions to be related to the surface activity of the black as indicated by the amount of bound rubber. Payne (91) used a model of the breaking and reforming of the "structure" of the particles to encompass part of the dynamic properties of carbon black-oil as well as those of carbon black–rubber mixtures. Voet (92) has connected both the persistent and transient particle to particle interaction with "structure" phenomena. Boonstra and Taylor (93) have noted that the restriction of swelling of carbon black vulcanizates can be correlated with the number of reactive sites on the carbon black surface. We are inclined to agree that rubber reinforcement is a complex phenomenon not well understood. Years ago, fine carbon black was shown empirically to be the best reinforcing pigment for rubber. Over the years, we have learned much by experiment regarding optimum particle size and structure. This knowledge has resulted in tremendous black reinforcement capabilities. So far, the theoretical speculations have contributed little to this progress.

As long as the particle size of the filler is sufficiently small, i.e., below 300 Å in diameter, fillers such as silica or calcium carbonate reinforce rubber like the more generally used carbon blacks. With some rubbers, as with the silicone elastomers, the silica pigments may be preferred. The activity of silica reinforcing fillers has been studied with respect to the reaction which may occur between the silica pigment and zinc oxide (94).

The stress-softening or Mullins effect of vulcanizates has previously been associated with the rubber to filler bond. Harwood, Mullins, and Payne (95) have recently shown that, when compared at the same degree of stress, the extent of softening is similar for both gum and filled vulcanizates.

The increasing production of elastomers in hydrocarbon solution has resulted in the development of solution masterbatching of filler and rubber. Excellent dispersions of filler in rubber have thus been obtained (96).

As stated previously it is often useful both for reasons of economy and processibility to use elastomer–oil blends. Oil extended polybutadiene may be used with normal levels of compounding ingredients, with only a small increase in the required amount of accelerator (97). Bruzzone has developed quantitative relationships between the composition of extender oils, the interaction parameter determined by swelling measurements and the physical properties of oil extended SBR compounds (98). The proper type and amount of oil to be added to rubber compound will depend upon the exact structure of the elastomer itself.

B. Vulcanization and Aging

The theoretical equibilibrium modulus of a rubber compound depends upon the number of crosslinking sites per cubic centimeter and is relatively independent of the nature of the crosslink itself. However, certain properties such as heat buildup, flex and age resistance of the vulcanizates do depend upon the type of bond which makes up the crosslink. The basic sulfur, accelerator, and antioxidant systems developed for natural rubber may generally be applied to the newly developed elastomers (99). Alliger and Sjothun (100) have edited a book, reviewing the history of the development and current state of the art of vulcanization.

The delayed action sulfenamide accelerated curing system has been reviewed by Morita (101). Studebaker (102) has shown that a larger proportion of monosulfide crosslinks result from the vulcanization of cis-1,4 polybutadiene than is the case with natural rubber. The reactivity of double bonds toward vulcanizing agents increases as we go from ethylene-propylene-dicyclopentadiene terpolymer, to isobutylene-isoprene copolymer, to ethylene-propylene-cyclooctadiene terpolymer (103).

The degree of crosslinking obtained through dicumyl peroxide varies greatly depending upon the chemical nature of the rubber. The overall crosslinking in terms of crosslinks formed per molecule of decomposed peroxide varies from 0.4/1 for EPDM, 1.0/1 for natural rubber, and 15.0/1 for polybutadiene. Loan (104) postulated that scission results from attack on tertiary hydrogen atoms and crosslinking from attack at secondary hydrogens. Van der Hoff (105) suggests that the polybutadienyl radical itself is capable of attacking the double bonds of polybutadiene, thus increasing the crosslinking efficiency of a peroxide radical.

Vulcanizates obtained with maleimide and its derivates are attractive relative to conventional sulfur cures in that short time–high temperature curing cycles are more feasible. The good aging characteristic of nonsulfur vulcanizates is observed, and the vulcanizates retain their ability to crystallize upon stretching (106).

In the case of high cis-1,4 polybutadiene or polyisoprene, the vulcanization reaction, either with dicumylperoxide or sulfur tends to isomerize the pure microstructure toward the equilibrium cis-trans mixture (107).

The nature and usefulness of antioxidants and antiozonants have been reviewed recently by Ambelang (108). Shelton (109) has elucidated the mechanism for antioxidant activity in *cis*-1,4 polybutadiene. Parks and Lorenz (110) have shown how the chemical nature of the network affects the rate of oxidation of the vulcanizates. Andrews and Braden (111) have proposed that the rubber surface exposed to ozone is protected by an ozonized layer 100 Å thick. Certain antiozonants protect only against crack initiation and others retard the rate of crack growth. Of course, the selection of a saturated as opposed to an unsaturated elastomer also results in vulcanizates which exhibit high resistance to oxygen or ozone degradation.

C. Rheology and Processing

As already noted in the discussion of polybutadienes (11), long chain branching or broadness in the molecular weight distribution of an elastomer results in non-Newtonian flow behavior. The incorporation of black and oil further increases the degree by which the apparent viscosity of a compound decreases with increasing shear rate. The rubber technologist should have at hand not merely a single Mooney viscosity value, but must also know the resistance to flow of polymers and compounds at all rates of flow from those of static storage to those of high speed tubing operations. Wolstenholme (112) has noted that at low rates of shear, solution SBR is softer than emulsion SBR, while at high rates of shear the relative apparent viscosities are reversed.

The processing characteristics of factory compounds are now being evaluated by observation of the degree of die swell and surface roughness of Garvey Die extrudates (113), extrusion data interpreted in terms of competitive degrees of breakdown and crosslinking (114), the time required for incorporation of black in a Braebender Plastograph (115), rheological results obtained in an oscillating disk rheometer, and capillary extrusion at varying shear rates (116).

Powder technology may put rubber mixing on a more competitive basis with plastics processing. All of the rubber ingredients are mixed in a Henschel mixer to form a powder with little work going into the mix (117). New coprecipitates of synthetic rubber and silicic acid are also usable in the form of fine dry powders (118).

Injection molding techniques are expected to be valuable in the processing of rubber mechanical goods. Gregory (119) however, suggests that there are still technical and economic problems which must be overcome in injection molding machinery design.

Elastomers form the basis of many adhesives. Kaelble (120) has analyzed the peel adhesion of polymers in terms of their bulk rheological behavior over a range of strain rates and temperatures. The use of resins in rubber has been reviewed by Powers (121).

IV. PHYSICAL TESTING AND PRODUCT PERFORMANCE

The quality of rubber vulcanizates is evaluated both by laboratory tests and product performance.   Juve has suggested that there is a tendency to run too many laboratory tests which may be either inadequate or not well understood (122).   Timm (123) has pointed out that the difficulty in finding more exact correlations between laboratory and service tests arises from the complicated macromolecular structure of polymers which permits different modes of molecular motion.   These relaxation processes are dependent upon both time and temperature. From the detailed mechanical spectral response of an elastomeric product the range of commercial utility of a given material can be evaluated. In addition the time and temperature performance characteristics of the rubber-cord composite that makes up the major portion of a tire is also required.

The single most important molecular parameter which governs the mechanical response of a polymer over a wide range of test speeds and temperatures is the T_g.   Boyer has comprehensively reviewed the chemical and other structural factors which affect both the glass and crystalline softening temperatures of high polymers (124).   Peticolas (125) has reviewed the molecular viscoelastic theory of high polymers.

Nielsen reviewed the use of the torsional pendulum, vibrating reed, and forced vibrators in the measurement of the dynamic properties of elastomers (126).   Similar information has been obtained by means of a rolling ball loss spectrometer (127).   A high speed tester for direct measurement of physical properties at very high strain rates has been developed (128).

Yin and Pariser have determined the time–temperature dependent mechanical properties of several elastomers to establish the most effective temperature–frequency region of vibration control for each elastomer (129).   Dunnom and de Decker (130) have observed that the optimum damping range of a single polymer is only about 90°F.   Therefore, to secure damping in the temperature range of −20 to 200°F, which is encountered in body mount applications, a blend of styrene-butadiene elastomers rather than a single polymer is required.   By varying the styrene content, these copolymers exhibit transition zones varying from −100 to 150°F.

Bassi (131) has found that among the various physical properties such as hardness, rebound, dynamic stiffness, dynamic loss, and damping factor, it is the latter that shows the best correlation with the coefficient of friction. In addition to the damping factor, surface phenomena such as adhesion between rubber and water also play an important part in wet traction. A high coefficient of friction rubber ("high $m\mu$"), an SBR with higher than normal styrene content, has been introduced in England for improved wet traction (132).

In relating the energy losses in a rolling tire to the physical properties of the basic rubber components, Collins (133) has shown that the bending of the tread (20% of the energy losses of the tire), carcass rubber (4%), sidewall (5%), and cord system (40%) are related to their respective loss moduli, while tread compression losses are proportional to the value of the loss modulus divided by the square of the complex modulus.

The failure characteristics of rubbers are also related to their basic viscoelastic properties. The stress and elongation at break of elastomers determined over a wide range of temperatures and test speeds have been encompassed by a failure envelope as described by Smith (134). Halpin and Bueche (135) proposed a failure theory to predict the time dependence of the tensile strength and ultimate elongation of both gum and filled vulcanizates. Failure was related to the propagation of tears and cracks within the viscoelastic body. Thomas has shown that the strength of an amorphous rubber is governed by the internal viscosity of the system (136).

The fatigue life of both SBR and natural rubber can be accounted for in terms of cut growth from small flaws initially present in the rubber. The cut growth of SBR is proportional to the fourth power of the tearing energy, whereas that of natural rubber is proportional to the second power (137). Lake and Lindley (138) have correlated the characteristics of cut growth at low tearing energies to the effects of polymer, vulcanizing system, oxygen, and filler on the fatigue limits of rubber vulcanizates.

Schallamach has divided tire wear into two patterns. Wear on sharp-pointed abrasives is dominated by tensile failure, while on more blunt surfaces abrasion losses are related to fatigue (139). Eckes has proposed an empirical equation which relates abrasion losses to the tensile strength at high speeds, dynamic properties, and coefficient of friction of the compound (140). Davison (141) has had some success in correlating laboratory abrasion results obtained under carefully controlled conditions of transmitted power, temperature, load, and velocity to wear results obtained on tires tested under road conditions encompassing a wide range of abrasive severities.

The radial ply tire has recently been introduced with promise of greater mileage, safety, and fuel economy; although with some sacrifice in ride qualities. Mulligan (142) has stated that the radial ply tire will probably not require any major change in the use of SBR, polybutadiene, and natural rubber although some of the polymer ratios may be altered. A tough rubber is needed in the sidewall where most of the flexing of the radial tire takes place.

It is difficult to cover comprehensively the range of nontire applications of rubber compounds. It is expected that with the increase in the diversity of elastomer and rubber compound designs and the development of the complete mechanical spectral analysis of the behavior of each class of compounds, the design engineer will be better equipped to select the best rubber for a given application. Davey and Payne (143) have noted that rubber is an excellent basic material for bridge bearings because of its ability to resist large compressive forces and yet allow for easy deformation under weak shear stresses. Gent (144) has studied the critical compressive characteristics of rubber springs subjected to a variety of loading conditions.

In the past two years, an upturn has occurred in the use of latex foam in furniture applications. Rogers (145) has reviewed the development of the Dunlop and Talalay methods for the production of latex foam.

A large percentage of elastomeric material is used in solid fuel rockets. Polysulfide, polyurethane, and modified polybutadiene rubbers serve as binders for the solid propellant, as secondary fuels, and they also provide the necessary mechanical and physical properties required to maintain the rocket's structural integrity. Inert components of the rocket such as insulators and liners also require rubber compositions able to withstand severe environmental conditions (146).

There is no doubt that the extensive body of viscoelastic data and theory that has been developed as part of the propellant program can be quickly applied toward the understanding of all the modes of rubbery behavior of interest to the rubber industry.

A final particular example of the direction of current research toward the more complete understanding of elastomer behavior is an excellent chapter, "Rupture of Amorphous Unfilled Polymers," by Landel and Fedors (147) reviewing the time, temperature, and crosslink dependence of the rupture process and the current molecular theories of uniaxial tensile strength.

Polymer chemists are developing the tools for the production of polymers of unique structures. The glass and melting temperatures, the macro-structure including molecular weight distribution and chain branching, solubility and chemical reactivity of the elastomer are distinguishing parameters which affect the end use. More extensive testing and evalua-tion of rubber over a wide range of test speeds and temperatures are providing data which can be used by the engineer to choose the appro-priate rubber, a low shear but high compression modulus material of construction.

References

1. G. Alliger, *Ind. Eng. Chem.*, **55**, (11), 52 (1963).
2. G. Alliger and F. C. Weissert, *Ind. Eng. Chem.*, **56**, (8), 28 (1964).
3. G. Alliger and F. C. Weissert, *Ind. Eng. Chem.*, **57**, (8), 61 (1965).
4. G. Alliger and F. C. Weissert, *Ind. Eng. Chem.*, **58**, (8), 36 (1966).
5. W. Cooper and N. S. Grace, *J. Polymer Sci.*, **C12**, 133 (1966).
6. B. Ke., *Polymer Rev.*, **6**, 722 (1964).
7. A. M. Borders and R. D. Juve, *Ind. Eng. Chem.*, **38**, 1066 (1946).
8. V. L. Hansley and G. Greenberg, *Rubber Chem. Technol.*, **38**, 103 (1965).
9. H. E. Adams, K. Farhat, and B. L. Johnson, *Ind. Eng. Chem. Prod. Res. Develop.*, **5**, 126 (1966).
10. B. M. E. van der Hoff, J. F. Henderson, and R. M. B. Small, *Rubber Plastics Age*, **46**, 821 (1965).
11. G. Kraus, R. P. Zelinski, C. F. Wofford, and J. T. Gruver, *Rubber Chem. Technol.*, **38**, 871, 881, 893, 903 (1965).
12. M. Bruzzone, G. Corradini, and F. Amato, *Rubber Plastics Age*, **46**, 278 (1965).
13. W. M. Saltman, F. S. Farson, and E. Schoenberg, *Rubber Plastics Age*, **46**, 302 (1965).
14. J. F. Kerscher, *Akron Rubber Group Tech. Symp.*, (1965–66).
15. L. H. Krol, *Rubber Plastics Age*, **45**, 1341 (1964).
16. S. T. Semegen, *Akron Rubber Group Tech. Symp.*, (1965–66).
17. M. Fleurot, *Rev. Gen. Caoutchouc*, **42**, 873 (1965).
18. C. G. Moore, K. E. Simpson, P. M. Swift, and M. A. Wheelans, *Rubber Age*, **97** (1), 61 (1965).
19. M. Abbott, *Rubber World*, **150** (1), 81 (1964).
20. G. Alliger and F. C. Weissert, *Intern. Rubber Conf., Paris* (1966).
21. D. V. Sarbach, *Rubber Age*, **89**, 283 (1961).
22. D. V. Sarbach and R. W. Hallman, *Rubber Plastics Age*, **46**, 1151, 1272 (1965).
23. F. C. Weissert and R. R. Cundiff, *Rubber Age*, **92**, 881 (1963).
24. J. J. Briggs, E. J. Hutchison, and R. C. Klingender, *Rubber World*, **150** (6), 41 (1964).
25. K. Vohwinkel, *Kautschuk Gummi*, **18**, 433 (1965).
26. J. F. Svetlik and E. F. Ross, *Rubber Age*, **96**, 570 (1965).
27. E. F. Engel, J. Schafer, and K. M. Kiepert, *Rubber Age*, **96**, 410 (1964).
28. P. Simmons, *Rubber Plastics Age*, **45**, 1347 (1964).
29. H. K. de Decker, C. A. McCall, and W. S. Bahary, *Rubber Plastics Age*, **46**, 286 (1965).

30. H. Blumel, *Rubber Chem. Technol.*, **37**, 408 (1964).
31. H. L. Hsieh, *Rubber Chem. Technol.*, **39**, 491 (1966).
32. F. W. Barlow, *Rubber J.*, **147** (9), 30 (1965).
33. J. R. Haws, *Rubber Plastics Age*, **46**, 1144 (1965).
34. F. C. Weissert and B. L. Johnson, *Preprint*, Div. of Rubber Chem., American Chemical Society, New York (1966).
35. F. Lepetti, *Rev. Gen. Caoutchouc*, **42**, 363 (1965).
36. W. Hofmann, *Rubber Chem. Technol.*, **37**, part 2, 1 (1964).
37. D. A. Booth, P. P. Brown, and L. Mayor, *Rubber Plastics Age*, **46**, 173 (1965).
38. P. Huot and P. Agius, *Rev. Gen. Caoutchouc*, **42**, 1276 (1965).
39. R. L. Zapp, *Rev. Gen. Caoutchouc*, **40**, 265 (1963).
40. C. W. Umland, *Mon. Summ. Automols. Eng. Lit.*, **9**, abs 13 (1963).
41. R. H. Dudley and A. J. Wallace, *Rubber World*, **152** (2), 66 (1965).
42. S. van der Burg and J. G. Manchette, *Rubber Age*, **93**, 594 (1963).
43. R. R. Garrett, *Rubber Plastics Age*, **46**, 915 (1965).
44. J. J. Mauer, *Rubber Chem. Technol.*, **38**, 979 (1965).
45. G. Sartori, A. Valvassori, and S. Faina, *Rubber Chem. Technol.*, **38**, 620 (1965).
46. R. F. McCabe, *Rubber Plastics Age*, **45**, 1492 (1964).
47. G. Crespi, G. Natta, A. Valvassori, and G. Sartori, *Rubber Plastics Age*, **45**, 1181 (1964).
48. R. M. White, *Rubber Age*, **94**, 897 (1964).
49. R. Natta, G. Crespi, A. Valvassori, and G. Sartori, *Rubber Chem. Technol.*, **36**, 1583 (1963).
50. M. E. Samuels, *Chem. Eng. Progr.*, **61** (4), 15 (1965).
51. R. F. McCabe, *Rubber Age*, **96**, 397 (1964).
52. J. H. Saunders, *Rubber Chem. Technol.*, **33**, 1259 (1960).
53. F. B. Testroet, *U.S. Gov't. Res. Rept.*, **39** (11), S-26 (1964).
54. J. J. Anderson, *Ind. Eng. Prod. Res. Develop.*, **2** (4), 260 (1963).
55. R. D. Whitman, J. A. Faucher, and F. P. Reding, *Rubber Plastics Age*, **44** (6), 683 (1963).
56. R. H. Harding and C. J. Hilado, *J. Appl. Polymer Sci.*, **8**, 2445 (1964).
57. A. Damusis, W. Ashe, and K. E. Frisch, *J. Appl. Polymer Sci.*, **9**, 2965 (1965).
58. R. J. Athey, *Rubber Age*, **96**, 705 (1965).
59. Thiokol Chem. Corp., Brit. Pat. 927,612 (May 29, 1963).
60. P. Kudriavetz, Jr., *Rubber Age*, **92**, 84 (1962).
61. P. Kurtz, *Chim. Ind. (Paris)*, **33** (3), 222 (1962).
62. C. B. Griffis and M. C. Henry, *Rubber Plastics Age*, **46**, 63 (1965).
63. R. L. Alexander, H. D. Anspon, F. E. Brown, B. N. Clampitt, and R. N. Hughes, *Polymer Eng. Sci.*, **6** (1), 5 (1966).
64. *Modern Plastics*, **43** (1), 84 (1965).
65. J. V. DelGatto, *Rubber World*, **152** (1), 95 (1965).
66. E. E. Gruber, D. A. Meyer, G. H. Swart, and K. V. Weinstock, *Rubber Age*, **94**, 921 (1964).
67. A. E. Gurgiolo, W. E. Prescott, and J. D. Hendrikson, *Rubber Age*, **93**, 101 (1963).
68. K. P. Perry, W. J. Jackson, Jr., and S. R. Caldwell, *J. Appl. Polymer Sci.*, **9**, 3451 (1965).
69. *Rubber Age*, **96**, 438 (1964).
70. D. J. Kelly, *Rubber World*, **150** (1), 82 (1964).

71. C. Houdret and M. Morin, *Rev. Gen. Caoutchouc*, **42**, 395 (1965).
72. H. J. Oswald and E. T. Kuber, *SPE Tech. Papers*, Paper 1 (1963).
73. J. T. Maynard and P. R. Johnson, *Rubber World*, **148** (1), 75 (1963).
74. E. J. Vanderberg, *Rubber Plastics Age*, **46**, 1139 (1965).
75. J. Lefrancois, *Rev. Gen. Caoutchouc*, **42**, 83 (1965).
76. D. K. Thomas, L. N. Phillips, A. S. Atkinson, and R. Sinnott, *Rubber Plastics Age*, **46**, 1020 (965).
77. J. C. Montermoso, *Rubber Chem. Technol.*, **34**, 1521 (1961).
78. F. M. Lewis, *Rubber Age*, **94**, 647 (1964).
79. K. C. Tsow, R. N. Bodley, and B. D. Halpern, *U.S. Govt. Res. Rept.*, **39** (2), 30 (1964).
80. D. A. George, P. Stone, L. A. Dunlap, and F. Roth, *Adhesive Age*, **6** (2), 32 (1963).
81. A. V. Tobolsky, *J. Colloid Sci.*, **18**, 353 (1963).
82. R. T. Woodhams, S. Adamek, and B. J. Wood, *Rubber Plastics Age*, **46**, 56 (1965).
83. K. R. Eilar and R. I. Wagner, *Chem. Eng. News*, **40** (33), 138 (1963).
84. Z. T. Ossefort and W. M. Veroeven, *Preprint*, Div. of Rubber Chem., American Chemical Society, New York (1966).
85. P. J. Corish, *Preprint*, Div. of Rubber Chem., American Chemical Society, New York (1966).
86. D. J. Angier and E. M. Fettes, *Rubber Chem. Technol.*, **38**, 1164 (1965).
87. G. Kraus, *Reinforcement of Elastomers*, Wiley, New York, 1965.
88. W. C. Wake, *Rubber Age*, **97** (6), 112 (1965).
89. E. H. Andrews, *Rubber Chem. Technol.*, **36**, 325 (1963).
90. J. J. Brennan and T. E. Jermyn, *J. Appl. Polymer Sci.*, **9**, 2749 (1965).
91. A. R. Payne, *Rubber Chem. Technol.*, **38**, 387 (1965).
92. A. Voet, *Rubber Chem. Technol.*, **38**, 677 (1965).
93. B. B. Boonstra and G. L. Taylor, *Rubber Chem. Technol.*, **38**, 943 (1965).
94. A. R. Payne, *Rev. Gen. Caoutchouc*, **41**, 507 (1964).
95. J. A. C. Harwood, L. Mullins, and A. R. Payne, *J. Appl. Polymer Sci.*, **9**, 3011 (1965).
96. K. A. Burgess, S. M. Hirschfield, and C. A. Stokes, *Rubber Age*, **93**, 588 (1963).
97. R. W. Hallman, *Rubber Age*, **95**, 791 (1964).
98. M. Bruzzone and G. Modini, *Rubber Chem. Technol.*, **37**, 451 (1964).
99. D. K. Roychaudhuri, *Rubber India*, **16** (5), 9 (1964).
100. G. Alliger and I. J. Sjothun, *Vulcanization of Elastomers*, Reinhold, New York, 1964.
101. E. Morita and E. J. Young, *Rubber Chem. Technol.*, **36**, 844 (1963).
102. M. L. Studebaker and L. G. Nabors, *Rubber Age*, **93**, 95 (1963).
103. G. Crespi and A. Arcozzi, *Rubber Chem. Technol.*, **38**, 590 (1965).
104. L. D. Loan, *Rubber Chem. Technol.*, **38**, 22 (1965).
105. B. M. E. van der Hoff, *Rubber Chem. Technol.*, **38**, 560 (1965).
106. P. O. Tawney, W. J. Wenisch, S. van der Burg, and D. I. Relyea, *Rubber Chem. Technol.*, **38**, 352 (1965).
107. D. Reichenbach, *Kautschuk Gummi*, **18**, 213 (1965).
108. J. C. Ambelang, R. H. Kline, O. M. Lorenz, C. R. Parks, and C. Wadelin, *Rubber Chem. Technol.*, **36**, 1497 (1965).
109. J. R. Shelton and D. N. Vincent, *J. Am. Chem. Soc.*, **85**, 2433 (1963).
110. C. R. Parks and O. Lorenz, *Rubber Age*, **93**, 95 (1963).

111. E. H. Andrews and M. Braden, *J. Appl. Polymer Sci.*, **7**, 1003 (1963).
112. W. E. Wolstenholme, *Rubber Chem. Technol.*, **38**, 769 (1965).
113. J. H. Macey, *Rubber Age*, **96**, 221 (1964).
114. W. E. Claxton and F. S. Conant, *Rubber Age*, **95**, 466 (1964).
115. A. Meder and M. May, *Rubber J.*, **146** (6), 39 (1964).
116. S. C. Einhorn and S. B. Turetzky, *J. Appl. Polymer Sci.*, **8**, 1257 (1964).
117. T. R. Goshorn and F. R. Wolfe, *Rubber Age*, **97** (8), 77 (1965).
118. F. Zeppernick, *Kautschuk Gummi*, **18**, 231 (1965).
119. C. H. Gregory, *Rubber J.*, **147** (4), 50 (1965).
120. D. H. Kaelble, *J. Colloid Sci.*, **19**, 413 (1964).
121. P. O. Powers, *Rubber Chem. Technol.*, **36**, 1542 (1963).
122. A. E. Juve, *Rubber Chem. Technol.*, **37**, XXIV (1964).
123. Th. Timm, *Kautschuk Gummi*, **18**, 801 (1965).
124. R. F. Boyer, *Rubber Chem. Technol.*, **36**, 1303 (1963).
125. W. L. Peticolas, *Rubber Chem. Technol.*, **36**, 1422 (1963).
126. L. E. Nielsen, *Mechanical Properties of Polymers*, Reinhold, New York, 1962.
127. I. C. Cheetham, *Rubber J.*, **146**, 77 (1964).
128. Plas-Tech Equipment Corp., *Rubber World*, **150** (4), 120 (1964).
129. T. P. Yin and R. Pariser, *J. Appl. Polymer Sci.*, **8**, 2427 (1964).
130. D. D. Dunnom and H. K. de Decker, *Rubber Age*, **97** (8), 85 (1965).
131. A. C. Bassi, *Rubber Chem. Technol.*, **38**, 112 (1965).
132. *Rubber Plastics Weekly*, **142**, 494 (1962).
133. J. M. Collins, W. L. Jackson, Jr., and W. L. Oubridge, *Rubber Chem. Technol.*, **38**, 400 (1965).
134. T. L. Smith, *J. Appl. Physics*, **35**, 27 (1964).
135. J. C. Halpin and F. Bueche, *Rubber Chem. Technol.*, **38**, 263, 278 (1965).
136. A. G. Thomas, *Rubber Plastics Age*, **45** (5), 548 (1964).
137. A. N. Gent, P. B. Lindley, and A. G. Thomas, *Rubber Chem. Technol.*, **38**, 292 (1965).
138. G. J. Lake and P. B. Lindley, *J. Appl. Polymer Sci.*, **9**, 1233 (1965).
139. A. Schallamach, *Rubber J.*, **145** (5), 34 (1964).
140. R. Ecker, *Rubber Chem. Technol.*, **39**, 823 (1966).
141. S. Davison, M. A. Deisz, D. J. Meier, R. J. Reynolds, and R. D. Cook, *Rubber Chem. Technol.*, **38**, 457 (1965).
142. W. Mulligan, *Rubber World*, **153** (2), 59 (1965).
143. A. B. Davey and A. R. Payne, *Rubber J.*, **147** (2), 24 (1965).
144. A. N. Gent, *Rubber Chem. Technol.*, **38**, 415 (1965).
145. T. H. Rogers, *Rev. Gen. Caoutchouc*, **42**, 1265 (1965).
146. D. C. Sayles, *Rubber World*, **153** (2), 89 (1965).
147. R. F. Landel and R. F. Fedors, in *Fracture Processes in Polymeric Solids*, B. Rosen, Ed., Wiley, New York, 1964.

ELASTOMERS BY
RADICAL AND REDOX MECHANISMS

A. Butadiene-Styrene Rubbers (SBR) and Rubbers from Substituted Butadienes and Styrenes

CARL A. URANECK

Phillips Petroleum Company, Bartlesville, Oklahoma

Contents

I. INTRODUCTION

The world production of synthetic rubber in 1941 was approximately 80,000 tons; three years later it was 900,000 tons. The increase was due chiefly to the creation of a synthetic rubber industry by the U.S. Government, which produced 760,000 tons in 1944 on a war emergency basis. After the war a downtrend in synthetic SBR (styrene-butadiene rubber) consumption started as natural rubber became available. The trend, however, was reversed in 1949 with the advent of "cold" SBR which together with furnace blacks yielded a tire rubber that was superior to natural rubber (1).

In mid-1950 the stereoregular diene rubbers were announced (2,3). The 75-year search for a synthetic natural rubber was essentially completed, and the high abrasion resistance of the newly discovered cis-polybutadiene portended further stiff competition for SBR. Although these stereorubbers slowed the expansion of the production facilities for the established rubbers, total consumption of SBR was not greatly changed over the ensuing years. The total U.S. consumption of SBR in 1965 was 1.08 million long tons, and the predicted value for 1975 is 1.13 million long tons (4). The latter amount indicates that SBR will be an important factor in the rubber industry for at least another decade. Research and technology should contribute to improved processes and products, and, as in the case of many other polymers in similar competitive situations, the improvements could result in enhanced rather than limited growth. Since the improvements will probably stem from fundamental studies of the emulsion polymerization process, this treatise will emphasize the fundamental aspects of the SBR process.

II. PREPARATION OF MONOMERS

A. Butadiene

1. Processes Based on Acetylene or Ethanol

The large scale development of a synthetic rubber capability naturally requires a corresponding production of the monomers. Normally economic factors largely dictate the manufacturing process but during a war the existing technology, availability of raw materials, and exigencies are controlling factors. In Germany the plan to become economically self-sufficient between the wars included the development of a synthetic rubber industry. There, lime and coal served as the raw materials for the production of acetylene, and two processes, the four-step process involving acetaldehyde as an intermediate and the Reppe-process, were the chief means of obtaining the necessary butadiene (5a). After the war, these processes could not compete with those based on petroleum.

In Russia the Lebedev process utilizing alcohol was the main source of butadiene, and in the U.S. over one-half of the diene in the first three years of the emergency was produced by a modified Ostromislensky process starting from ethanol (5b). Like the acetylene based processes, those based on ethanol could not compete with the manufacture of butadiene from petroleum.

2. Processes Involving Dehydrogenation of Petroleum Fractions

a. **Thermal Cracking of Hydrocarbons.** Thermal cracking of petroleum fractions during the War was of importance as a part of the Synthetic Rubber Program. This provided a rapid means of obtaining butadiene in existing refinery and gas producing equipment. This approach saved steel and provided butadiene before the more complex large plants specifically designed for butadiene could be put in production. Although these cracking plants served their purpose of producing butadiene quickly, as well as valuable by-products, they later lost out to more efficient plants.

However, in recent times, developments in the petrochemical industry have significantly altered the importance of thermal cracking as a source of butadiene. The great demand for ethylene has resulted in new production facilities for cracking of naphthas. By careful temperature control the yield of ethylene from naphtha can be increased to about 30%. A by-product of this cracking process is a large amount of butadiene. An estimate shows that more butadiene will be available in Germany by 1970 as a by-product of naphtha cracking than will be produced by the existing and planned butadiene plants (6).

b. **Dehydrogenation of Butanes or Butenes.** The most important process for production of butadiene is based on the dehydrogenation of butenes. A process was originally developed by Standard Oil Development Company (7). The original "1707" catalyst used in this process was altered considerably and one modification introduced by Shell in form of a high iron oxide-chromium oxide composition containing potassium

TABLE I

Operating Conditions and Results of Various Butadiene Processes

| | | | | Phillips | |
| | Jersey- | | | | |
Process	Shell	Dow	Houdry	Step 1	Step 2
Feed	Butene	Butene	Butane	Butane	Butene
Catalyst	Fe_2O_3-	$CaNiPO_4$	Cr_2O_3-	Cr_2O_3-	Fe_2O_3-
	K_2CO_3-		Al_2O_3	Na_2O-	K_2CO_3-
	Cr_2O_3			Al_2O_3	Cr_2O_3
Space velocity, $V/V/hr$	200–500	100–200	225–450	500–1000	250–450
Temp., F	1100–1250	900–1200	1100–1250	1000–1100	1075–1150
Steam/hydro	8–18	20	—[a]	0	10
Conversion, %	20–30	40–50	25	30–40	25–30
Selectivity, %	70–80	90	80	85–90	75
Regeneration of catalyst	1/day	100 min	10 min	60 min	None

[a] Reduced pressure obtained by vacuum.

oxide came into extensive use (8). Dow also developed an improved catalyst for dehydrogenation of butenes (9). The shortage of butenes during the war warranted Phillips Petroleum Company's development of a two-step process starting with butane (10), and caused Houdry (11) to originate a one-step process based on a feed of butane. Some of the operating conditions and results of the processes are given in Table I.

The dehydrogenation of butanes or butenes is an endothermic equilibrium reaction. The formation of butadiene is favored by heat, low pressure, and a rapid quench of the reaction mixture. The operating conditions in the tabulation represent optimization for each process which also takes into account factors such as catalyst life, by-product formation, and others.

c. **Oxidative Dehydrogenation of Butanes and Butenes.** According to thermodynamic equilibrium theory, the endothermic dehydrogenation process is favored by reduced pressure and high temperature. In the oxidative dehydrogenation process an oxidant is used which reacts with hydrogen by one of several mechanisms to drive the reaction to completion. The oxidation is an exothermic reaction furnishing heat which can lower the overall heat requirement. Furthermore, conversion and selectivity are improved so that claims to yields per pass as high as 80% have been made; whereas the yields for conventional dehydrogenation are in the 20–45% range.

Two types of oxidative dehydrogenation of hydrocarbons are revealed in the literature: those involving oxygen and a catalyst (12) and those involving a halogen or a halogen compound as well (13). In the latter type of reaction, the halogen abstracts the hydrogen to form a double bond and hydrogen halide which is oxidized to regenerate the halogen and water. The regeneration may be accomplished concurrently with the dehydrogenation, in which case oxygen is added to the reactor, or in a separate step. In the halogen-free process, the oxygen reacts with hydrogen to form water and the unsaturated hydrocarbon. At present, two plants are thought to be producing butadiene by oxidative dehydrogenation: Shell has reached the startup stage for a plant near Marseille, France, that uses iodine, and Petro-Tex is operating a plant at Houston, Texas, to make butadiene from butenes (14). Although not proven commercially, these new procedures are supposed to be more economical than the older ones because of the lower requirements with respect to plant, recycle, and utilities which follow from the higher yields.

3. Purification of Butadiene

About 80% of the butadiene manufactured in the U.S. comes from catalytic dehydrogenation of butenes or n-butane. The light fractions are readily separated from the effluent from the dehydrogenation reactor, while the separation of the butadiene from the crude butadiene fraction is more complicated. One of three extractive distillation procedures are normally used for this purpose: Esso Research and Engineering's cuprous ammonium acetate process, Phillips Petroleum's furfural process, or Shell's acetonitrile process. In Germany, Badische Anilin und Soda-Fabrik has an N-methylpyrrolidone based process and most recently Geon of Japan has announced a process based on dimethylformamide (15). In the conventional butadiene plant, about two-thirds of the capital investment and operating costs are associated with the butadiene purification. For one oxidative dehydrogenation process, butadiene of 98.5% purity after fractionation without extractive distillation has been claimed (16). Considerable savings in plant investment can be expected for such a case.

B. Styrene

During World War II, several processes were developed for the production of styrene. The most important involves catalytic dehydrogenation of ethylbenzene prepared from benzene and ethylene with the aid of a Friedel-Crafts catalyst:

$$C_2H_4 + C_6H_6 \underset{}{\overset{AlCl_3}{\rightleftharpoons}} C_6H_5—CH_2—CH_3 + \text{by-products}$$

$$C_6H_5—CH_2—CH_3 \xrightarrow[\text{Catalyst}]{} H_2 + C_6H_5—CH=CH_2 + \text{by-products}$$

This process requires extensive fractionation equipment to yield high purity monomer.

Another process, developed by Carbide and Carbon (5c), was based on the oxidation of ethylbenzene to a mixture of α-phenylethanol, acetophenone, and by-products. The acetophenone was hydrogenated to the alcohol and the combined phenylethanol stream was dehydrated to styrene. The purification problem in this process was much simpler, but the economics depended on the utilization of by-products and intermediates.

The oxidative dehydrogenation of ethylbenzene to styrene is also disclosed in the patent literature (17). Although the improvements for styrene are not as dramatic as those for butadiene, the production of styrene by the oxidative dehydrogenation route should also receive considerable attention.

"Polymerization grade" butadiene and styrene are manufactured to exacting specifications which generally require higher than 98 mole % monomer. The economical production of these high purity products is a noteworthy accomplishment of the chemical industry. The general and fundamental problems associated with the conversion of these monomers into elastomers is the concern of the following sections of this chapter.

III. GENERAL ASPECTS OF SBR POLYMERIZATIONS

SBR is a styrene-butadiene copolymer prepared in an emulsion system by a free radical mechanism. The major portion of SBR manufactured in the United States comprises a 23:77 styrene-butadiene ratio, and the largest volume use of this rubber is in tire production. The monomer units are randomly distributed in the copolymer and the configuration of the butadiene unit in a typical polymer exists in a 68:14:18/ *trans*:*cis*:vinyl ratio. A carbon skeletal representation of the various structural units, but not in proportion, follows:

Temperature exerts a small effect on the diene configuration and the monomer ratios are largely predictable by copolymer composition laws which are discussed briefly in Section VII of this chapter and in greater detail in Chapter 4.B on nitrile rubber.

Most of the SBR is produced by means of two basic systems in which the ingredients are determined largely by the temperature of the polymerization. "Hot" rubber, made at 50°C, was the basic rubber produced during World War II. "Cold" rubber, made at 5°C, is a second generation rubber of improved properties. Typical recipes for "hot" and "cold" emulsion polymerizations are listed in Table II.

Each ingredient in a recipe has one or more functions. The water is the reaction medium solubilizing salts and also serves as a heat transfer agent. The KCl as an electrolyte enters into the colloidal system and helps to stabilize the emulsion, and this is also one of the functions of the $K_2S_2O_8$ in the SBR 1000 recipe. At least three functions are served by

TABLE II
Typical Recipes for Preparing "Hot" and "Cold" SBR

	Weight by parts	
	SBR 1000[a] ("Hot")	SBR 1500[a] ("Cold")
Butadiene	75	72
Styrene	25	28
Water	180	180
Fatty acid soap[b]	4.5	4.5
KCl	—	0.3
Auxiliary surface active agent	—	0.3
t-Dodecyl mercaptan	0.28	0.20
$K_2S_2O_8$	0.3	—
p-Menthane hydroperoxide	—	0.063
$FeSO_4 \cdot 7H_2O$	—	0.010
Ethylenediamine tetraacetic acid salt	—	0.050
Sodium formaldehyde sulfoxylate	—	0.050
Temperature of polymerization, °C	50	5
Conversion, %	72	60
Polymerization stopper	HQ[c]	DMDT[c]
Antioxidant	BLE[c]	BLE

[a] These numbers are designations for types of rubber given by an IISRP committee.
[b] Fatty acid soaps are used alone and in mixtures with rosin soaps.
[c] HQ = hydroquinone, DMDT = N,N-dimethyldithiocarbamate, BLE = diphenylamine-acetone condensation product.

the soap: to emulsify the monomers, to solubilize the monomers in the micelles which are the loci of initiation at the beginning of the polymerization, and to stabilize the latex particles during the polymerization and subsequent steps. The auxiliary surface active agent is a dispersant added to help stabilize the emulsion. Free radicals for initiation in the SBR 1000 recipe are formed by thermal decomposition of the persulfate. In the SBR 1500 system a p-menthyloxy radical is formed by a redox reaction of a complexed ferrous ion with the hydroperoxide and the sulfoxylate regenerates the ferrous ion from the ferric state. Primary or tertiary dodecyl mercaptans are used to control the molecular weight of the polymer. In the SBR 1000 system the mercaptan also acts as a "promoter" of initiation. All of these functions are expressed in equations or mechanisms which are presented in the following sections of this chapter.

The existence of individual latex particles is the distinguishing feature of emulsion polymerization systems. The particles serve as tiny reaction sites in which initiation, propagation, chain transfer, and termination can occur without competition with or interference from identical reactions occurring in neighboring particles. The simultaneous growth of many isolated polymer chains and the absence of termination by combination leads to high rates of polymerization and high molecular weights. The mechanism of the formation of the particles and the kinetic expressions related to their formation and the subsequent propagation are known as the Harkins-Smith-Ewart theory of emulsion polymerization.

Polymerizations are conducted in a straightforward manner. The monomers and hydrocarbon-soluble ingredients are charged as one phase and the water and water-soluble ingredients as another, while one component of the initiation system is usually withheld and added after the hydrocarbon phase is emulsified and the system has attained the proper temperature. In the laboratory these solutions are added to beverage bottles which are provided with a self-sealing gasket through which ingredients can be added or the reaction mixture withdrawn. For precise kinetic and mechanistic studies, high vacuum line techniques are employed principally to eliminate oxygen. In commercial plants the ingredients are usually added continuously in three streams to a series of six or more 3700 gallon autoclave reactors.

SBR polymerizations are terminated at conversions of between 60–72% by addition of a free radical inhibitor such as hydroquinone or a thiocarbamate. In plant operation, the unreacted monomers are recycled. The rubber is recovered by destroying the emulsion by means of a brine–acid mixture. Before the latter step is performed, an antioxidant, such as BLE, is added to the latex. The rubber crumb is composed of about 93% rubber hydrocarbon, 4% organic acid, and 1.5% antioxidant, while the remainder is salt and other impurities occluded during the coagulation step. Plasticizers and/or carbon black can be incorporated into the rubber crumb by co-coagulating the latex with an emulsion of a viscous petroleum based oil or a dispersion of carbon black. These mixtures are known as masterbatches. Currently over 90 types of SBR are produced which differ in recipe composition, in manufacturing steps, in plasticizer and/or a carbon black content, or in combinations of these variables. The different types are designated by a four digit number which specifies the variables.

Contributions chiefly by Harkins and by Smith and Ewart placed the complicated emulsion polymerization process on a theoretical basis. In the following two sections the fundamental aspects of emulsion polymerization are presented.

IV. HARKINS-SMITH-EWART THEORY OF EMULSION POLYMERIZATION

Harkins (18,19) developed a theory of emulsion polymerization based on the phases in an emulsion system and their participation as the loci of initiation and propagation. On the basis of these ideas Smith and Ewart (20) transformed the phenomena of emulsion polymerization systems into kinetic expressions. The following description, which combines these ideas, is being referred to in the literature as the Harkins-Smith-Ewart (H-S-E) theory of emulsion polymerization. Although these men based their theories predominantly on experimental evidence for the emulsion polymerization of styrene, the description will be broadened to incorporate the ingredients of an SBR polymerization system, such as those given in the SBR 1500 recipe of the preceding section. Following Harkins then, the hydrocarbon phase consists of emulsifier stabilized droplets containing most of the monomers, the alkaryl hydroperoxide, and the modifier. In the water phase are dissolved the salts and trace amounts of monomers, peroxide, and modifier. Also in the water phase is the soap in a concentration high enough to exceed the critical micelle concentration. The micelles, which have solubilized a small amount of the monomers, peroxide, and modifier, comprise the third phase. The various ingredients may be in equilibrium distribution between the phases depending mainly on thermodynamic and colloid chemical factors. When an activator is added to this system, free radicals are generated.

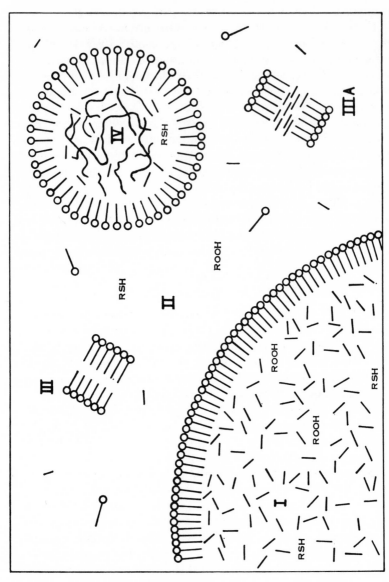

Fig. 1. Schematic representation of an emulsion polymerization system (after Fryling (23f.)). I, hydrocarbon phase contains monomers, —, hydroperoxide, and mercaptan; II, aqueous phase contains salts, soap—o, and trace amounts of monomers, hydroperoxide, and mercaptan; III, soap micelles; IIIA, soap micelles containing a small amount of solubilized hydrocarbon phase; IV, monomer-polymer particle containing monomers, polymer, and mercaptan,

The free radicals are captured preferentially by the micelles because of the relatively large surface offered by the latter as a consequence of their large number and very small size. As monomer solubilized within a micelle begins to polymerize, the growing polymer chain imbibes more monomer to form a swollen monomer–polymer (M/P) particle. A schematic representation of the emulsion polymerization system is shown in Fig. 1.

Polymerization within the particles creates more interface onto which emulsifier is adsorbed from the water phase. The formation of new particles and stabilization of growing particles continues until all the micelles are adsorbed on either the droplets of the hydrocarbon phase or the M/P particles. The disappearance of the micelles represents a discontinuity in the emulsion polymerization system and marks the end of the particle forming stage, which is followed by the steady-state stage.

The latter is characterized by a constant number of M/P particles, a constant rate of polymerization, and a constant monomer concentration or a constant function involving various ingredients in the polymerization system. An emulsion polymerization system exhibiting this constancy is

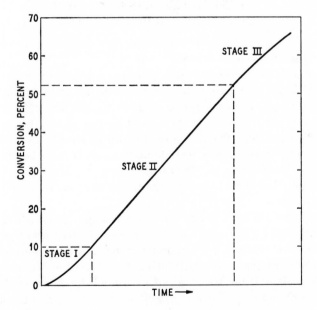

Fig. 2. Conversion–time curve for an "ideal" emulsion polymerization. Stage 1: M/P particle formation; Stage II: steady-state; Stage III: solution polymerization.

sometimes called an "ideal" system and was first characterized by Smith and Ewart (20). In an SBR system, this steady-state stage covers an interval from approximately 10 to 55% conversion. During this stage, the particles increase in size while the monomer phase decreases. The steady-state stage ends when this phase disappears. This is a second discontinuity since the monomer concentration in the particles now changes with conversion, and the kinetics of the polymerization have the characteristics of a solution system.

The third or solution polymerization stage constitutes only a short interval in SBR systems terminated at 60% conversion. Since transitions between the stages are not abrupt, the bulk of the polymer in SBR systems is formed during the second stage. The average physical properties are dominated by the polymer formed during this interval.

An emulsion polymerization system fulfilling the conditions of the H-S-E theory displays a characteristic conversion-time curve (Fig. 2). Although many emulsion polymerization systems, including SBR systems, display the phenomena described in this introduction, the situation becomes more complicated when kinetic expressions are sought.

V. KINETIC EXPRESSIONS FOR EMULSION POLYMERIZATION

Harkins (18,19) established that the micelles are the loci of initiation at the beginning of the polymerization and that the M/P particles become the principal loci of polymerization after the micelles have disappeared. Smith and Ewart (20) accepted these ideas and derived a kinetic expression for the rate of polymerization. However, the derivation was necessarily restricted to certain conditions that were either defined or were obvious from the derivation. Three important conditions were (1) the monomers were only slightly soluble in water as is the case for styrene, butadiene, or their mixtures, (2) the polymer was soluble in the monomer as in bulk polymerization, and (3) the initiating radicals were generated in the water phase.

Noncompliance of some emulsion systems with the H-S-E theory can be attributed to a violation of one of these conditions; however, in some instances a clear reason for noncompliance has not been established. In the following treatise conformity with and deviation from the H-S-E theory for SBR will be discussed. Since much of the theoretical work has been carried out with monovinyl monomers, work with the latter will be included to illustrate some of the alternative theories and kinetic expressions.

Since in emulsion systems propagation occurs within the M/P particle, the polymerization is restricted to the confines of a colloidal particle. This confinement confers unique possibilities on emulsion polymerization not available to bulk or solution systems. A large number of isolated particles in the presence of a limited concentration of free radicals make the concentration of radicals per particle of primary importance. The chief factors determining the rate of polymerization in an emulsion system, restricted to the preceding conditions, are the number of particles, the concentration of the monomers in the particle, and the number of free radicals per particle.

From mathematical derivation and deduction, Smith and Ewart (20) concluded that for a polymerization system where radicals enter particles one at a time and the kinetic life of a single radical is long compared with that of two simultaneous radicals in the same particle, the number of propagating radicals per particle is on the average one for half the time. Under these conditions the rate is

$$- \frac{d[\text{M}]}{dt} = k_p \, [\text{M}]N/2 \tag{1}$$

where k_p is the rate constant for solution polymerization, [M] is the monomer concentration in the particle, N the number of M/P particles. This elegantly simple equation has become the subject of numerous investigations. Most of the studies have been concerned with homopolymerization of monovinyl monomers and a number of excellent reviews have been written (21,22).

Research on synthetic rubber conducted under U.S. Government control was reviewed in mid-1950 (23a,24a). Work with dienes and diene copolymerization has been relatively sparse since that time; some of the more recent research has been reviewed in the German literature (22,25). In the following sections the emulsion polymerization of dienes and diene–styrene mixtures will be emphasized, but the studies of styrene homopolymerization will also be included.

A. Stage I. Particle Formation

1. Effect of Soap on N

Harkins and coworkers presented experimental evidence for the existence of the phases and processes comprising the first stage of an emulsion polymerization. A 3% solution of potassium dodecanoate contains approximately 10^{18} micelles per cm³. Only a fraction of these micelles form particles, and the final latex contains some 10^{15} M/P particles per cm³ (19).

On the basis of several simplifying assumptions Smith and Ewart (20) derived an expression for the number of particles per cm³ at the end of the particle forming stage:

$$N = k(\rho/\mu)^{2/5}(a_sS)^{3/5} \qquad (2)$$

where ρ is the rate of entrance of free radicals into the particles, μ is the rate of volume increase of a particle, a_s is the interfacial area occupied per unit amount of soap, and S is a unit amount of soap in excess of the critical micelle concentration. The constant k lies between 0.53 and 0.37. Another simplifying assumption is that the amount of free radicals captured by the particles is proportional to the number of radicals generated by the initiating system in the aqueous phase.

On the basis of these assumptions, calculations compared favorably with the number of particles in some of the emulsion polymerization systems. Smith (26) calculated N to be approximately 10^{15} for styrene polymerization initiated by persulfate and emulsified with 1% fatty acid soap. The count from Smith's graph (26) of the number of particles versus soap concentration gave a value of approximately 8×10^{14} for the 1% soap concentration. French (27) replotted Smith's data and showed that $N = f(\text{soap})^{3/5}$. In another study of styrene polymerization the graph at 1% potassium laurate solution (28) showed the calculated and experimental values of N per gram of monomer to be approximately 3.6×10^{15} and 2×10^{15}, respectively. Likewise, Bartholome et al. (29), in an examination of the dependence of N on the amount of soap, found that the slope of a log N − log (soap) plot was $3/5$ power of the amount of soap as required by Eq. 2.

In many studies related to the H-S-E theory, the number of particles was determined even though this was not the primary purpose of the research. Some values of N in emulsion polymerizations of dienes and the experimental conditions are listed in Table III.

Although the first set of data in Table III shows a decrease in the number of particles with conversion, Morton believed this trend could be attributed to a problem associated with the measurement of the particle diameter by the soap titration procedure (30). For the other three sets of data in Table III, the number of particles did not vary with conversion. The polymerizations conducted with the polyamine activator complied with one of the requirements of the H-S-E theory: the rate did not increase when more initiator was added during the steady-state stage.

TABLE III

Number of Particles and Experimental Conditions for Emulsion
Polymerization of Dienes

Mono-mers	Soap		Initiation		Polymeri-zation		$N \times 10^{-15}$, cm^{-3}	Ref.
	Type	PHM[a]	Type	PHM[a]	%	Temp, °C		
Bd[a]	ORR[a]	2.5	$S_2O_8{}^{2-}$	0.3	33[b]	50	7.0	30
					77[b]		4.4	30
Bd	ORR	5.0	$S_2O_8{}^{2-}$	0.3	29[b]	50	9.2	30
					65[b]		6.7	30
Bd	ORR	10.0	$S_2O_8{}^{2-}$	0.3	63	50	8.8	30
Bd	ORR	5.0	DIP–TETA[a]	0.05[d]	30–66	10	5.1[c]	31
Bd	ORR	5.0	DIP–TETA	0.20[d]	34–65	10	8.8[c]	31
Bd	ORR	5.0	DIP–TEPA[a]	0.10[d]	58–64	10	6.9[c]	31
Bd	ORR	5.0	DIP–TEPA	0.05[d]	31–74	5	6.2[c]	31
Iso[a]	ORR	5.0	DIP–TEPA	0.1	57	5	11.9	32
Iso	ORR	5.0	DIP–TEPA	0.4	64	15	15.5	32
DMBd[a]	ORR	5.0	$S_2O_8{}^{2-}$	0.3	62	50	3.2	33
DMBd	ORR	5.0	DIP–TEPA	0.2	61–78	5	3.3	33

[a] PHM = parts per 100 monomer, Bd = butadiene, ORR = Office of Rubber Reserve Soap, DIP = diisopropylbenzene hydroperoxide, TETA = triethylenetetramine, TEPA = tetraethylene pentamine, Iso = isoprene, DMBd = dimethylbutadiene.

[b] Same polymerization at two conversions.

[c] Average values within range indicated.

[d] Initiator in initial charge.

Furthermore, a plot of the first set of experimental data to check the equation $N_s = kC_E^{-2/5}$ showed that the exponent was closer to -0.5 than to the predicted -0.4 when N_s is the number of particles per gram of soap. Morton and coworkers did not consider the -0.5 value conclusive but rather considered this as a general confirmation of the theoretical value (30).

Klevens (34) determined the particle diameter for the emulsion polymerization of a 75:25 isoprene–styrene mixture with persulfate initiator at 50°C in a 2.5% soap solution. The diameters ranged from 310 to 460 Å at 17.6–55.5% conversion, respectively. Using these values and a density of 0.9 for the copolymer, an N value of 4×10^{15} can be calculated for the system. Klevens refrained from calculating any values because of uncertainty about the interpretation of soap titration data. However, this value of N falls in the range calculated and found for diene emulsion polymerizations.

2. Effect of Initiator on N

Smith (26) in the original study of the polymerization of styrene with persulfate at three concentrations found that the experimentally determined relative number of particles corresponded to the relative number predicted by Eq. 2. Bartholome and coworkers (29) showed that N was proportional to the 0.4 power of the amount of initiator for the polymerization of styrene.

Comparable studies of the effect of initiator on the number of particles in emulsion polymerization of dienes have not been made. The use of some of the available data, such as that in Table III, for calculations is not justified because of the small number of experiments and the uncertainty as to the amount of initiator actually employed. However, other tests of the H-S-E theory have been made for emulsion polymerization of dienes and these will be presented in later sections.

Besides the S-E expression (2) for N as a function of soap and initiator, several investigators have found experimental support for other relationships. Bakker (35) derived an expression, $N = kC_I^{1/2}C_E^{1/2}$ where C_I and C_E are initiator and soap concentration, respectively, and gave data for polymerization of styrene that supported the expression. Bakker's relationship has been questioned (36) because of the simplifying boundary condition used in the derivation. Brodnyan et al. (37) found that in emulsion polymerization of methyl and butyl methylacrylate the exponent depended on the particle concentration:

$$N \propto C_E^{0.5} \qquad N < 3 \times 10^{14}$$

$$N \propto C_E^{3.0} \qquad N > 3 \times 10^{14}$$

These investigators suggested that the surface and the interior of the particles may represent two different loci of polymerization with the predominant one determined by the ratio of the surface to volume for the particles. French (27) in polymerizing vinyl acetate emulsified with a nonionic ethylene oxide–propylene oxide agent found $N = f(\text{soap})^3$. However, the M/P ratio for vinyl acetate was approximately 6.5 at 14% conversion so that a solution polymerization condition existed early in the polymerization. This deviates from one of the conditions assumed by Smith and Ewart. Recently a more fundamental approach was taken by Parts et al. to find a relation between the variables, such as initiator and soap content, and $N/2$ at the end of stage one (38). This seems to be an important point, since the number of particles is several orders of magnitude less than the number of micelles. Although these authors did not derive a satisfactory expression, their theoretical analysis indicated some phenomenon such as coalescence of latex particles during the first stage of polymerization to be needed to get the necessary decrease in the particle number.

To summarize, conditions exist for the polymerization of styrene where good agreement is obtained between experimental results and those predicted by the S-E Eq. 2. For polymerization of dienes, the experimental evidence supporting Eq. 2 is sparse, but some of the information in the following section offers more convincing evidence that to a first approximation the H-S-E theory can be used as a good starting point.

B. Stage II. Steady State

1. Rate of Homopolymerization

The steady stage is the one most characteristic of emulsion polymerization systems. Following a short first stage, the rate of polymerization is constant for an appreciable part of the polymerization (see Fig. 2). Equation 1 provides a ready explanation for this constant rate. If the M/P ratio and N are constant, then the right side of the equation is constant. Thus, one of the first tests of an "ideal" emulsion system is a measure of the constancy of the rate after the initial nonlinear stage. This constancy of rate has been demonstrated so many times that it is an accepted characteristic of emulsion polymerization systems. Butadiene-styrene copolymerization in many SBR systems also exhibits a constant rate for an appreciable period of the polymerization. This is a more complicated case which will be treated in some detail in Section V-B-2.

Another test of the H-S-E theory is the relation of the rate of polymerization to the soap and initiator concentration. This follows from a combination of Eqs. 1 and 2:

$$\text{Rate} \propto N \propto C_E^{0.6} C_I^{0.4} \tag{3}$$

The exponents found by various investigators for some typical styrene and diene emulsion homopolymerizations are listed in Table IV together with the emulsifiers and initiator systems used.

The experimental evidence in the previous section for the number of particles and the rate data in Table IV fairly well verify that the H-S-E

TABLE IV
Test of $R \propto C_E^\alpha C_I^\beta$ for Styrene and Diene Emulsion Homopolymerization

Monomer	Emulsifier	Initiator system	Polymerization temp., °C	Exponent for		Ref.
				C_E	C_I	
Styrene	ORR[a]	$S_2O_8{}^{2-}$	50	—	0.40	24b
Styrene	ORR	$S_2O_8{}^{2-}$	50	0.5	—	24c
Styrene	Amphoseife[a]	$S_2O_8{}^{2-}$	45	0.6	0.40	29
Butadiene	ORR	$S_2O_8{}^{2-}$	50	—	ind.[a]	24d

[a] ORR = Office of Rubber Reserve Soap, Amphoseife = potassium salt of sulfonated C_{18}-olefin, ind. = independent.

theory holds for emulsion polymerization of styrene under the conditions used. Similar experimental evidence for emulsion polymerization of dienes is lacking. Studies of the latter type were probably discouraged early because the rates of polymerization of butadiene in emulsion are independent of the persulfate concentration. This anomaly will be discussed in Section VI-A.

Although Morton and coworkers did a considerable amount of testing of the H-S-E theory for the polymerization of dienes, data for checking Eq. 3 were not obtained. However, they did show that one criterion of the theory was obeyed for the polymerization of butadiene, isoprene, and dimethylbutadiene by means of the ROOH-amine initiator couple; the rate for the three monomers was not affected by the addition of increments of initiator during the steady-state stage but was responsive to temperature.

2. Rate of Copolymerization

The H-S-E theory should also apply to copolymerization of water insoluble monomers if the necessary conditions prevail. Mayo and Walling (39) presented the simplified rate equation for the copolymerization of monomers in an ideal emulsion system:

$$- \frac{d([M_1] + [M_2])}{dt} = kC_I^{2/5} C_E^{3/5} V^{3/5} \tag{4}$$

$$- \frac{d([M_1] + [M_2])}{dt} = V \frac{N}{2} \tag{5}$$

where V is a copolymerization function:

$$V = \frac{k_{11}k_{22}[r_1[M_1]^2 + 2[M_1][M_2] + r_2[M_2]^2]}{k_{22}r_1[M_1] + k_{11}r_2[M_2]} \quad (6)$$

(For explanation of symbols, see Chapter 4.B by W. Hofmann, p. 193.)

The rate curves of monomers polymerizing in conformity with Eq. 5 will resemble those of a homopolymerization obeying Eq. 1 if in the former case V is a constant. Calculations with the help of Eq. 6 are presented in Table V. The values used are either presented in other tables in this chapter with references or their source is identified in the table.

For a conversion rate of 6% per hour and an average V of 38.91, a value for N of 2.3×10^{15} per ml of latex is obtained by means of Eq. 5.

Although some of the values in Table V are uncertain, this has little bearing on the test of the constancy of V. Uniformity of V might be expected since these monomers form a nearly ideal pair (41) in solution copolymerization, i.e., $r_1 \cdot r_2 = 1.08$. These data offer a reasonable explanation for the constant rate for the copolymerization of butadiene and styrene in emulsion systems. In Table VI are listed experimental results from the literature and some unpublished data for testing Eq. 4 for some typical emulsion copolymerizations of butadiene and styrene. Rate curves for emulsion polymerization in the laboratory with the SBR-1500

TABLE V

Calculation of Copolymerization Function, V, for Emulsion
Polymerization of 72:28 Butadiene–Styrene Mixture at 5°C

	Experimental Data		
	Styrene, M_1	Butadiene, M_2	Polymer
k_p, 1 mol^{-1} sec^{-1}	17.35	6.22	
r	0.64	1.38	
d, g/ml	0.923	0.64	0.942 (23b)
% M at 20% (40)	22.3	77.7	
% M at 52% (40)	22.8	77.2	
M/P ratio			0.92

Calculation for V, at Conversion

	20%	52%
V, sec^{-1}	39.17	38.65

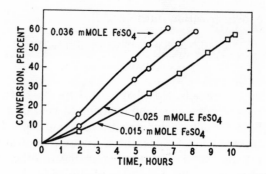

Fig. 3. Conversion–time curves for laboratory polymerization with the SBR 1500 system at variable iron level (47).

system are presented in Fig. 3, and some rate-iron level curves in Fig. 4. The latter were used for calculating data in the last two rows of Table VI. Each point in curve 3, Fig. 4, represents an average of from 10 to 50 individual runs.

The data in Table VI and Fig. 4 show some instances of compliance with Eq. 4 whereas some show little compliance. The independence of the rate of polymerization of dienes of the persulfate concentration (24d,30) is particularly puzzling, since the same initiator shows excellent

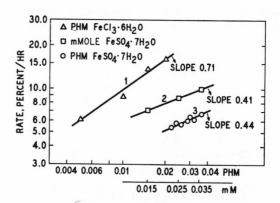

Fig. 4. Rate of polymerization vs. iron level: curve 1, PHM FeCl₃·6H₂O (Fig. 1, Ref. 46); curve 2, mM FeSO₄·7H₂O, bottle polymerization (Ref. 47); curve 3, PHM FeSO₄·7H₂O plant polymerization (Ref. 47). △, PHM FeCl₃·6H₂O; □, mM FeSO₄· 7H₂O; ○, PHM FeSO₄·7H₂O.

TABLE VI
Test for $R \propto C_E{}^{\alpha}C_I{}^{\beta}$ for Emulsion Copolymerization of
Butadiene–Styrene Mixtures

Ratio Bd/S	Emulsifier	Initiator system	Polymer-ization temp, °C	Exponent for C_E	Exponent for C_I	Ref.
75/25	ORR[a]	$S_2O_8{}^-$	50	0.7[b]	ind.[a]	24d
75/25	KO_2CC_{13}	ROOH–FePP[a]	30	0.75	0.4[c]	42
75/25	KO_2CC_{13}	ROOH–FePP	0	—	0.4–0.5[c]	43
75/25	K–FA[a]	ROOH–N_2H_4[a]	5	0.7	ind.	44
75/25	ORR	ΦN_2OAc[a]	25	—	ind.	45
70/30	ORR	SFS[a]	5	—	0.71[d]	46
70/30	1500[a]	SFS	5	—	0.41[d]	47
72/28	1502[a]	SFS	5	—	0.44[d]	47

[a] ORR = Office of Rubber Reserve, ROOH–FePP = hydroperoxide–ferrous pyrophosphate initiator, ind. = independent, K–FA = potassium fatty acid soap, ROOH–N_2H_4 = hydroperoxide–hydrazine initiator, ΦN_2OAc = N-nitrosoacetanilide, SFS = sodium formaldehyde sulfoxylate, 1500 = SBR 1500 recipe, 1502 = SBR 1502 recipe.

[b] Based on data of Kolthoff cited in Ref. 19.

[c] Based on consumption of hydroperoxide.

[d] Based on iron level.

confirmation of the H-S-E theory for styrene polymerization. (See Section VI-A for further discussion.)

Others have proposed different equations for the rate as a function of the soap and initiator. However, the application of these equations to dienes has been infrequent. Bakker (35) and Medvedev (48) both proposed similar modifications of Eq. 3:

$$\text{Rate} \propto C_E{}^{\frac{1}{2}}C_I{}^{\frac{1}{2}} \qquad (7)$$

albeit, for different reasons. Brodnyan and coworkers (37) attempted to distinguish between Eqs. 3 and 7 by using kinetic data for emulsion polymerization of methyl and n-butyl methacrylates and concluded that both equations fitted the data equally well. The same conclusion was reached by Kolthoff and Medalia (43) in one instance of copolymerization of butadiene and styrene. The different investigators (37,48) proposing Eq. 7 for the rate expression for some emulsion polymerizations believe that adsorption layers are involved in initiation and propagation.

Rysanek (36) derived an expression, based on the H-S-E theory, for

TABLE VII
Effect of Water Solubility of Monomers on Exponents
of the Rate Equation 3

	H$_2$O solubility, %	C_E	C_I
		Exponent for	
Vinyl caproate	0.004	0.6	—
Acrylonitrile	9	0.3	0.4
Acrolein	20.8	0.2	0.4

emulsion polymerization in the steady-state stage in the presence of inhibitors, and experimental verification was offered for several monomers.

Cherdron (49) found the exponent for the rate expression 3 related to the water solubility of the monomers. The data for three monomers initiated with a persulfate-silver couple were as shown in Table VII.

Significantly when the emulsion polymerization of styrene was carried out in 50% methanol, the resulting S-E equation had an exponent of 0.2 for C_E (50). Other special cases of emulsion polymerization will not be considered in detail here. Bamford et al. (51) reviewed heterogeneous polymerization in which the polymer is insoluble in the polymerization medium. Finally several groups (52,53) have successfully polymerized special combinations of monomers in soap-free systems. These special emulsion systems contain at least one water soluble monomer. Excellent rates have been obtained for a 70:30:5 butadiene–styrene–acrylonitrile mixture in 500 parts of water at 30°C.

In summary, for the second polymerization stage, styrene has been shown to obey the S-E equation for persulfate initiated polymerization under conditions which are close to those used in SBR-type polymerizations. However, for dienes in the same system, the rate has been independent of the persulfate concentration. For some iron-pyrophosphate, hydrazine, and sulfoxylate initiator systems, partial or near compliance has been found for the copolymerization of butadiene and styrene. For a well-balanced sulfoxylate system the rate was approximately proportional to $(Fe)^{0.4}$ in both laboratory and plant polymerizations.

3. Absolute Rate Constant

Equation 1 became attractive to polymerization researchers since it apparently offered a means for determining the propagation constant for monomers without the use of the tricks or assumptions employed in solution polymerization measurements. Again most of the work in this area has been done with styrene and has been adequately reviewed (21,22,54). Burnett and Lehrle (54) gave an average value for k_p determined for solution polymerization of styrene, but they refrained from averaging values for emulsion polymerization because of the uncertainty in the determination of some of the values. Uncertainties were concerned with the determination of the number of particles, the concentration of styrene in the particles, and even the density of the polystyrene in the styrene monomer. The number of radicals per particle also becomes important in the use of Eq. 1 as is discussed in the following section.

Van der Hoff (21) and Gerrens (22) also used restraint in obtaining an Arrhenius equation from log k_p vs $1/T$ plots. Despite the hesitation shown by investigators to use the rate constant for emulsion polymerization of styrene, the values for k_p obtained by Morton and coworkers for polymerization of butadiene (31), isoprene (42), and dimethylbutadiene (33) are cited by many reviewers and in most tabulations of rate constants:

$$\text{for butadiene, } k_p = 1.2 \times 10^8 \exp\left(-9300/RT\right)$$

$$\text{for isoprene, } k_p = 1.2 \times 10^8 \exp\left(-9800/RT\right)$$

$$\text{for dimethylbutadiene, } k_p = 8.9 \times 10^7 \exp\left(-9000/RT\right)$$

This method of obtaining k_p is apparently the only good one available for these dienes, since they polymerize poorly in solution systems (55a,56).

In summary, the use of Eq. 1 for determining k_p for styrene has been questioned because of uncertainties regarding certain determinations. However, the method has proven useful for dienes and further research could provide a valuable method for studying polymerization kinetics.

4. Number of Radicals per Particle

Smith and Ewart derived a recursion formula for the steady state distribution of free radicals among the emulsion particles. These authors showed that, for a limiting case which occurred for very small particles, the total number of free radicals was approximately equal to $N/2$. Roe and Brass (57) and Gerrens (22) obtained limited solutions and Stockmayer (58) derived a general solution for the recursion formula for the calculation of the average number of free radicals per particle, $\bar{n}$. Van der

Hoff (59) expressed the pertinent equations as follows:

$$\frac{\bar{n}}{a/4} = \frac{I_0(a)}{I_1(a)} \quad \text{or} \quad \bar{n} = \frac{a}{4}\frac{I_0(a)}{I_1(a)} \tag{8}$$

$$a = 4(R_i/2k_t)^{0.5}N_A v \tag{9}$$

$$R_p = k_p(M)\bar{n}N/N_A \tag{10}$$

where $I_0(a)$ and $I_1(a)$ are hyperbolic Bessel functions, k_t denotes the rate of termination, v is the volume of a particle, N_A is Avogadro's number, R_i is the rate of initiation, and the other terms are the same as used previously; $\bar{n}$ is a function of a, Eq. 8, and when less than 1, the value of $\bar{n}$ rapidly approaches 0.5. The parameter a, Eq. 9, will be small when k_t is large, and R_i and v are small. Haward (60) calculated that the critical size associated with one active center in the bulk polymerization of various monomers is approximately 0.1–1.0 μ when the rate of polymerization is 10% per hour and the rate constant is approximately that of styrene. Gerrens (61) devised a means of determining the change in $\bar{n}$ with conversion for styrene. For particles less than 900 Å, $\bar{n}$ was close to 0.5 up to 60% conversion. At higher conversion, $\bar{n}$ became larger, especially for larger particles. The latter trend could be attributed to a smaller rate of termination because of increased viscosity of the polymerization solution within the particles (Trommsdorff gel effect (62)). Van der Hoff found similar values for $\bar{n}$ for the persulfate and cumene hydroperoxide thermal initiation of styrene for particle diameters less than 0.12 μ (28,63).

In SBR polymerizations for preparing solid elastomers, the soap concentration is at a level where the number of particles is large and the average diameter is generally less than 600 Å. Only in the preparation of high solids latexes, where an effort is made to obtain large particles, is the average diameter over 1000 Å. Furthermore, by using Haward's method for calculating the critical droplet size for a butadiene polymerization at 5°C in which the rate of monomer conversion is 10% per hour, one obtains a value of approximately 1000 Å. These values and the high molecular weights obtained in the copolymerization of butadiene–styrene mixtures in the absence of modifiers suggest that the $\bar{n}$ in the M/P particles of standard SBR systems is low and probably less than 1.

For most practical purposes the value of $\bar{n}$ is of little import as long as it remains constant. For theoretical studies, the absolute value of $\bar{n}$ is critical and part of the discrepancy for k_p determined by the solution and emulsion procedures could be attributed to the uncertainty in the determination of $\bar{n}$.

The "ideal" emulsion polymerization systems considered to this point were based on the assumption that radicals captured by particles are not desorbed. Smith and Ewart also considered the case for free radicals transferring out of particles in which the average number of free radicals is much less than unity. Termination could take place in either the water phase or the polymer particles. The rate laws would be similar to these for a solution polymerization except that factors are needed to account for the ratio of radicals in the two phases or the probability of escape of radicals from the particles. Stockmayer's general solution for the recursion equation was also based on the assumption that the particles did not lose activity. Recently O'Toole (64) extended the theory to include a finite rate of radical loss from the particles. The expression for the average number of free radicals per particle is

$$\bar{n} \approx k_a/(k_d + 2k_a)$$

where k_a and k_d are the rate constants for absorption and desorption, respectively; hence, when the loss of radicals to the aqueous phase is negligible, $n = \frac{1}{2}$ as for the "ideal" emulsion. Different cases of desorption and of initiation within the particles were also considered (64) but they will not be presented here.

In summary, for styrene under normal bulk polymerization conditions the critical size of a droplet associated with one radical site has been calculated to be from 0.1 to 1.0 μ, and since the simultaneous presence of two radicals within a sphere of critical size leads to rapid combination, an average $\bar{n}$ of 0.5 should be found. Experiments established that in emulsion polymerization of styrene where the particle size is less than 0.1 μ, $\bar{n}$ is close to 0.5 when conversion is kept below 60%. Although a similar treatment was not given to butadiene or SBR systems, the similarity of polymerization characteristics suggests that in the latter systems $\bar{n}$ is also less than 1.0.

VI. INITIATION SYSTEMS FOR EMULSION POLYMERIZATION

A. Persulfate

The general reviews (21,22) of emulsion polymerization present data showing the conditions under which the H-S-E theory holds for the polymerization of styrene. Various kinetic studies of the thermal decomposition of persulfate in aqueous solution and of the emulsion polymerization of styrene show that in the presence of emulsifiers the rate of initiation of styrene is equal to the rate of thermal dissociation of the persulfate (65).

The dissociation and initiation equations are:

$$S_2O_8{}^{2-} \xrightarrow{k_d} 2SO_4{}^-\cdot$$
$$SO_4{}^-\cdot + M \xrightarrow{k_i} {}^-SO_4M\cdot$$

Furthermore, the rate of polymerization of styrene is not affected by the presence of mercaptan nor is the amount of sulfate per gram of polymer significantly changed. In the emulsion polymerization of styrene mercaptan acts chiefly as a chain transfer agent.

The simple picture for styrene does not hold for the persulfate initiated polymerization of dienes. Kolthoff and Harris (66) showed that the polymerization of butadiene and butadiene–styrene mixtures in persulfate initiated systems depended on the presence of mercaptans. A minimum amount (0.005 phm) of n-dodecyl mercaptan was required for satisfactory rates, but above this amount the rates were practically independent of the concentration (66). This "promotion" effect can also be demonstrated by the use of a low molecular weight mercaptan that is rapidly depleted resulting in cessation of the reaction which can be restarted by incremental addition of the same mercaptan (Fig. 5).

Another anomaly of the persulfate initiation of dienes is the lack of

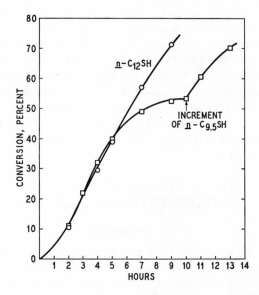

Fig. 5. Effect of increment of n-$C_{9.5}SH$ on persulfate initiated polymerization of a 70/30 butadiene–styrene system in comparison to a n-$C_{12}SH$ control (47).

dependence of the rate on the persulfate concentration as indicated in Table IV. The dependence of the rate on the presence of mercaptan suggests the mercaptyl radical as the initiating species, but this suggestion does not conform with the preceding data that the sulfate content of the polymer is independent of the mercaptan level. Reducing agents other than mercaptans do not significantly improve the initiation efficiency of persulfate; although other monomers respond favorably to redox couples involving persulfate (25). The mechanism of initiation of dienes by persulfate remains a puzzle.

B. Oxidants for Redox Initiating Systems

The demonstration that low temperature SBR produced superior tires (1) led to a concentrated search for practical initiation systems. An excellent review exists of the various commercially feasible redox systems developed under the Government program (24). The initiator systems were generally designated by the reductant. The approximate chronological order of the development was diazothioether, high sugar–iron–pyrophosphate, low sugar–iron–pyrophosphate (Custom recipe), sugar-free–iron–pyrophosphate, polyamine, and finally the sodium formaldehyde sulfoxylate system (SFS).

For the original development of redox initiator systems, cumene hydroperoxide was the best oxidant commercially available. Extensive search for other peroxides (67) showed that a considerable number of such compounds gave rates of polymerization in the Custom recipe that were severalfold higher than that obtained with cumene hydroperoxide. However, only a few could be produced economically and these were investigated more extensively. In these studies the rate constant for the reaction of ferrous iron with the different hydroperoxides was partially correlated with the rate of polymerization with the same peroxides in the Custom recipe, Table VIII.

With cumene, diisopropylbenzene, and t-butylisopropylbenzene hydroperoxides the activation energies decrease and the rates of polymerization increase in opposite order (70). However, the energy of activation is not the only criterion since phenylcyclohexane hydroperoxide appears out of order. Furthermore, the redox rate constant was determined in a low pH medium, whereas the polymerizations took place in a basic solution (24e). The solubility of the peroxides has been suggested as another factor in these reactions. Data correlating the solubility in different aqueous polymerization media and the rate of polymerization are shown in Table IX.

TABLE VIII
Rate Constants for the Expression $d(Fe^{+2})/dt = -k_1(Fe^{+2})(Ox)$ and the Rate of Polymerization

Hydroperoxide	k_1, 1 mol^{-1} sec^{-1}[a]	Polymerization rate, %/hr[b,c]	Ref.
Cumene	1.1×10^{10} exp $(-12,000/RT)$	5.8	68
p-Methane	6.3×10^9 exp $(-11,000/RT)$	8.0[d]	69
p-Diisopropylbenzene	4×10^9 exp $(-10,800/RT)$	8.8	70
p-t-Butylisopropylbenzene	1.8×10^9 exp $(-9,900/RT)$	13.1	70
Phenylcyclohexane	2.4×10^9 exp $(-10,600/RT)$	14.3	69

[a] Rate constants determined in acid solution.
[b] Rates for Custom recipe (67).
[c] The peroxides used in the polymerizations were technical grade.
[d] Unreported data from Phillips Petroleum Company.

TABLE IX
Solubility of Hydroperoxide in Aqueous Phases and Rates of Polymerization

Hydroperoxide	ROOH, in aqueous phase, %		Conversion, % in 7 hr[a]	Custom recipe, rate, %/hr[b]	H$_2$O/HOAc conversion, % in 2 hr[c]
	H$_2$O/MeOH[a]	H$_2$O/MeOH/soap[a]			
Cumene	17	40	19	5.8	<30[d]
Diisopropylbenzene	10	21	42	8.8	34
Triisopropylbenzene	6	16	53	10.2	—
t-Butylisopropylbenzene	2	10	47	13.1	56
Dodecylisopropylbenzene	2	10	41	<5[d]	72
Chloroisopropylbenzene	—	—	—	15.8	59

[a] Experiments at $-10°C$ (71).
[b] Rate in Custom recipe (67).
[c] Ferrous-pyrophosphate activator, polymerization at 5°C (72).
[d] Unpublished data from Phillips Petroleum Company.

Different hydroperoxides give maximum rates in different recipes. Polymerizations with cumene hydroperoxide tend to be inefficient; good initial rates are followed by premature cessation (67,71). This is attributed to a high solubility of the peroxide in the aqueous phase where the peroxy free radicals are destroyed in side reactions before being able to initiate polymerization.

For the series cumene, diisopropyl- and t-butylcumene hydroperoxide, it was noted in some emulsion systems at lower than 0°C that the rate of polymerization increased as the electropositive nature of the substituent increased (73). However, this generalization does not apply to chloro-cumene hydroperoxide, Table IX, which is a more effective initiator than cumene hydroperoxide. Apparently solubility overrides other effects such as electronegativity of substituents in the cumyl radical.

C. Sodium Formaldehyde Sulfoxylate (SFS) System

The search for a better initiator system for low temperature polymerization led to the development of the SFS recipe (46), which has been widely adopted in the synthetic rubber industry. The chief advantages of this recipe over previous systems are the ease of preparation, low cost, reproducibility, and responsiveness of polymerization rate to initiator level.

The composition of a typical SFS initiator system is shown in Table X.

This combination as the initiator for an SBR 1502 polymerization gave the rate curves shown in Fig. 3. A simplified mechanism for this system is:

$$\text{ROOH} + \text{FeY}^{-2} \xrightarrow{k_1} \text{RO} \cdot + \text{FeY}^{-1} + \text{OH}^- \tag{11}$$

$$2\text{FeY}^{-1} + \text{HSO}_2^{-} \cdot \text{HCHO} + 3\text{OH}^- \rightleftharpoons 2\text{FeY}^{-2} + \text{SO}_3^{-2} \cdot \text{HCHO} + 2\text{H}_2\text{O} \tag{12}$$

$$\text{RO} \cdot + \text{M} \xrightarrow{k_i} \text{ROM} \cdot \tag{13}$$

(Y = EDTA ligand).

Reynolds and Kolthoff (74) studied the kinetics of Eq. 11 and found the rate constant, $k_1 = 5 \times 10^{10} \exp(-10{,}400/RT)$, independent of pH in the range 3.7–10.3, and the simplified equation representative even though the iron complex at higher pH values is of the type $\text{Fe(OH)}_m\text{Y}^{-(2+m)}$. The stoichiometry of reaction 12 is uncertain. In a recent study of persulfate–iron–sulfoxylate kinetics (75), a formal and tentative solution was formulated as:

$$\text{FeY}^{-1} + \text{SFS} \xrightleftharpoons{k_2} \text{FeY}^{-2} + \text{X}$$

TABLE X
Ingredients for an SFS Initiating System

	PHM	mmole
p-Menthane hydroperoxide	0.063	0.366
FeSO$_4$·7H$_2$O	0.010	0.036
EDTA Na$_4$·4H$_2$O[a]	0.050	0.111
HOCH$_2$SO$_2$Na·2H$_2$O	0.050	0.333

[a] Ethylenediamine tetraacetic acid sodium salt.

where X represents an unknown oxidation product of SFS. Reaction 13 must be considered only as indicative because in the study of the kinetics of the cumene hydroperoxide–iron–hydrazine redox cycle, cumyl peroxy radicals apparently did not initiate polymerization of acrylate (76), since stoichiometric amounts of acetophenone were found under the conditions of the experiment. However, Eq. 13 is favored for SBR systems in view of the results obtained by Orr and Williams (77) who found that the peroxy radical was attached to the polybutadiene molecule when the oxidant was the dihydroperoxide of diisopropylbenzene.

Chelating agents for iron were originally investigated in connection with the SFS system (46). A more extensive examination devoted to iron chelating agents was published recently (78). Probably the best complexing agent for SBR systems is EDTA, although various modifications of this type of agent have been found satisfactory. Ratios of EDTA/Fe greater than 1:1 do not affect the rate in SBR systems. In plant practice, an excess of the chelate is generally used to aid in stabilizing the latex, complexing adventitious polyvalent ions, improving color, etc.

The success of the SFS system probably accounts for the scarcity of recent publications on initiator systems for the preparation of SBR, although much work is being reported on initiator systems for emulsion polymerization of other monomers and monomer mixtures. Some examples of interesting initiator systems used for emulsion polymerization of dienes follow.

Initiator systems simply having the ability of starting the polymerization as their only feature offer little attraction for production of SBR. Kolthoff and Meehan (79) developed a ferric-EDTA-dithionite-hydroperoxide initiator that has many of the characteristics of the SFS system. In comparison with SFS, the dithionite system gives higher rates of polymerization. One drawback seems to be a low stability of the dithionite solution in air.

The use of diisopropylbenzene dihydroperoxide for the emulsion polymerization of butadiene by means of an iron-pyrophosphate complex enabled Orr and Williams (77) to prepare a hydroperoxy-telechelic polybutadiene which initiated the polymerization of styrene to form a block copolymer. The preparation of the hydroperoxy-telechelic polymer was attributed to the transfer of the —ROOH group into the M/P particle where further reaction with the ferrous ion was prevented.

Diazothioethers, ZArN=NSArZ, were proposed as initiator-modifiers for preparing telechelic polymers of butadiene-styrene mixtures in emulsion systems (80). The Z groups can be any functional group that does not interfere with either free radicals or the transfer of the radical into the M/P particle.

Other initiators that have been used to introduce functional groups into polymers are aliphatic diazo compounds, acyl peroxides containing functional groups and cyclic hydroperoxides which rearrange into carboxylic groups as free radicals. Evidence for the presence of many other functional groups attached to polymers arising from different initiating systems has been obtained by a dye partition test (81). Another means of introducing functional endgroups into SBR type polymers will be discussed in Section VIII-B on modification.

VII. COPOLYMERIZATION

The rate equation for copolymerization in emulsion systems was presented in Section V-B-2 and will be discussed no further here. The rate expression is more complicated in solution polymerization where different modes of termination must be considered (82).

For emulsion polymerization, the equation describing the copolymer composition is modified to allow for propagation in a phase where the monomer concentrations are $[M_1']$ and $[M_2']$ while the overall concentrations are $[M_1]$ and $[M_2]$. For an emulsion where $[M_1']/[M_2'] = k[M_1]/[M_2]$, the composition equation is:

$$\frac{d[M_1]}{d[M_2]} = \frac{[M_1]}{[M_2]} \cdot \frac{r_1'[M_1] + [M_2]}{[M_1] + r_2'[M_2]} \tag{14}$$

where $r_1' = kr_1$ and $r_2' = r_2/k$ (83). (For definition of symbols, see Chapter 4.B by W. Hofmann, p. 193.) Under these circumstances, the reactivity ratios for solution and emulsion copolymerizations should differ. Most of the reactivity ratios presented in the literature for the copolymerization of butadiene and styrene in solution and in emulsion are listed in Tables XI and XII.

TABLE XI

Butadiene (r_1)-Styrene (r_2) Reactivity Ratios in Solution Copolymerization

r_1	r_2	r_1r_2	T, °C	Ref.
1.39 ± 0.03	0.78 ± 0.01	1.08	60	41
1.48 ± 0.08	0.23 ± 0.07	0.34	50	83
1.35 ± 0.12	0.58 ± 0.15	0.78	50	85
1.39 ± 0.15	0.83 ± 0.09	1.15	30	86

TABLE XII
Butadiene (r_1')-Styrene (r_2') Reactivity Ratios in Emulsion Copolymerization

r_1'	r_2'	T, °C	Ref.
1.4 ± 0.2	0.5 ± 0.1	50	87
1.59 ± 0.05	0.44 ± 0.3	50	85
1.8 ± 0.4	0.6 ± 0.1	45	88
1.83	0.65	45	89
1.30 ± 0.1	0.01 ± 0.01	43	90
1.38	0.64	5	91
1.37	0.38	−18	92
1.40	0.44	—[a]	93

[a] Values calculated from polymerization conducted at 5 and 50°C.

While the agreement among the first five pairs of values in Table XII is not very good, average values for the monomer reactivity ratios at about 50°C were given as $r_1' = 1.6$ and $r_2' = 0.5$ (24f). The same authors point out that the copolymer composition-conversion curve changes unexpectedly as the temperature of polymerization is lowered. The composition of the copolymer changed much less with conversion at 55°C than at 50°C, and at −18°C, the combined monomer-conversion curve

TABLE XIII
Reactivity Ratios for Copolymerization of Butadiene with Vinyl Aromatic Compounds or Other Dienes in Emulsion and Solution Systems

Butadiene, r_1	Comonomer	r_2	T, °C	Systems	Ref.
0.78 ± 0.12	Benzalacetophenone	-0.03 ± 0.05	60	emul or bulk	94
1.07	p-Chlorostyrene	0.42	50	emul	83
0.65 ± 0.1	2,5-Dichlorostyrene	0.2 ± 0.04	70	bulk	95
0.46 ± 0.01	2,5-Dichlorostyrene	0.46 ± 0.01	50	emul	83
1.26	Dimethylbutadiene	0.78	−18	emul	96
0.85	2,3-Dimethylbutadiene	0.63	5	emul	91
0.75	Isoprene	0.85	5	emul	91
0.94	Isoprene	1.06	−18	emul	96
1.07	Methyl 2-chlorocinnamate	−0.02	60	bulk	94
1.20	Methyl 2-chlorocinnamate	−0.03	80	bulk	94
2.7	Methyl 4-chlorocinnamate	0.0	80	bulk	94
1.08	α-Methylstyrene	0.57	60	emul	97
1.20	α-Methylstyrene	0.18	45	emul	97
1.55	α-Methylstyrene	0.01	12.8	emul	97

showed a minimum. This violates the monotonic shape predicted by the copolymerization Eq. 14. The minimum in the low temperature composition curve cannot be explained on the basis of any theory of copolymerization and is attributed to a difference between the rates of solution of the two monomers in the polymer particles. However, when the polymerization conditions permit an equilibrium distribution of the monomers between the phases, the r values for solution and emulsion polymerization, with several exceptions, compare fairly well, Tables XI and XII.

A tabulation of r values for copolymerization of butadiene with various vinylaromatic compounds is given in Table XIII, and r values for different diene derivatives with vinylaromatics or dienes are listed in Table XIV.

Du Pont investigators studied the polymerization in emulsion systems of 214 different unsaturated compounds (99), which were polymerized either alone or in mixtures with dienes. This study established that a wide variety of unsaturated compounds are capable of copolymerizing with dienes to give vulcanizable elastomers. Many of the experiments were conducted before the emulsion process was well understood and the results are not definitive as far as the value of these compounds as comonomers is concerned. The work, done after the introduction of newer initiators, modifiers, and processes is more quantitative and has been reviewed elsewhere (23c). Much of the latter research was a part of the Government rubber program.

TABLE XIV
Reactivity Ratios for Copolymerization of Substituted Dienes with
Other Dienes or Vinylaromatic Monomers

Substituted diene	r_1	Comonomer	r_2	T, °C	System	Ref.
Isoprene	2.05 ± 0.45	Styrene	1.38 ± 0.54	50	—	85
Isoprene	1.68 ± 0.0	Styrene	0.8 ± 0.0	50	emul	85
Isoprene	1.30 ± 0.02	Styrene	0.48 ± 0.01	−18	emul	96
Isoprene	2.02	Styrene	0.42	—[a]	bulk	98
Isoprene	1.98	Styrene	0.44	—[a]	emul	98
Isoprene	1.18	Dimethyl-butadiene	0.84	−18	—[b]	96
Dimethyl-butadiene	0.92 ± 0.02	Styrene	0.42 ± 0.02	−18	emul	96

[a] Temperature not given in *Chemical Abstracts*.
[b] Calculated from Q and e values.

Some of the polymers made from monomers other than styrene and butadiene were superior to SBR types, especially in tire applications, but when cost of monomers, monomer-recovery, compatibility with other elastomers, overall balance of physical properties, etc., were considered, the results were not favorable in comparison with the SBR controls. However, large scale production of a butadiene-α-methylstyrene copolymer (SKSM type), as well as a considerable interest in polypiperylene, exists in Russia. These polymers have been examined in the U.S. without being commercialized.

The heats of homo- and copolymerization of butadiene and styrene at 25°C obtained from measurement of heats of combustion are (in kcal/mole of monomer)

Styrene (100)	Butadiene (101)	SBR (101)
16.7	17.4	17.1–17.7

The copolymers ranged in styrene content from 8 to 55 wt % and were prepared at 5 and 50°C. No appreciable difference was found between the heats of copolymerization for polymers made at 5 and 50°C. A means of determining the thermodynamic values for the heterocopolymerization reactions in the copolymerization of butadiene and styrene was reported (93). The heat of polymerization and entropy for the heterocopolymerization of butadiene and styrene at 30°C are 34.6 kcal/mole and 50 eu/mole.

VIII. MOLECULAR WEIGHT AND MOLECULAR WEIGHT DISTRBIUTION (MWD) OF SBR

A. Standard Procedure

Two sets of equations in the literature show the relationships between different variables and transfer agents for free radical polymerizations conducted in solution and emulsion, respectively. The differences arise from treating the monomer concentration either as a variable as in solution polymerization or as a constant as in case of an "ideal" emulsion polymerization. The two sets of equations describing modifier depletion and molecular weights are shown in Table XV for the two systems where the principal termination reaction is by chain transfer with a growing polymer radical.

At low conversion in emulsion polymerizations, the equations derived for solution systems are sometimes used in modifier studies (102,105,106),

TABLE XV

Equations Involving Transfer Agents for Solution and Emulsion Free
Radical Systems

No.	Expression for	Solution $(24g,102)$[a]	Emulsion $(103,104)$[a]
(15)	Modifier depletion, S	$S_0(1 - X)^c$	$S_0 e^{-RX}$
(16)	No. average degree of polymerization, $\bar{P}_n$	$\dfrac{M_0 X}{S_0[1 - (1 - X)^c]}$	$\dfrac{M_0 X}{S_0[1 - e^{-RX}]}$
(17)	Wt. average degree of polymerization, $\bar{P}_w$	$\dfrac{2}{c}\left(\dfrac{M_0}{S_0}\right)\left[\dfrac{1 - (1 - X)^{2-c}}{(2 - c)X}\right]$	$2\left(\dfrac{M_0}{S_0}\right)\dfrac{(e^{RX} - 1)}{(R^2 X)}$
(18)	Vis. average degree of polymerization, $\bar{P}_v$	$\dfrac{k_s}{c}\left(\dfrac{M_0}{S_0}\right)\left[\dfrac{(1 - X)^{-\beta} - 1}{X\beta}\right]^{1/\alpha}$	$k_e\left(\dfrac{M_0}{S_0}\right)\left[\dfrac{e^{\alpha RX} - 1}{\alpha X R^{1+\alpha}}\right]^{1/\alpha}$

[a] $-\beta = \alpha(1 - c) + 1$, α = exponent in the Mark-Houwink equation, M_0 and S_0 are original monomer and modifier concentrations, c = transfer costant, R = regulating index, X = fractional conversion.

and Meehan and coauthors (107) point out that at zero conversion the transfer constants for mercaptans in emulsion polymerization of butadiene or styrene are essentially the same as those determined in solution polymerization. The experimental data from many studies of the kinetics of emulsion polymerization, copolymer composition, and modifier depletion indicate, however, that the initial solution polymerization behavior becomes dominated by *emulsion* polymerization phenomena at a relatively low conversion.

For the second stage in emulsion systems, the equations of the second column of Table XV are best for predicting the behavior of modifiers and molecular weight values. For the third stage in the polymerization as well as for seeding experiments where a separate monomer phase does not exist, the solution polymerization equations are used (107).

Smith (106) derived an expression for the modifier reaction in copolymerization systems, which in terms of the symbols previously used is:

$$\frac{d \ln S}{dX} = -k\left(\frac{C_1 r_1 M_1 + C_2 r_2 M_2}{r_1 M_1^2 + 2M_1 M_2 + r_2 M_2^2}\right) \tag{19}$$

where $[M_1]$ and $[M_2]$ are the mole fractions of the two monomers and the expression in parentheses is called a modification function. Application of Eq. 19 to the kinetics of modifier reactions will be simplified if the modifier function is constant or varies in a uniform manner. By using the reactivity ratios and other data presented in Table V for monomer concentration calculations and values of 1.4 and 1.3 as transfer constants for butadiene and styrene, respectively, the modification function at three conversions can be calculated as follows:

Conversion, %	0	20	52
Modification function	1.38	1.31	1.30

The value varies only 5.3% in the range from zero to 52% conversion, and in the steady-state stage the modification function for this system is essentially constant.

The treatment of the data as presented in Refs. 108 and 110 followed from the excellent correspondence of the experimental data with the theoretical curves. The rationale lies in an assumed constant monomer concentration over the main polymerization interval.

The equations in Table XV, admittedly inexact, are useful for studying modification in SBR systems. Simulated modifier depletion curves, Fig. 6, calculated by means of Eq. 15 show a broad range in effectiveness as the regulating index varies. Experimental depletion curves, Fig. 7, depart from linearity after 50% conversion for some modifiers, which coincides with the disappearance of the monomer phase, i.e., the end of stage II. The selection of a modifier is determined largely by the physical properties desired in the polymer with certain limitations.

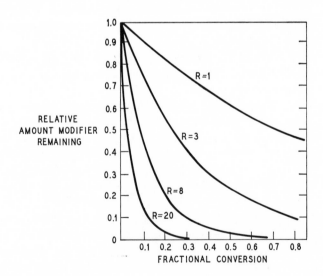

Fig. 6. Effect of regulating index, R, on depletion of modifier (108).

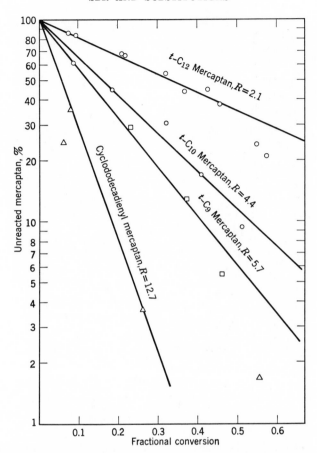

Fig. 7. Modifier depletion curves for various mercaptans in an SBR 1502 system (108).

Simulated curves calculated by means of Eq. 18 for $\bar{P}_v$, or for some property which is a simple function of $\bar{P}_v$, are of help in selecting the most efficient modifier for polymerizing to a specific Mooney value, Fig. 8.

The curves in Fig. 8 indicate that for each terminal conversion there is an optimum regulating index which gives a minimum molecular weight. The $\bar{P}_v$ Eq. 18 is differentiable and the solution after equating to zero is:

$$e^{\alpha RX}[\alpha RX - (1 + \alpha)] = -(1 + \alpha) \qquad (20)$$

With Eq. 20, the optimum regulating indexes (R values) calculated for

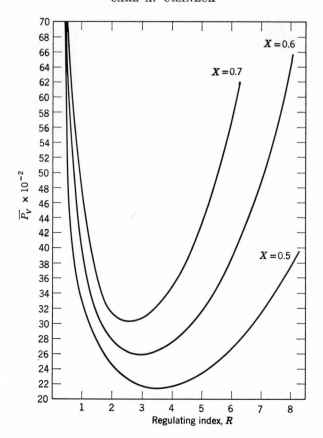

Fig. 8. Effect of regulating index on $\bar{P}_v$.   Terminated at different conversions, X (108).

polymerizations terminated at 50, 60, and 70% conversions are 3.50, 2.91, and 2.50, respectively.   Significantly, the amount of modifier remaining at each conversion is 17.4%.   This value signifies that the most efficient modifier in an ideal emulsion system is one which remains at a small but measurable concentration at the terminal point.   Zero modifier content at terminal conversion is not the most efficient condition.

The amounts of some mercaptans with different regulating indexes needed for a constant Mooney viscosity in the SBR 1502 system are shown in Table XVI.   The amounts of a t-C_{12} and of a t-$C_{10.5}$ mercaptan required in three different SBR systems are shown in Table XVII.   All regulating index-mole of mercaptan values of Table XVI and XVII with

TABLE XVI
Effect of Regulating Index on Modification in SBR 1502

| | RSH | | | Polymerization | | Viscosity | |
| | | | | Time, | Conversion, | | |
t-RSH	PHM	$M \times 10^3$	R	hr	%	Inh.	ML-4
C_{12}^a	0.21	1.04	2.1	6.2	60	2.19	51
C_9^a	0.18	1.12	5.7	6.8	61	2.15	46
C_7^a	0.24	1.97	7.0	14.9	61	1.93	45
C_{10}^b	0.17	0.93	4.3	6.6	60	2.00	49
C_9^b	0.18	1.12	6.7	7.1	60	2.30	51
C_8^b	0.24	1.65	7.4	8.5	60	2.10	36
sec-$C_{12}^{b,c}$	0.46	2.36	12.7	6.8	61	2.64	52

[a] Commercial mercaptans from Phillips Petroleum Co.
[b] Experimental mercaptans.
[c] Experimental cyclododecadienyl mercaptan.

TABLE XVII
Regulating Index—RSH Requirement for t-C_{12} SH and
t-$C_{10.5}$ SH in SBR 1500, 1502, and 1503

| | 1500 | | | 1502 | | | 1503 | | |
	R	PHM	$M \times 10^3$	R	PHM	$M \times 10^3$	R	PHM	$M \times 10^3$
t-C_{12}	2.9	0.187	0.927	2.1	0.21	1.04	2.0	0.235	1.15
t-$C_{10.5}$	5.4	0.176	0.968	4.4	0.157	0.864	3.3	0.155	0.835

the exception of the value for the mercaptan from cyclododecatriene are plotted in Fig. 9. The shape of this curve resembles the shape of those of Fig. 8, and the minimum falls between R of 3 and 4. This agrees well with the theoretical value of approximately 3.

Mooney viscosity vs. R curves for polymerizations conducted at a constant mercaptan level, but varying regulating indexes, are shown in Fig. 10. The curves in Figs. 9 and 10 establish fairly well that for SBR systems the most efficient modification is obtained with the modifier having the optimum regulating index.

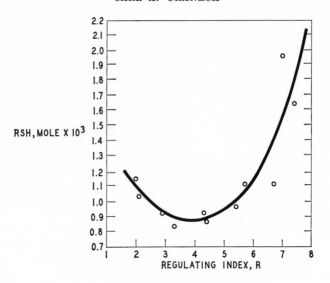

Fig. 9. Effect of regulating index, R, on modifier requirements, mole of RSH, for 50 Mooney polymer made in SBR systems (108).

B. Incremental Addition of Modifier

The control of molecular weight distribution by incremental addition of modifier has long been an unsolved problem in SBR production (23d,109). Recently, a theoretical and practical approach to the problem has been offered (110). A model equation for the viscosity average degree of polymerization, $\bar{P}_v$, of a polymer prepared by the addition of one increment is:

$$\bar{P}_v = k \left[\frac{M_0}{S_0} \right] \left(\frac{1}{X_2 \alpha R^{1+\alpha}} \right)^{1/\alpha} \left\{ (e^{\alpha R X_1} - 1) \right.$$
$$\left. + \left(\frac{b}{b + e^{X_1 R}} \right)^{\alpha} (e^{\alpha R X_2} - e^{\alpha R X_1}) \right\}^{1/\alpha} \quad (21)$$

where X_1 is the fractional conversion at which the increment is added, X_2 is the final conversion, b is the ratio of initial to final increment, and the other terms have been previously defined. The derivation of this and other molecular weight equations has been published (108).

Three sets of curves obtained with the help of Eq. 21 are $\bar{P}_v$ vs. X_1 for different values of R, Fig. 11; $\bar{P}_v$ vs. X_1 for different values of b, Fig. 12; and $\bar{P}_v$ vs. X_1 for different values of X_2, Fig. 13. Experimental verification of the effect of R on Mooney viscosity vs. X_1 curves is shown in

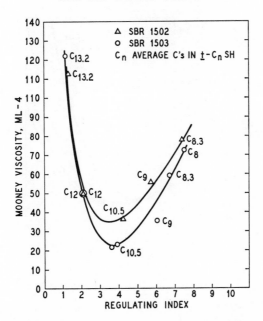

Fig. 10. Mooney-R curves for SBR systems at constant modifier level: SBR 1502 at 1.04 mmole and 1503 at 1.19 mmole/100 monomers.

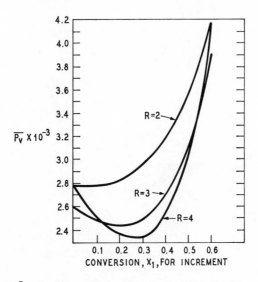

Fig 11. Effect on $\bar{P}_v$ of adding the increment at different conversions, $R = 2, 3,$ 4 (108).

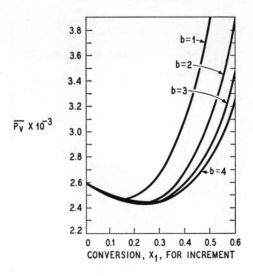

Fig. 12. Effect of variable b on $\bar{P}_v - X_1$ curves, $b = 1, 2, 3, 4$.   (108).

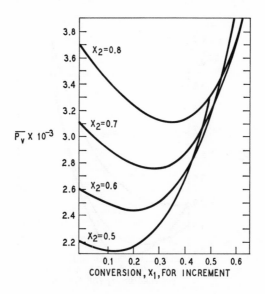

Fig. 13. Effect of variable X_2 on $\bar{P}_v - X_1$ curves (108).

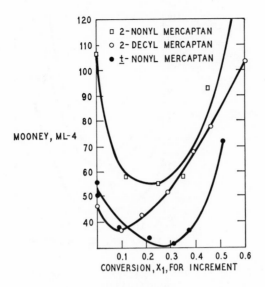

Fig. 14. Mooney viscosity vs. X_1 for various mercaptans in SBR 1502 (108).

Fig. 14, and the effect of b on the same relationship is shown in Fig. 15. Recently the incremental addition of bisisopropyl xanthogen disulfide (Dixie) has been reported as a means of improving the physical properties of the vulcanizate (111).

The curves in Fig. 13 indicate that incremental addition of modifier to polymerizations taken to a higher conversion can give the same molecular weight as a control polymerization stopped at a lower conversion. This is readily seen by the point at which the $X_2 = 0.8$ and $X_2 = 0.5$ curves meet.

Not all modifiers can be used advantageously in the incremental addition technique (23d,109,110). Requirements for successful application of this technique (110) in SBR production are:

1. Adding increments in such a manner that formation of gel during the interval of additions is avoided.
2. Using modifiers that are capable of being distributed between phases in a reasonable time.
3. Finding an optimum division of the amount of modifier between the original charge and the increments as well as an optimum conversion for each addition so that the greatest efficiency in utilization of the modifier is attained.

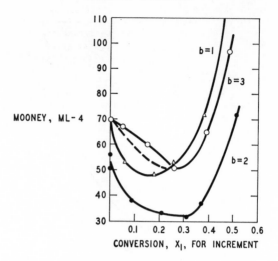

Fig. 15. Effect of variable b on Mooney-X_1 curves for t-nonyl mercaptan in SBR 1502 system (108).

When the proper conditions are fulfilled, incremental addition of mercaptans to SBR systems results in uniform polymer viscosities during the polymerization, changes in MWD, and more efficient utilization of the mercaptan without apparent sacrifice of raw or vulcanizate properties in standard formulations.

C. Molecular Weight of SBR Polymers

Over the years the molecular weights and MWD of SBR have been determined by various procedures. Table XVIII lists some reported values.

The following molecular weight relations are for three SBR polymers fractionated with the Water's Gel Permeation apparatus (123) at 130°C using trichlorobenzene solvent; one 10^6, two 10^4, and one 10^3 Å columns; and a polybutadiene calibration curve with a factor of 1.2 (47):

		Mol wt $\times$ 10^{-3}		
SBR	η_{inh}	$\bar{M}_n$	$\bar{M}_w$	$\bar{M}_w/\bar{M}_n$
1006	1.94	48	365	7.6
1500	1.98	66	275	4.2
1502	1.95	46	275	6.0

TABLE XVIII
Molecular Weight Data for SBR Polymers

SBR type	RSH type	Viscosity	Mol wt × 10⁻³ $\bar{M}_n$	$\bar{M}_w$	$\bar{M}_v$	HI[a]	Characterization procedure[a]	Ref.
1000	n-C₁₂	2.36	96.5	—	330[b]	3.42[b]	o	112
1000	n-C₁₂	2.68	116[c]	—	—	—	o	113
1000	n-C₁₂	2.37	92	—	—	—	o	114
1000[d]	n-C₁₂	2.43	—	—	—	5.9	f and v	115
1000	n-C₁₂	2.25	—	—	—	3.1	f and v	116
1000	t-C₁₂	2.24	—	—	—	4.3	f and v	116
at 10°C[e]	—	1.90	112	800	—	7.2	o and l	115
at 10°C[e]	—	1.86	84.1	740	—	8.7	o and l	115
1500	t-C₁₂	1.95	80	276	246	3.5[f]	o, l, and v	117
1500	t-C₁₂	2.02	95	—	246	2.6	o and v	109
1500	t-C₁₂	2.16	—	—	287	—	v	118
1500	t-C₁₂	2.00	88	—	—	3.9–9.3[g]	o and l	119
1500	t-C₁₂	2.14	113	—	282	2.50	o and v	120
1502	t-C₁₂	—	99	—	261	2.6	o and v	109
1507	t-C₁₂	—	127[h]	—	—	—	u and v	121
1502	t-C₁₂	2.1	—	—	—	2.1	f and v	116
1502	sec-C₁₂	2.6	—	—	—	4.3	f and v	116
1502	t-C₉	2.3	—	—	—	3.0	f and v	116

[a] HI = heterogeneity index: $\bar{M}_w/\bar{M}_n$ or $\bar{M}_v/\bar{M}_n$; o = osmometry; f = fractionation; v = viscometry; l = light scattering; u = ultracentrifuge.

[b] Calculated by Mochel and Nichols (120) from data in Ref. (112).

[c] Reprecipitated polymer.

[d] A 70/30 copolymer prepared at 50°C in a standard system.

[e] 75/25 copolymers prepared at 10°C in a standard system. The $\bar{M}_w/\bar{M}_n$ calculated from the viscosity of fractionated polymers were 2.3 and 2.4 for the two polymers.

[f] $\bar{M}_w/\bar{M}_n$.

[g] For untreated polymer or polymer treated to remove or breakdown microgel.

[h] Value taken from Fig. 5 and calculated with Eq. 7 of Ref. (121).

The range of values in Table XVIII for the same type of polymer reflects the procedure used and the history, the polydispersity, and the non-uniformity of the polymers. Treatment of the polymer solution, SBR or diene, before a measurement can have a profound effect on the molecular weight value, especially the one for weight average by light scattering (115,119,124). Bahary and Bsharah (119) suggest two measurements for SBR polymers, one on the raw polymer and another on treated polymer from which microgel has been removed. Since microgel is a component of many diene polymers and influences their properties, some measurement related to this gel should be of value.

Comparison of values determined by different procedures probably have little significance. However, fairly consistent values are obtained when the same procedure is used on the same type of polymer by different experimenters. For example, the osmotic $\bar{M}_n$ values in Table XVIII show reasonable agreement. In SBR polymerization, where the molecular weight is controlled by the use of mercaptans, the number average molecular weight has been found to be close to that based on the mercaptan consumed (33,101,122a,132). Again, when determinations are made in parallel, the relative values show the proper order as is the case for the last three rows of data in Table XVIII (116). Others have reported different procedures and different aspects of the determination of molecular weight and MWD of SBR polymers (122a–f). At this stage of development, Water's Gel Permeation apparatus is attractive, but recent experience indicates a thorough study of many details of the procedure will be needed before meaningful comparative and absolute values can be obtained.

During the steady-state stage of "ideal" emulsion systems, the rate of polymerization is independent of the initiator concentration and also of the rate of initiation. However, the molecular weight during the second stage is dependent on the rate of initiation and also on that of termination if the latter occurs principally by radical chain combination. When the initiation and termination rates are constant, the molecular weight should be constant. The molecular weight of polystyrene formed in an emulsion system under continuous uniform latex polymerization conditions has been measured recently. The observed variation in molecular weight with particle size was interpreted as departure from the H-S-E theory (125). In a previous study of the molecular weight of polymethyl methacrylate formed under polydisperse particle formation conditions, the variation in molecular weight with conversion was attributed to a change in initiation efficiency and in the termination rate constant (126). Low efficiency of primary radicals in starting polymer chains was found by van der Hoff (63) in similar persulfate initiated systems.

IX. FINE STRUCTURE OF SBR MOLECULES

A. Endgroup Problem

The problem of free chain ends in the rubber network structure was first pointed out by Flory (127) and was studied as a part of the Government research program (128). The preparation of elastomeric molecules with groups capable of being incorporated into the network would be desirable. Such molecules could be prepared by using initiators possessing functional groups and then relying on combination of growing radicals as the principal termination reaction. Examples of this type were included in Section VI-C.

Another means of preparing telechelic polymers is the use of bis-type modifiers for which the transfer process can be represented as (129):

$$\underset{\substack{\text{Growing} \\ \text{polymer} \\ \text{chain}}}{\text{ZRXP·}} + \underset{\substack{\text{Bis-type} \\ \text{modifier}}}{\text{ZRXXRZ}} \rightarrow \underset{\substack{\text{Telechelic} \\ \text{polymer}}}{\text{ZRXPXRZ}} + \underset{\substack{\text{Radical} \\ \text{from} \\ \text{modifier}}}{\text{ZRX·}} \xrightarrow{M} \underset{\text{New chain}}{\text{ZRXM·}}$$

This procedure is of limited value because only a few bis-type modifiers are sufficiently reactive in SBR systems to offer a general method of preparing telechelic polymers. Pierson and coworkers have reported on the use of Dixie which should produce a xanthate-telechelic polymer. Modification with Dixie is well known in the synthetic rubber industry but wide scale use has never been attained. The retardation obtained with Dixie in low temperature systems is a deterrent to its use in commercial production.

The use of a diazothioether as an initiator-modifier as mentioned in Section VI-C, presents probably the most general method of preparing telechelic polymers in SBR systems. Telechelic butadiene-styrene copolymers prepared with an acetyl-diazothioether had vulcanizate properties that were superior to those of a control prepared in a conventional system. The improvement was attributed to the incorporation of free chain ends into the vulcanizate network structure (80). These results support the belief (128) that improved SBR can be obtained with telechelic polymers possessing endgroups capable of reacting to eliminate free chain ends.

B. Branching and Crosslinking

Flory distinguished between branching and crosslinking in diene polymerization by pointing out that the former linkage arises from a chain transfer process, whereas the latter arises from polymer molecules participating in a propagation reaction in place of monomer (130). Branching by a chain transfer process gives a trifunctional unit and, in the absence of other processes, will not lead to an infinite network structure. Crosslinking, on the other hand, gives a tetrafunctional linkage, and when a critical amount is exceeded, gel begins to form. Polydienes possessing

TABLE XIX
Crosslinking and Ratio of Rate Constants for Butadiene,
Isoprene, and Dimethylbutadiene

Monomer	Crosslinking k_x, 1 mole^{-1} sec^{-1}	At 60°C $k_x/k_p \times 10^5$	Ref.
Butadiene	$1.9 \times 10^9 \exp{(-16,800/RT)}$	20	131
Isoprene	$8.5 \times 10^6 \exp{(-14,800/RT)}$	3.4	132
Dimethylbutadiene	$3.0 \times 10^6 \exp{(-14,400/RT)}$	0.92	33

butenyl units, $-CR_2C(R)\!=\!C(R)CR_2-$ where R may be hydrogen, are susceptible to attack by free radicals. Dienes polymerized in SBR systems in the absence of chain transfer agents form gel at low conversions, and the products are unsuitable for most end uses. An upper limiting molecular weight is usually encountered in diene polymerizations because of the crosslinking reaction. Modifiers are used in these cases to keep the average molecular weight at a safe value below this upper limit.

A kinetic expression, derived by Flory (130) for the rate of crosslinking in a polymerization, was adapted by Morton and coworkers (131) to the determination of the crosslinking constants for dienes in emulsion polymerization of butadiene, isoprene, and dimethylbutadiene in SBR systems. Some data are listed in Table XIX.

Morton and Gibbs (33) note that the Arrhenius expressions, Section V-B-3, for the propagation reaction for the three monomers are quite similar. The differences in the crosslinking expressions, Table XIX, are attributed to the shielding of the double bonds of the dienyl units in polymers from isoprene and dimethylbutadiene provided by the (one or two) methyl groups.

Another method of measuring the extent of crosslinking during the emulsion polymerization of butadiene has been reported by Howland et al. (133) who used radioactive modifiers containing S^{35} in order to measure the number of end groups and then related these to osmotic molecular weights. The following is a comparison of the crosslinking density, ρ, the fraction of the polymerized units involved in crosslinks, determined by the two groups of workers:

$\rho \times 10^4$ in Ref.

$\bar{P}_n$	(132)	(133)
1370	3.1	5.2 (Fig. 1)
1700	3.3	5.3 (Fig. 1)

The data of Morton and of Howland and their coworkers are noteworthy in showing that ρ is rather constant over an appreciable conversion range, especially at a low polymerization temperature. However, Flory's equation for solution polymerization predicts a considerable increase in ρ as conversion increases. The constant value of ρ for emulsion polymerization is consistent with predictions based on H-S-E concepts.

The value of the exponent in the Mark-Houwink equation: $[\eta] = KM^{\alpha}$ can also be used to detect branching in polymers. Zimm and Kilb (134) concluded that α should decrease as the density of crosslinking becomes greater. Johnson and Wolfangel (135) found that α did vary for the products of emulsion polymerizations of butadiene taken to various conversions at different temperatures.

Still another method of determining branching by viscometric measurements is based on the theoretical derivations of Zimm and Stockmayer (136) who define a function g as a ratio of some characteristic of a branched to a linear polymer of the same molecular weight. Pollock et al. (137) applied a simplified expression of g to viscometric measurements and found ρ for a polybutadiene prepared at 50 and 5°C to be 1.5×10^{-4} and 1×10^{-4}, respectively.

Some empirical relationships exist between the Mooney viscosity of polybutadiene and the Mooney viscosity average molecular weight, $\overline{M}_m$, or intrinsic viscosity (138):

$$(ML) = K_m(\overline{M}_m)^{\beta}$$

$$(ML\text{-}4) = K(\eta)^{\varphi}$$

In the latter expression, the constants would be expected to change as the extent of branching of the polymer varies. Expressions for Mooney viscosity of extended polymers and of mixtures of polymers were also presented.

The extent of branching formed in emulsion polymerization in comparison with that formed in solution or bulk was proposed as evidence for deviation from the H-S-E theory (139). Polymerization at a surface, where a high concentration of monomer relative to polymer exists, should produce less branching than polymerization throughout an M/P particle where a relatively high concentration of polymer should exist. The results obtained for the homopolymerization of styrene and methyl methacrylate in an emulsion system indicated, but did not prove conclusively, that less branching occurred in emulsion than in bulk polymerization of the same monomers (139). If Medvedev's mechanism is valid, then the basic equations such as (1) and (10) would be the same, if the monomer concentration is constant in the zone where propagation occurs.

C. Stereochemistry

The stereochemistry of the butenyl unit in SBR has played a minor part in the development of this elastomer microstructure because of the difficulty in changing the configuration significantly by any of the standard procedures.   The amounts and kinds of emulsifier, initiator system, electrolyte, organic diluent, inhibitor, modifier and the conversion have little or no effect on the microstructure.   Styrene as a comonomer has a measurable effect, and other comonomers such as acrylonitrile exhibit a marked penultimate effect (140).

Temperature exerts a considerable influence on the 1,4/1,2 as well as on the *cis/trans* ratio.   Condon (141) handled the effect of temperature on the molecular structure by treating the formation of *cis*, *trans*, and vinyl units as three simultaneous competing reactions and obtained the curves shown in Fig. 16.

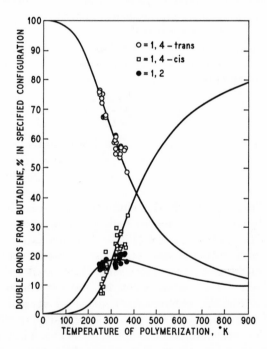

Fig. 16. Distribution of double bonds in polybutadiene as a function of polymerization temperature (after Condon 141).   O = 1,4-*trans*; □ = 1,4-*cis*; ● = 1,2.

Flory (130) originally suggested that the 1,4 and 1,2 units arise because of allylic resonance of the terminal unit:

$$\sim C - C = C - C \cdot \; \rightleftharpoons \; \sim C - C \cdot$$
$$\underset{\underset{C}{\parallel}}{\overset{\mid}{C}}$$

and that the structure was not fixed until the next butadiene molecule was added. However, this does not account for the *cis/trans* ratios. Walling (55b) proposed a combination of equilibrium and resonance steps to account for the structures found in polybutadiene. More definitive data are needed before Condon's proposal can be verified.

Figure 16 shows the limitation to obtaining a highly stereoregular polybutadiene by an emulsion, free radical mechanism. Polybutadiene prepared at $-40°C$ still has enough vinyl units to keep the polymer in a rubbery state at room temperature. Crystallinity can be detected in polybutadiene prepared at 20°C or below and the amount increases as the temperature of polymerization is decreased (142). Stereoregular polydienes have been obtained in emulsion systems by the use of transition metal initiators (see Chapter 8.B, Part II).

This brief examination of the current research in SBR systems indicates that control of MWD, attachment of endgroups to polymer molecules, and preparation of block polymers are possible; and with the use of transition metal initiators, even stereospecific polymerization can be accomplished in emulsion systems.

X. POLYMERIZATION PROCEDURES

The batch process was originally adopted for SBR production because of the many unknowns in the theory and the production and in the properties of the product. As the relationships between polymerization variables and polymer quality became clarified for batch polymerizations, some attention was directed to continuous processes. The developments in this area up to mid-1950 is reviewed in the literature (23e). Various theoretical studies were made during the Government program and several others have been reported since then. Denbigh (143) concluded that a narrower MWD would be obtained in a single continuous stirred reactor if the life of a growing chain was considerably less than the residence time in the reactor. For a continuous emulsion copolymerization in one reactor in which a constant (steady-state) monomer ratio was maintained, Wall and coworkers (144) found that compositional homogeneity was improved over that of a copolymer prepared in a batch process. Significantly, the copolymer composition expression developed was identical with that of Eq. (14) except that for the continuous process $[M_1]$ and $[M_2]$ are the steady state concentrations of the monomers.

General theoretical analyses have been reported recently for homo- and copolymerizations in both continuous and batch processes and for a number of kinetic models reported in the literature. In one approach to the solution of the various polymerization expressions, simultaneous first order differential equations for the concentrations of the different species were solved numerically by a digital computer (145). In another approach, only the term for the first derivative in the Taylor series expansion of the expression for the polymer radical concentration was retained (146). These approaches give estimates of the radical and the terminated molecule concentrations, as well as of the MWD. The analyses were conducted with idealized models and the approach was considered tentative and introductory for the method. This work was recently extended to an analytical study of the polymerization kinetics in batch reactors (147).

Computer control of SBR plants has been announced by at least four companies (148). Two reports (149) give qualitative descriptions of the principles on which the control models are based, and they present a limited number of the model equations used either for simulated calculations or as an operational expression. Both analog and digital computers are being used. Computer control of a process as complicated as SBR production is indicative of a high level of achievement for this complex plant operation.

References

1. W. A. Schulze, W. B. Reynolds, C. F. Fryling, L. R. Sperberg, and J. E. Troyan, *India Rubber World*, **117**, 739 (1948); W. H. Shearon, J. P. McKenzie, and M. E. Samuels, *Ind. Eng. Chem.*, **40**, 769 (1948).
2. *Chem. Eng. News*, **34**, 2223 (1956).
3. *Chem. Eng. News*, **33**, 3472 (1955); *ibid*, **33**, 3553 (1955); *ibid*, **33**, 4518 (1955).
4. R. E. Workman, *Rubber Age*, **97** (5), 52 (1965).
5. *a*. G. S. Whitby, Ed., *Synthetic Rubber*, Wiley, New York, 1954, Chap. 5; *b*. Chap. 4; *c*. Chap. 6.
6. W. Graulich, *Kautschuk Gummi Kunststoff*, **18**, 491 (1965).
7. L. H. Beckberger and K. M. Watson, *Chem. Eng. Prog.*, **44**, 229 (1948); K. K. Kearby, *Ind. Eng. Chem.*, **42**, 295 (1950); K. K. Kearby, U.S. Pat. 2,395,875 and -876 (Mar. 5, 1946) (to Jasco, Inc.); K. K. Kearby, U.S. Pat. 2,426,829 (Sept. 2, 1947) (to Standard Oil Development Co.); S. D. Sumerford, U.S. Pat. 2,436,616 (Feb. 24, 1948) (to Standard Oil Development Co.).

8. C. L. Gutzeit, U.S. Pat. 2,408,140 (Sept. 24, 1946) (to Shell Development Co.); U.S. Pat. 2,408,139 (Sept. 24, 1946) (to Shell Development Co.); F. T. Eggertsen and H. H. Voge, U.S. Pat. 2,414,585 (Jan. 21, 1947) (to Shell Development Co.); W. E. Armstrong and C. Z. Morgan, U.S. Pat. 2,916,531 (Dec. 8, 1959) (to Shell Development Co.); U.S. Pat. 2,870,228 (Jan. 20, 1959) (to Shell Development Co.).

9. E. C. Britton, A. J. Dietzler, and C. R. Noddings, *Ind. Eng. Chem.*, **43**, 2871 (1951); P. M. Reilly, *Chem. Can.*, **5**, 41 (1953).

10. G. H. Hanson and H. L. Hays, *Chem. Eng. Prog.*, **44**, 431 (1948): J. R. Owen, U.S. Pat. 2,606,159 (Aug. 5, 1952) (to Phillips Petroleum Co.); E. W. Pitzer, U.S. Pat. 2,866,790 (Dec. 30, 1958) (to Phillips Petroleum Co.); K. H. Hackmuth, U.S. Pat. 2,386,310 (Oct. 9, 1945) (to Phillips Petroleum Co.).

11. J. P. O'Donnell, *Oil Gas J.*, **41**, No. 29, 38 (1942); C. H. Thayer, R. C. Lassiat, and E. R. Lederer, *Chem. Met. Eng.* **49**, No. 11, 116 (1942).

12. C. R. Adams, H. H. Voge, C. Z. Morgan, and W. E. Armstrong, *J. Catalysis*, **3**, 379 (1964); PH. A. Batist, B. C. Lippens, and G. C. Schuit, *ibid.*, **5**, 55 (1966); H. H. Voge, W. E. Armstrong, L. B. Ryland, U.S. Pat. 3,110,746 (Nov. 12, 1963) (to Shell Oil Co.); H. H. Voge, U.S. Pat. 3,067,272 (Dec. 4, 1962) (to Shell Oil Co.); B. Masterton and J. A. Langton, U.S. Pat. 3,050,572 (Aug. 21, 1962) (to Shell Oil Co.); G. W. Hearne and K. E. Furman, U.S. Pat. 2,991,320 (July 4, 1961) (to Shell Oil Co.), *Chem. Abstr.*, **56**, 3351 (1962); H. H. Voge and C. R. Adams, U.S. Pat. 2,991,321, Appl. (Dec. 29, 1959) (to Shell Oil Co.), *Chem. Abstr.*, **66**, 6706 (1962); W. E. Armstrong, H. H. Voge, and C. R. Adams, U.S. Pat. 2,991,322, Appl. (Dec. 29, 1959) (to Shell Oil Co.), *Chem. Abstr.*, **56**, 3351 (1962); J. L. Callahan, B. Gertisser, and G. Grasselli U.S. Pat. 3,260,768, (July 12, 1966) (to Ohio Standard Oil Co.); U.S. Pat. 3,257,474 (July 21, 1966) (to Ohio Standard Oil Co.); L. Bajars, U.S. Pat. 3,260,767 (July 12, 1966) (to Petro-Tex Chemical Corp.).

13. J. H. Raley, R. D. Mullineaux, and C. W. Bittner, *J. Am. Chem. Soc.*, **85**, 3174 (1963); R. J. Gay, U.S. Pat. 3,207,805 (Sept. 21, 1965) (to Petro-Tex Chem. Corp.); L. Bajars, U.S. Pat. 3,207,806 (Sept. 21, 1965) (to Petro-Tex Chem. Corp.); L. Bajars and R. M. Mantell, U.S. Pat. 3,207,807 (Sept. 21, 1965) (to Petro-Tex Chem. Corp.); L. Bajars, U.S. Pat. 3,207,808-811 (Sept. 21, 1965) (to Petro-Tex Chem. Corp.); L. Bajars and B. M. Mantell, U.S. Pat. 3,210,436 (Oct. 5, 1965) (to Petro-Tex Chem. Corp.); R. D. Mullineaux and J. H. Raley, U.S. Pat. 2,890,253 (Aug. 9, 1959) (to Shell Development Co.); J. H. Raley, R. D. Mullineaux, and S. A. Ballard, U.S. Pat. 2,880,249 (Mar. 31, 1959) (to Shell Development Co.), *Chem. Abstr.*, **53**, 15548 (1959); J. H. Raley, R. D. Mullineaux, and S. A. Ballard, U.S. Pat. 2,898,386 (Aug. 4, 1959) (to Shell Development Co.), *Chem. Abstr.*, **54**, 10298 (1960); R. L. Magovern, U.S. Pat. 2,921,101 (Jan. 12, 1960) (to Shell Development Co.), *Chem. Abstr.*, **54**, 12569 (1960); M. Nager, U.S. Pat. 3,080,435 (Mar. 5, 1963) (to Shell Development Co.); F. Wattimena and W. F. Engel, U.S. Pat. 3,205,280 (Sept. 7, 1965) (to Shell Development Co.); R. P. Arganbright, U.S. Pat. 3,028,440 (April 3, 1962) (to Monsanto); C. C. Kennedy and C. R. Russell, U.S. Pat. 2,224,155 (Dec. 10, 1940) (to Dow Chemical Company), *Chem. Abstr.*, **35**, 1808 (1941); H. Bahr and W. Deiters, U.S. Pat. 2,259,195 (Oct. 14, 1941) (to Jasco, Inc.), *Chem. Abstr.*, **36**, 491 (1942); A. Cantzler and H. Krekeler, U.S. Pat. 2,274,358 (Feb. 12, 1942) (to Jasco, Inc.), *Chem. Abstr.*, **36**, 4138 (1942).

14. C. N. Wolf, R. I. Bergman, and M. Sittig, *Chem. Week*, **98**, No. 22, 113 (1966); *Eur. Chem. News*, **3**, No. 47, 25 (1963).
15. *Chem. Eng. News*, **44**, No. 10, 54 (1966).
16. L. Bajars and R. M. Mantell, Belg. Pat. 617,893 (Nov. 21, 1962) (to Petro-Tex Chem. Corp.).
17. A. M. Dadasheva, T. G. Alkhazor, and M. S. Belenkii, *Izv. Vysshekh. Uchebn. Zavedenii, Neft i Gaz.*, **8**, No. 11, 8 (1965); G. H. Kalb, U.S. Pat. 2,719,171 (Sept. 27, 1955) (to du Pont); E. H. Lee, U.S. Pat. 3,179,707 (April 20, 1965) (to Monsanto).
18. W. D. Harkins, *J. Am. Chem. Soc.*, **69**, 1428 (1947).
19. W. D. Harkins, *J. Polymer Sci.*, **5**, 217 (1950).
20. W. V. Smith and R. H. Ewart, *J. Chem. Phys.*, **16**, 592 (1948).
21. B. M. E. van der Hoff, *Advan. Chem.*, **34**, 6 (1962).
22. H. Gerrens, *Fortschr. Hochpolymer.-Forsch.*, **1**, 234 (1959).
23. *a*. G. S. Whitby, Ed., *Synthetic Rubber*, Wiley, New York, 1954; *b*. p. 350; *c*. Chap. 21; *d*. p. 253; *e*. Chap. 7 and Chap. 26; *f*. p. 235.
24. *a*. F. A. Bovey, I. M. Kolthoff, A. I. Medalia, and E. J. Meehan, *Emulsion Polymerization*, Interscience, New York, 1955; *b*. p. 196; *c*. p. 201; *d*. p. 345; *e*. p. 390; *f*. p. 257; *g*. p. 132.
25. Houben-Weyl, *Methoden Der Organischen Chemie*, Georg Thieme Verlag, Stuttgart, Vol. 14/1, 1961.
26. W. V. Smith, *J. Am. Chem. Soc.*, **70**, 3695 (1948).
27. D. M. French, *J. Polymer Sci.*, **32**, 395 (1958).
28. B. M. E. van der Hoff, *J. Polymer Sci.*, **44**, 241 (1960).
29. E. Bartholome, H. Gerrens, R. Herbeck, and H. M. Weitz, *Z. Elektrochem.*, **60**, 334 (1956).
30. M. Morton, P. P. Salatiello, and H. Landfield, *J. Polymer Sci.*, **8**, 111 (1952).
31. M. Morton, P. P. Salatiello, and H. Landfield, *J. Polymer Sci.*, **8**, 215 (1952).
32. M. Morton, P. P. Salatiello, and H. Landfield, *J. Polymer Sci.*, **8**, 279 (1952).
33. M. Morton and W. E. Gibbs, *J. Polymer Sci.*, **1A**, 2679 (1963).
34. H. B. Klevens, *J. Colloid. Sci.*, **2**, 365 (1947).
35. J. Bakker, *Philips Res. Rept.*, **7**, 344 (1952).
36. A. Rysanek, *J. Polymer Sci.*, **29**, 557 (1958); *ibid.*, **52**, 91 (1961).
37. J. G. Brodnyan, J. A. Cala, T. Konen, and E. L. Kelley, *J. Colloid. Sci.*, **18**, 73 (1963).
38. A. G. Parts, D. E. Moore, and J. G. Watterson, *Makromol. Chem.*, **89**, 156 (1965).
39. F. R. Mayo and C. Walling, *Chem. Rev.*, **46**, 191 (1950).
40. C. M. Tucker and J. E. Troyan, *Rubber World*, **121**, 67, 190 (1949).
41. F. M. Lewis, C. Walling, W. Cummings, E. R. Briggs, and W. T. Wenisch, *J. Am. Chem. Soc.*, **70**, 1527 (1948).
42. I. M. Kolthoff and A. I. Medalia, *J. Polymer Sci.*, **5**, 391 (1950).
43. I. M. Kolthoff and A. I. Medalia, *J. Polymer Sci.*, **6**, 195 (1951).
44. I. M. Kolthoff and E. J. Meehan, *J. Polymer Sci.*, **9**, 355 (1952).
45. I. M. Kolthoff, D. S. Brackman, and E. J. Meehan, *J. Polymer Sci.*, **21**, 529 (1956).
46. R. W. Brown, C. V. Bawn, E. B. Hansen, and L. H. Howland, *Ind. Eng. Chem.*, **46**, 1073 (1954).
47. Unpublished data from Phillips Petroleum Company.

48. S. S. Medvedev, *International Symposium on Macromolecular Chemistry*, Pergamon Press, New York, 1959, p. 174.
49. H. Cherdron, *Kunststoffe*, **50**, 568 (1960).
50. S. Okamura and T. Motoyama, *J. Polymer Sci.*, **58**, 221 (1962).
51. C. H. Bamford, W. G. Barb, A. D. Jenkins, and P. F. Onyon, *The Kinetics of Vinyl Polymerization by Radical Mechanism*, Academic Press, New York, 1958.
52. G. S. Whitby, M. D. Gross, J. R. Miller, and A. J. Costanza, *J. Polymer Sci.*, **16**, 549 (1955).
53. J. E. Pritchard, M. H. Opheim, and P. H. Moyer, *Ind. Eng. Chem.*, **47**, 863 (1955).
54. G. M. Burnett and R. S. Lehrle, *Proc. Roy. Soc. (London)*, **253A**, 331 (1959).
55. a. C. Walling, *Free Radicals in Solution*, Wiley, New York, 1957, p. 208; b. p. 229.
56. R. H. Gobran, M. G. Berenbaum, and A. V. Tobolsky, *J. Polymer Sci.*, **46**, 431 (1960).
57. C. P. Roe and P. D. Brass, *J. Polymer Sci.*, **24**, 401 (1957).
58. W. H. Stockmayer, *J. Polymer Sci.*, **24**, 314 (1957).
59. B. M. E. van der Hoff, *J. Polymer Sci.*, **33**, 487 (1958).
60. R. N. Haward, *J. Polymer Sci.*, **4**, 273 (1949).
61. H. Gerrens, *Z. Elektrochem.*, **60**, 400 (1956).
62. E. Trommsdorff, H. Kohle, and P. Lagally, *Makromol. Chem.*, **1**, 169 (1948).
63. B. M. E. van der Hoff, *J. Polymer Sci.*, **48**, 175 (1960).
64. J. T. O'Toole, *J. Appl. Polymer Sci.*, **9**, 1291 (1965).
65. I. M. Kolthoff, P. R. O'Connor, and J. L. Hansen, *J. Polymer Sci.*, **15**, 459 (1955).
66. I. M. Kolthoff and W. E. Harris, *J. Polymer Sci.*, **2**, 41 (1947).
67. J. E. Wicklatz, T. J. Kennedy, and W. B. Reynolds, *J. Polymer Sci.*, **6**, 45 (1951).
68. J. W. L. Fordham and H. L. Williams, *J. Am. Chem. Soc.*, **72**, 4465 (1950); *ibid.*, **73**, 1634 (1951).
69. R. J. Orr and H. L. Williams, *J. Phys. Chem.*, **57**, 925 (1953).
70. R. J. Orr and H. L. Williams, *Can. J. Chem.*, **30**, 985 (1952).
71. C. F. Fryling and A. E. Follett, *J. Polymer Sci.*, **6**, 59 (1951).
72. C. A. Uraneck and R. J. Sonnenfeld, *Ind. Eng. Chem.*, **52**, 790 (1960).
73. R. J. Orr and H. L. Williams, *Can. J. Technol.*, **29**, 29 (1951).
74. W. L. Reynolds and I. M. Kolthoff, *J. Phys. Chem.*, **60**, 996 (1956).
75. H. M. Andersen and S. I. Proctor, *J. Polymer Sci.*, **3A**, 2343 (1965).
76. E. J. Meehan, I. M. Kolthoff, C. Auerbach, and H. Minato, *J. Am. Chem. Soc.*, **83**, 2232 (1961).
77. R. J. Orr and H. L. Williams, *J. Am. Chem. Soc.*, **79**, 3137 (1957).
78. A. K. Prince and R. D. Spitz, *Ind. Eng. Chem.*, **52**, 235 (1960).
79. I. M. Kolthoff and E. J. Meehan, *J. Appl. Polymer Sci.*, **1**, 200 (1959).
80. C. A. Uraneck, H. L. Hsieh, and O. G. Buck, *J. Polymer Sci.*, **46**, 535 (1960).
81. P. Ghosh, S. C. Chadha, and S. R. Palit, *J. Polymer Sci.*, **2A**, 4441 (1964), and preceding publications of senior author S. R. Palit.
82. H. W. Melville, B. Noble, and W. F. Watson, *J. Polymer Sci.*, **2**, 229 (1947).
83. F. T. Wall, R. W. Powers, G. D. Sands, and G. S. Stent, *J. Am. Chem. Soc.*, **70**, 1031 (1948).
84. R. Simha and L. A. Wall, *J. Res. Natl. Bur. Std.*, **41**, 521 (1948). From data of K. R. Henery-Logan and R. V. V. Nicholls.
85. F. R. Mayo and C. Walling, *Chem. Rev.*, **46**, 191 (1950). From data of K. R. Henery-Logan and R. V. V. Nicholls.

86. L. Crescentini, G. B. Gechele, and A. Zanella, *J, Appl. Polymer Sci.*, **9**, 1323 (1965).

87. E. J. Meehan, *J. Polymer Sci.*, **1**, 318 (1946).

88. J. M. Mitchell and H. L. Williams, *Can. J. Res.*, **27F**, 35 (1949).

89. R. D. Gilbert and H₄ L. Williams, *J. Am. Chem. Soc.*, **74**, 4114 (1952). See Ref. 75 for the values.

90. N. Ashikari, *Bull. Chem. Soc. Japan*, **32**, 1060 (1954). Ref. given in G. E. Ham, *Copolymerization*, Interscience, New York, 1964, Appendix A.

91. R. D. Gilbert and H. L. Williams, *J. Am. Chem. Soc.*, **74**, 4114 (1952).

92. R. J. Orr and H. L. Williams, *Can. J. Chem.*, **29**, 270 (1951).

93. R. J. Orr, *Polymer*, **2**, 79 (1961).

94. G. P. Scott, *J. Org. Chem.*, **20**, 736 (1955).

95. T. Alfrey, A. I. Goldberg, and W. P. Hohenstein, *J. Am. Chem. Soc.*, **68**, 2464 (1946).

96. R. J. Orr and H. L. Williams, *Can. J. Chem.*, **30**, 108 (1952).

97. J. W. L. Fordham and H. L. Williams, *J. Phys. Chem.*, **57**, 346 (1953).

98. M. P. Zverev and M. F. Margaritovs, *Ukrain. Khim. Zh.*, **24**, 626 (1958), through *Chem. Abstr.*, **53**, 10823 (1959).

99. H. W. Starkweather, R. O. Barl, H. S. Carter, F. B. Hill, V. R. Hurka, C. J. Mighton, P. A. Sanders, H. W. Walker, and M. A. Youker, *Ind. Eng. Chem.*, **39**, 210 (1947).

100. D. E. Roberts, W. W. Walton, and R. S. Jessup, *J. Polymer Sci.*, **2**, 420 (1947).

101. R. A. Nelson, R. S. Jessup, and D. E. Roberts, *J. Res. Natl. Bur. Std.*, **48**, 275 (1952).

102. W. V. Smith, *J. Am. Chem. Soc.*, **68**, 2059 (1946).

103. J. Bardwell and C. A. Winkler, *Can. J. Res.*, **27B**, 116, 128 (1949); R. H. Ewart, W. V. Smith, and G. E. Hulse, War Production Board, CR-73, June 24, 1943.

104. D. S. Montgomery and C. A. Winkler, *Can. J. Res.*, **28B**, 416 (1950).

105. W. V. Smith, *J. Am. Chem. Soc.*, **68**, 2064 (1946).

106. W. V. Smith, *J. Am. Chem. Soc.*, **68**, 2069 (1946).

107. E. J. Meehan, I. M. Kolthoff, and P. R. Sinha, *J. Polymer Sci.*, **2A**, 4911 (1964).

108. C. A. Uraneck and J. E. Burleigh, *Kautschuk Gummi Kunststoffe*, **19**, 532 (1965).

109. C. Booth, L. R. Beason, and J. T. Bailey, *J. Appl. Polymer Sci.*, **5**, 116 (1961).

110. C. A. Uraneck and J. E. Burleigh, *J. Appl. Polymer Sci.*, **9**, 1273 (1965).

111. I. Y. Poddubnyi and M. A. Rabinerzon, *J. Appl. Polymer Sci.*, **9**, 2527 (1965).

112. J. A. Yanko, *J. Polymer Sci.*, **3**, 584 (1948).

113. G. D. Sands and B. L. Johnson, *Ind. Eng. Chem., Anal. Ed.*, **19**, 261 (1947).

114. D. M. French and R. H. Ewart, *Ind. Eng. Chem., Anal. Ed.*, **19**, 165 (1947).

115. W. Cooper, G. Vaughan, D. E. Eaves, and R. W. Madden, *J. Polymer Sci.*, **50**, 159 (1961).

116. C. A. Uraneck, M. G. Barker, and W. D. Johnson, *Rubber Chem. Technol.*, **38**, 809 (1965).

117. C. Booth and L. R. Beason, *J. Polymer Sci.*, **42**, 93 (1960).

118. T. Homma, K. Kawakara, and H. Fujita, *J. Appl. Polymer Sci.*, **8**, 2853 (1964).

119. W. S. Bahary and L. Bsharah, *Polymer Preprints*, **5**, (1) 1 (1964).

120. W. E. Mochel and J. B. Nichols, *Ind. Eng. Chem.*, **43**, 154 (1951).

121. T. Homma and H. Fujita, *J. Appl. Polymer Sci.*, **9**, 1701 (1965).

122. *a.* E. J. Meehan, *J. Polymer Sci.*, **6**, 255 (1949); *b.* R. L. Scott, W. C. Carter, and M. Magat, *J. Am. Chem. Soc.*, **71**, 220 (1946); *c.* L. H. Cragg and J. E. Simkins, *Can. J. Res.*, **27B**, 961 (1949); *d.* B. L. Johnson, *Ind. Eng. Chem.*, **40**, 351 (1948); *e.* A. W. Meyer, *Ind. Eng. Chem.*, **41**, 1570 (1949); *f.* M. A. Golub, *J. Polymer Sci.*, **11**, 583 (1953).

123. J. C. Moore, *J. Polymer Sci.*, **2A**, 835 (1964).

124. W. Cooper, D. E. Eaves, and G. Vaughan, *J. Polymer Sci.*, **59**, 241 (1962).

125. D. J. Williams and E. G. Bobalek, *Polymer Preprints*, **5**, (2) 688 (1964).

126. W. S. Zimmt, *J. Appl. Polymer Sci.*, **1**, 323 (1959).

127. P. J. Flory, *Ind. Eng. Chem.*, **38**, 417 (1946).

128. R. M. Pierson, "Formation of Polymer Net-Works Through the Use of Reactive End Groups. I. Theory and Discussion," private communication to Office of Synthetic Rubber, R. F. C., 1951.

129. R. M. Pierson, A. J. Costanza, and A. H. Weinstein, *J. Polymer Sci.*, **17**, 221 (1955); A. J. Costanza, R. J. Coleman, R. M. Pierson, C. S. Marvel, and C. King. *ibid.*, **17**, 319 (1955).

130. P. J. Flory, *Principles of Polymer Chemistry*, Cornell University Press, Ithaca, N.Y., 1953.

131. M. Morton and P. P. Salatiello, *J. Polymer Sci.*, **6**, 225 (1951).

132. M. Morton, J. A. Cala, and I. Piirma, *J. Polymer Sci.*, **15**, 167 (1955).

133. L. H. Howland, A. Nisonoff, L. E. Dannals, V. S. Chambers, *J. Polymer Sci.*, **27**, 115 (1958).

134. B. H. Zimm and R. W. Kilb, *J. Polymer Sci.*, **37**, 19 (1959).

135. B. L. Johnson and R. D. Wolfangel, *Ind. Eng. Chem.*, **41**, 1580 (1949); *ibid.*, **44**, 752 (1952).

136. B. H. Zimm and W. H. Stockmayer, *J. Chem. Phys.*, **17**, 1301 (1949).

137. D. J. Pollock, L. J. Elyash, and T. W. DeWitt, *J. Polymer Sci.*, **15**, 335 (1955).

138. B. M. E. van der Hoff, J. F. Henderson, and R. M. B. Small, *Rubber Plastics Age*, **46**, 821 (1965).

139. J. G. Brodnyan, E. Cohn-Ginsberg, and T. Konen, *J. Polymer Sci.*, **3A**, 2392 (1965).

140. F. C. Foster and J. L. Binder, *J. Am. Chem. Soc.*, **75**, 2910 (1953).

141. F. E. Condon, *J. Polymer Sci.*, **11**, 139 (1953).

142. K. E. Beu, W. B. Reynolds, C. F. Fryling, and H. L. McMurry, *J. Polymer Sci.*, **3**, 465 (1948).

143. K. G. Denbigh, *Trans. Faraday Soc.*, **40**, 352 (1944); *ibid.*, **43**, 648 (1947).

144. F. T. Wall, C. J. Delberq, and R. E. Florin, *J. Polymer Sci.*, **9**, 177 (1952).

145. Shean-Lin Liu and N. R. Amundson, *Rubber Chem. Technol.*, **34**, 995 (1961).

146. R. Zeman and N. R. Amundson, *A.I.Ch.E. J.*, **9**, 297 (1963).

147. R. J. Zeman and N. R. Amundson, *Chem. Eng. Sci.*, **20**, 637 (1965).

148. *Chem. Eng. News*, **40**, Oct. 22, 66 (1962); *Chem. Week*, **95**, Nov. 7, 95 (1964).

149. K. G. Roquemore and E. E. Eddy, *Chem. Eng. Progr.*, **57** (9) 35 (1961); C. Vanderslice, *Oil Gas J.*, Aug. 9, (1965), p. 85.

CHAPTER 4

ELASTOMERS BY RADICAL AND REDOX MECHANISMS

B. Nitrile Rubber and Other Nitrogen-Containing Rubbers Except Polyurethanes*

WERNER HOFMANN

Farbenfabriken Bayer A. G., Leverkusen-Bayerwerk

Contents

* Translated by J. P. Kennedy and E. G. M. Törnqvist.

185

I. THE HISTORY OF NITRILE RUBBER (1–3)

Late in the 1920s, in connection with the development of synthetic rubber in Germany (from butadiene), it was found that the processibility of the rubber could be significantly improved by copolymerizing butadiene with a monovinyl compound. Although these monovinyl compounds give thermoplastics when homopolymerized, they cause a certain softening of the copolymers at the temperatures used for blending and processing. During a systematic search in 1929–30 in Leverkusen, styrene, acrylonitrile, and vinylpyridine were singled out among a large number of potential monovinyl components.

When the deteriorating economic situation in the world shortly after these discoveries made a rapid commercialization of styrene-butadiene copolymers impossible for reasons of cost, more emphasis was put on the development of acrylonitrile-butadiene copolymers (nitrile rubber) because of their gasoline, oil, heat, aging, and abrasion resistance. In spite of an appreciable price disadvantage for nitrile rubber relative to natural rubber and in spite of differences in processing behavior, the first small-scale manufacturing unit was started in 1934 in Leverkusen. The high price of rubber products manufactured from this specialty material was more than compensated for by their excellent durability.

The first large quantities of nitrile rubber were shipped to the U.S.A. in 1937 under the trade name Perbunan. Because of the importance which this rubber had gained, the Standard Oil Development Company received a request from Jasco* to start production of nitrile rubber after the outbreak of World War II. This production was based on licensing negotiations conducted with I.G. Farbenindustrie A.G. since 1932. The B.F. Goodrich Co., Firestone Tire & Rubber Co., Goodyear Tire and Rubber Co., and U.S. Rubber Co., and later Polymer Corp. (Sarnia, Canada), also got involved in this development.

After World War II, nitrile rubber production was also started in other countries, i.e., in England by the British Geon, Ltd. and Imperial Chemical Industries, Ltd., in France by Ugine S.A., in Italy by Montecatini and Anic, in Japan by the Japanese Geon, and in Russia.

II. PREPARATION OF MONOMERS (2–5)

A. Preparation of Butadiene

The preparation of butadiene is discussed in Chapter 4.A.

B. Preparation of Acrylonitrile

1. Economic Situation

Acrylonitrile manufacture is closely connected with the production of acrylic fibers, since about two-thirds of all acrylonitrile produced is consumed by fiber producers. Because of the rapidly growing demand for acrylontrile, novel economical manufacturing routes have recently been explored. The classical "Bayer process" developed by Kurtz, which uses acetylene and hydrogen cyanide together with an aqueous catalyst, proved too expensive in the long run. A certain reduction in the price of acrylontrile could be realized by adding hydrogen cyanide to the more readily available ethylene oxide or acetaldehyde according to the cyano-

---

*Joint American Study Company established by Standard Oil Company (N.J.) and I.G. Farbenindustrie A.G. (cf. Chapter 2, p. 76).

hydrin synthesis and dehydrating the adduct. The acetaldehyde–hydrogen cyanide process (Knapsack) is presently so economical that the Knapsack-Griesheim A.G. is granting licenses.

However, even this acrylonitrile process is based on intermediates made from petrochemical base products and is, as a consequence, dependent on processes which produce these basic materials. The simplest possible starting materials for acrylonitrile production are propylene, ammonia, and oxygen (air). In fact, as early as 1947, Cosby demonstrated the feasibility of synthesizing acrylonitrile from these raw materials. The process was developed commercially in 1959 by the Standard Oil Company of Ohio (Sohio). At about the same time, the Distillers Co. and Farbenfabriken Bayer A.G. also succeeded in developing a "propylene–ammonia–air process."

Another modern process has been developed by the du Pont Company. This process converts propylene with nitric oxides into acrylonitrile.

As a consequence of the commercialization of these economical processes, cheap acrylonitrile has been supplied by U.S. firms to the world markets since 1961. This has forced earlier acrylonitrile producers to terminate their production or to switch to the new propylene–ammonia–air process. Since the older processes have been replaced by the newer ones, they will not be further described.

2. Preparation of Acrylonitrile from Acetaldehyde and Hydrogen Cyanide (Knapsack Process)

Acetaldehyde became readily available through a new process developed by Farbwerke Hoechst and Wacker Chemie based on a patent to the Konsortium für Elektrochemie, Munich. In this process, ethylene is oxidized with palladium(II) salts to acetaldehyde, whereupon the metallic palladium is oxidized with atmospheric oxygen in the presence of copper salts.

The very cheap acetaldehyde thus obtained is then combined with hydrogen cyanide in a cyanohydrin synthesis in an aqueous acid medium. α-Oxypropionic nitrile or lactic nitrile is obtained (Eq. 1):

$$CH_3—CHO + HCN \rightleftharpoons CH_3—\underset{\underset{OH}{|}}{CH}—CN \tag{1}$$

which on dehydration gives acrylonitrile (Eq. 2):

$$CH_3—\underset{\underset{OH}{|}}{CH}—CN \rightarrow CH_2{=}CH—CN + H_2O \tag{2}$$

3. Direct Catalytic Preparation of Acrylonitrile from Propylene, Ammonia, and Air

The following reaction is the basis for the preparation of acrylonitrile from propylene, ammonia, and air (Eq. 3):

$$CH_2{=}CH{-}CH_3 + NH_3 + \tfrac{3}{2}O_2 \rightarrow CH_2{=}CH{-}CN + 3H_2O \qquad (3)$$

A large number of patent applications have been published. At least three companies have erected or will construct production facilities in which various catalysts will be used.

a. Bismuth Phosphomolybdate Catalyst (Sohio). Significantly higher yields can be obtained with bismuth molybdate on a silica gel carrier than by the so-called Cosby process from which this process originated.

According to this process, propylene, ammonia, water vapor, and air in the average mole ratio of 1:1:1:7.5 are passed over a fluidized bed of finely dispersed catalyst with a few seconds of contact time at 450–500°C and 1.35–3.11 atm and in the presence of diluents. The reaction products, acrylonitrile, hydrogen cyanide, and acetonitrile, are removed from the gas stream by washing with water and worked up by known methods.

Several acrylonitrile plants have already been constructed for synthesis according to this process which yields an acrylonitrile of such high purity (>99%) that it is excellently suitable for polymerization purposes.

A disadvantage of the Sohio process is the large amount of acetonitrile formed (9 kg on 101 kg acrylonitrile). It was hoped originally that large quantities of acetonitrile could be sold for solvent purposes, for example, in the selective absorption of butadiene, or as an intermediate. However, these plans have been realized only partially so that this process is encumbered by this side product.

b. Catalysis with Cobalt, Molybdenum, Antimony, and Tin Compounds (Distillers Co. Process). According to a new process by the Distillers Co., mixtures of cobalt and molybdenum oxides as well as mixtures of antimony and tin oxides are good catalysts for the manufacture of acrylonitrile from propylene, ammonia, and oxygen.

According to data in the patents, 75% of the propylene charge is converted to acrylonitrile. No precise information is available as to the amount of by-products and operational details.

The plant built by Ugilor is probably operated with such a catalyst.

c. Catalysis with Solid Phosphoric Acid and Activators (Farben-fabriken Bayer A.G. Process). The process developed by Farben-fabriken Bayer A.G. for the preparation of acrylonitrile uses so-called solid phosphoric acid (e.g., boron phosphate or titanium phosphate) activated by small amounts of metal compounds.

Mixtures of propylene, ammonia, air, and water vapor at an average mole ratio of $1:1:7:1$ to 5 are passed through a zone of fluidized catalyst. The contact time is 0.5–2.0 sec and the reaction temperature 480–540°C.

The propylene conversion is up to 85%. The products are acrylonitrile and hydrogen cyanide. Unreacted propylene and ammonia are removed from the reaction mixture and recycled. This process has a good space-time yield.

Thus these three processes for the manufacture of acrylonitrile from propylene, ammonia, and air differ mainly in the nature of the catalyst employed and presumably also in the individual process steps and in the composition of the crude products.

4. Preparation of Acrylonitrile by Direct Catalyzed Reaction between Propylene and Nitric Oxide (du Pont Process) and Addition of Hydrogen Cyanide to Acetylene with an Anhydrous Catalyst

The reaction upon which the first process is based is shown by Eq. 4:

$$4CH_2{=}CH{-}CH_3 + 6NO \rightarrow 4CH_2{=}CH{-}CN + 6H_2O + N_2 \qquad (4)$$

A mixture of propylene and nitric oxide is reacted in the presence of a dehydrogenation catalyst by an exothermic reaction. Besides other dehydrogenation catalysts, silver on an inert support, i.e., charcoal or silica gel, has been found particularly effective. The conversion is carried out by heterogeneous catalysis at 450–550° and normal or elevated pressures.

The du Pont Company constructed a 40,000-ton plant based on this process. Since this process yields large amounts of hydrogen cyanide, this plant operates in conjunction with the older process based on acety-lene and hydrogen cyanide.

In contrast to the Bayer process, the du Pont process uses an anhydrous copper chloride (Niewland) catalyst in organic nitriles. The addition of small amounts of hydrogen chloride enhances the activity of the catalyst and improves the final yield. A relatively rapid gas through-put and good yields are realized. The acrylonitrile produced by this process is purer than that obtained with aqueous catalysts.

TABLE I
Typical Analysis of Acrylonitrile
Prepared by the Sohio Process

Cyanobutadiene	Not detectable
Divinylacetylene	Not detectable
Methyl vinyl ketone	Not detectable
Peroxides	Less than 0.1 ppm
Iron	Less than 0.3 ppm
Copper	Less than 0.3 ppm
Hydrogen cyanide	Less than 5 ppm
Aldehydes	Less than 10 ppm
Acetonitrile	Less than 500 ppm
Nonvolatiles	Less than 60 ppm
pH Value	6.5–7.5
Color APHA	Less than 5
Oxygen bomb stability	More than 4 hr

5. Properties of Acrylonitrile

Acrylonitrile is a colorless, toxic liquid which boils at 77.3°C at atmospheric pressure. It is sparingly soluble in most organic solvents.

A number of acetylene derivatives which may be present, depending on the particular manufacturing process used, may affect the rate of polymerization of acrylonitrile. The presence of acetylene derivatives should be expected when the acetylene–hydrogen cyanide process is employed.

The composition of a particularly pure acrylonitrile is shown in Table I.

Acrylonitrile of such purity is eminently suitable for polymerization, since polymerization poisons are absent.

III. PREPARATION OF NITRILE RUBBER (1–3,6–9)

The preparation of nitrile rubber is usually carried out by emulsion copolymerization of butadiene and acrylonitrile. The ratio of the two monomers will vary within wide limits depending upon the type of nitrile rubber desired. The most commonly used nitrile rubbers contain 18–48%, preferentially 26–40% acrylonitrile (y in Eq. 5). The monomer

$$\begin{bmatrix} H & H & H & H \\ | & | & | & | \\ C=C-C=C \\ | & & & | \\ H & & & H \end{bmatrix}_x + \begin{bmatrix} H & H \\ | & | \\ C=C \\ | & | \\ H & C\equiv N \end{bmatrix}_y \rightarrow \begin{bmatrix} H & H & H & H \\ | & | & | & | \\ C-C=C-C \\ | & & & | \\ H & & & H \end{bmatrix}_x \begin{bmatrix} H & H \\ | & | \\ C-C \\ | & | \\ H & C\equiv N \end{bmatrix}_y \quad (5)$$

units are distributed statistically in the copolymer (Eq. 5). Besides giving rise to the desired linear or 1,4 enchainment butadiene may also enter the chain in the 1,2 fashion, i.e. (Eq. 6):

$$
\begin{bmatrix} H & H & H & H \\ & | & | & | & | \\ C{=}C{-}C{=}C \\ | & & & | \\ H & & & H \end{bmatrix}_x + \begin{bmatrix} H & H \\ | & | \\ C{=}C \\ | & | \\ H & C{\equiv}N \end{bmatrix}_y \rightarrow
$$

$$
\begin{bmatrix} H & H & H & H \\ | & | & | & | \\ C{-}C{=}C{-}C \\ | & & & | \\ H & & & H \end{bmatrix}_{x-z} \begin{bmatrix} H & H \\ | & | \\ C{-}C \\ | & | \\ H & C{\equiv}N \end{bmatrix}_y \begin{bmatrix} H & H \\ | & | \\ C{-}C \\ | & | \\ H & C{-}H \\ & & || \\ & H{-}C{-}H \end{bmatrix}_z \tag{6}
$$

The pendant vinyl groups may also participate in the polymerization and thus cause branching.

Readers interested in the fundamentals of emulsion copolymerization and the problems associated therewith are referred to the chapter on SBR in this book (Chapter 4.A). Only the pecularities of the butadiene-acrylonitrile copolymerization specifically caused by the presence of acrylonitrile will be described below.

A. Characteristics of the Butadiene-Acrylonitrile Copolymerization

Only in special cases will the mole ratio of the monomers in the copolymer be the same as in the charge. During the copolymerization, the incoming monomer unit can add to a growing chain end formed either from the same monomer or from the other one. Equations 7–10 show these possibilities:

$$M_1{}^* + M_1 \rightarrow M_1{}^* \quad \text{rate constant} \quad k_{11} \tag{7}$$

$$M_1{}^* + M_2 \rightarrow M_2{}^* \quad \text{rate constant} \quad k_{12} \tag{8}$$

$$M_2{}^* + M_2 \rightarrow M_2{}^* \quad \text{rate constant} \quad k_{22} \tag{9}$$

$$M_2{}^* + M_1 \rightarrow M_1{}^* \quad \text{rate constant} \quad k_{21} \tag{10}$$

where M_1 and M_2 denote the two monomers in the charge and $M_1{}^*$ and $M_2{}^*$ stand for the propagating chain ends derived from the M_1 and M_2 molecules, respectively.

The following assumptions are made:

1. The reactivity of the growing macromolecule is determined only by the active endgroup and is unaffected by the length and composition of the chain.

2. The average chain length of the macromolecules formed is sufficiently great so that the consumption of monomer by initiation, termination, and transfer reactions can be disregarded when compared with that by chain growth.

3. The life of the growing polymer chains is short compared to the total reaction time so that a quasi steady state can be assumed for the concentration of active chain ends. Copolymerization Eq. 11 can then be derived from Eqs. 7–10:

$$\frac{d[M_1]}{d[M_2]} = \frac{[M_1]}{[M_2]} \cdot \frac{r_1[M_1] + [M_2]}{[M_1] + r_2[M_2]} \tag{11}$$

where $[M]$ = molar concentration of M, $r_1 = k_{11}/k_{12}$, and $r_2 = k_{22}/k_{21}$. The correctness of this equation has often been demonstrated. Experience has also shown that the reactivity ratios r_1 and r_2 are only slightly affected by temperature and other factors such as the polymerization medium, catalysts, etc.

Similarly, the nature of the polymerization medium, i.e., solvent or emulsion, has no significant influence on the reactivity ratios, provided the monomer composition in the aqueous and organic phases, i.e., the relative solubility of the monomers in the two phases, is not very different. The latter is, however, not the case in the copolymerization of acrylonitrile and butadiene because of the relatively high water solubility of acrylonitrile.

Copolymerization Eq. 11 relates the molar amounts of the monomers incorporated into the copolymer to the molar amounts of the monomers in the charge. The initial monomer concentrations can be used in the copolymerization equation provided the reaction is stopped at such a low conversion (below about 10%) that the charge composition has not yet changed significantly. More complicated equations are available for higher conversions. The molar ratio of the monomers in the copolymer obtained from Eq. 11 or by an analytical method is, of course, only an average value reflecting overall copolymer composition and does not apply to individual macromolecules.

When a copolymer of a certain composition is desired, the required mole ratio of the monomers in the charge can be calculated if r_1 and r_2 are known. Thus, if one writes $d[M_1]/d[M_2] = X$ and rearranges Eq. 11

accordingly, one obtains

$$\frac{[M_1]}{[M_2]} = \frac{1}{2r_1} \cdot \{(X - 1) + [(X - 1)^2 + 4r_1r_2X]^{1/2}\} \qquad (12)$$

The following reactivity ratios were determined for butadiene and acrylonitrile in aqueous emulsion polymerizations at 50°C: r_1 (butadiene) $= 0.35 \pm 0.08$ and r_2 (acrylonitrile) $= 0.0 \pm 0.04$.

The possibly negative value for acrylonitrile is due to the water solubility of this monomer. The product of the reactivity ratios $r_1 \cdot r_2$ is <0.016.

The reactivity ratios at 5°C were found to be r_1 (butadiene) $= 0.28$ and r_2 (acrylonitrile) $= 0.02$.

Depending upon the magnitude of the reactivity ratios, *three extreme cases* arise in copolymerizations in general.

The *first case* occurs when the product of the reactivity ratios equals unity ($r_1 \cdot r_2 = 1$), i.e., when the relative reactivities of both monomers, M_1 and M_2, are the same toward either of the two possible growing chain ends M_1^* and M_2^*. This is an ideal copolymerization.

The SBR copolymerization (see Chapter 4.A) approaches the ideal ($r_1 \cdot r_2 = 1.08$; $r_1 \approx r_2 \approx 1$). Thus the relative molar concentration of the monomers in the copolymer and in the emulsion mixture are almost the same for all monomer ratios. This is not the case in nitrile rubber manufacture because this copolymerization is an example of the *second case*, which occurs when the product of the reactivity ratios is very small, $r_1 \cdot r_2 \ll 1$ (r_2 for acrylonitrile is close to 0). In this case, one of the monomers polymerizes much faster than the other.

In monomer mixtures with low acrylonitrile concentrations, the acrylonitrile will polymerize first, whereas in charges with high acrylonitrile concentrations, the butadiene will polymerize first. However, a so-called azeotropic mixture occurs when the monomer ratio corresponds to

$$\frac{[M_1]}{[M_2]} = \frac{r_2 - 1}{r_1 - 1} \qquad (13)$$

This is the case for butadiene-acrylonitrile copolymerizations when the acrylonitrile/butadiene ratio is 37:63 (25°C). At this monomer ratio, no change in monomer composition will occur during the polymerization and the polymer will contain the same ratio of the two monomers.

When the mole fraction of acrylonitrile is lower than 37%, more acrylonitrile will be incorporated in the copolymer initially than is

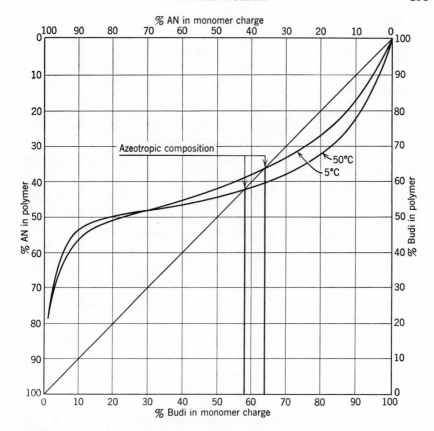

Fig. 1. Composition of butadiene-acrylonitrile copolymers as a function of initial monomer concentration at 5°C ($r_1 = 0.28$, $r_2 = 0.02$) and at 50°C ($r_1 = 0.42$, $r_2 = 0.04$).

present in the charge. On the other hand, when the acrylonitrile concentration is higher than 37 mole % the copolymer initially formed will contain less acrylonitrile than is present in the emulsion charge, i.e., butadiene will be preferentially polymerized.

This situation is shown in Fig. 1 where the relative monomer concentration in the charge is plotted against the copolymer composition at 5°C. The azeotropic composition is indicated where the differential curves cross the diagonal.

Although these curves are valid only for solution copolymerizations and only when monomer consumption by initiation, termination, and

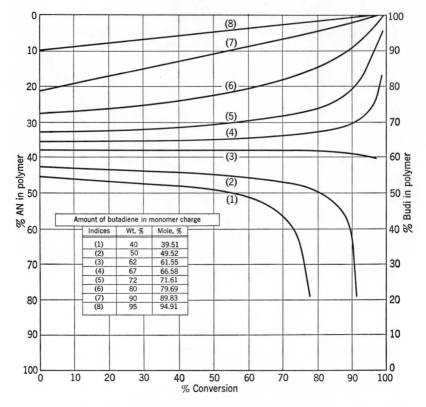

Fig. 2. Momentary composition of butadiene-acrylonitrile copolymers as a function of conversion for various monomer compositions at the start of polymerization
(differential curves). Temperature 50°C ($r_1 = 0.42$, $r_2 = 0.04$).

transfer can be ignored, they can be used to the first approximation for
true emulsion polymerizations. According to Fig. 1 (e.g., at 50°C), the
initial copolymer formed in a charge containing 30% acrylonitrile and
70% butadiene contains 38% acrylonitrile. Thus the concentration of
acrylonitrile in the emulsion will decrease and the composition of the
next increment of copolymer formed will contain a proportionally lesser
amount of acrylonitrile as indicated by the curve. The situation is
analogous above the azeotropic composition.

Another method of illustrating these relationships is shown in Figs.
2 and 3 where the instantaneous copolymer composition is given as a
function of conversion (Fig. 2 for 50°C, Fig 3 for 5°C). These differential

curves show the composition of the copolymers at the moment of their formation for various acrylonitrile concentrations (in mole %) in the charge. Since the molecular weights of butadiene (54) and acrylonitrile (53) are almost identical, there is not much difference between mole and weight percents for this monomer pair. Accordingly, these curves can be used approximately for charges expressed in weight %.

Figures 4 and 5 present in a similar manner so-called integral curves and show the overall composition of nitrile rubber as a function of conversion. These curves indicate the copolymer compositions obtained

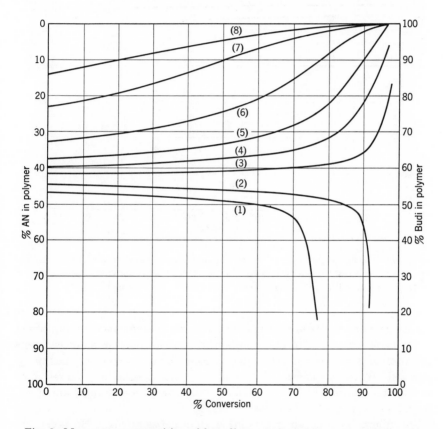

Fig. 3. Momentary composition of butadiene-acrylonitrile copolymers as a function of conversion for various monomer compositions at the start of polymerization (differential curves). Temperature 5°C ($r_1 = 0.28$, $r_2 = 0.02$). See legend in Fig. 2.

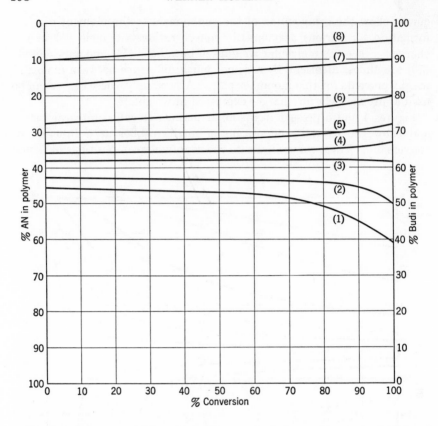

Fig. 4. Average composition of butadiene-acrylonitrile copolymers as a function of conversion for various monomer compositions at the start of polymerization (integral curves). Temperature 50°C ($r_1 = 0.42$, $r_2 = 0.04$). See legend in Fig. 2.

for various acrylonitrile concentrations in the charge as a function of the overall conversion. The end points of the integral curves show the composition of the original monomer mixture since everything has polymerized at these points. By contrast, the differential curves may terminate before 100% conversion, i.e., when one component (e.g., butadiene) has been consumed.

Butadiene homopolymerizes slower than acrylonitrile. However, the rate of polymerization of butadiene is accelerated by the presence of acrylonitrile. For azeotropic mixtures, the monomer units are incorporated into the copolymer in a predominantly alternating fashion. For other compositions, butadiene and acrylonitrile incorporation is essentially statistical. A truly alternating structure is obtained only in the special case when both reactivity ratios are practically zero, i.e., when only $M_1^* + M_2 \rightarrow M_1M_2^*$ or $M_2^* + M_1 \rightarrow M_2M_1^*$ can proceed. Under certain conditions, the butadiene-acrylonitrile copolymerization system approaches this ideal situation without reaching it.

The *third case*, which occurs when the product of the reactivity ratios is much larger than one $(r_1 \cdot r_2 \gg 1)$, has not yet been experimentally observed. In this case the individual growing chain ends react preferen-

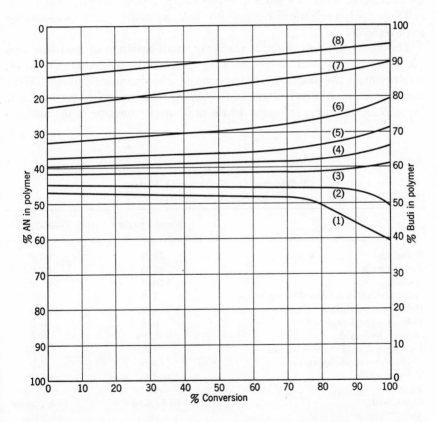

Fig. 5. Average composition of butadiene-acrylonitrile copolymers as a function of conversion for various monomer compositions at the start of polymerization (integral curves). Temperature 5°C ($r_1 = 0.28$, $r_2 = 0.02$). See legend in Fig. 2.

tially with their own monomer, i.e., $M_1^* + M_1 \rightarrow M_1M_1^*$ or $M_2^* + M_2 \rightarrow M_2M_2^*$. Under these conditions the two homopolymerizations proceed independently and no copolymerization occurs. However, if the product of the reactivity ratios is only somewhat greater than $1 (r_1 \cdot r_2 > 1)$ then the monomer units are incorporated into the polymer in a block fashion.

B. Effect of the Components in the Charge

1. Polymerization Charges

The fundamentals involved in composing a polymerization charge will be illustrated with two older German nitrile rubber recipes (Table II). These polymerization recipes do not normally meet present-day requirements.

The above charges contain relatively small amounts of modifiers and yield relatively hard and difficultly processible polymers, therefore. Furthermore, the products contain strongly discoloring stabilizers. With larger amounts of acrylonitrile in the charge, somewhat less modifier can be used for obtaining the same degree of plasticity because of the modify-

TABLE II
Polymerization Charges with Various
Amounts of Acrylonitrile

	Weight parts	Weight parts
Butadiene	73.0	60.0
Acrylonitrile	27.0	40.0
Water	150.0	200.0
Sodium dibutyl naphthaline sulfonate	3.6	3.6
Potassium persulfate	0.2	0.2
Sodium pyrophosphate	0.3	0.3
Sodium hydroxide	0.08	0.1
Diisopropyl xanthogen disulfide	0.03	0.27
N-Phenyl-β-naphthylamine	3.0	3.0
Temperature	∼29°C	∼23°C
Time	∼28 hr	∼28 hr
Conversion	10 kg/hr-m³	10 kg/hr-m³
Yield	∼76%	∼80%
Defo hardness	∼2800	∼2800
Incorporated acrylonitrile	∼26%	∼38%

ing effect of acrylonitrile. As a consequence of the faster polymerization rate of acrylonitrile, charges rich in this monomer can be run at somewhat lower temperatures or for a shorter length of time at conventional temperatures.

2. The Monomers and Their Relative Proportions

When butadiene and acrylonitrile are heated together in the absence of a radical catalyst, a cyclic codimer, cyanocyclohexene, is formed by a thermal reaction (Eq. 14).

$$\tag{14}$$

This product will not participate in subsequent polymerization reactions. In the presence of conventional emulsifying agents, this codimerization is substantially retarded, although not completely eliminated. The main reaction is the desired copolymerization. However, with increasing acrylonitrile content in the charge, this side reaction becomes more pronounced as evidenced by the odor of the copolymer.

Instead of butadiene, other conjugated dienes, e.g., isoprene, piperylene, or 2,3-dimethylbutadiene, can also be used in nitrile rubber manufacture. However, these substituted dienes yield primarily leathery copolymers and are of no practical significance, therefore. The elasticity of the copolymer diminishes as the number of alkyl substitutents on the butadiene increases.

Polymers with improved plasticity and solution properties are often obtained when methacrylonitrile or ethacrylonitrile are used instead of acrylonitrile. Larger amounts of such homologs in the rubber cause higher hysteresis, lower low-temperature flexibility and higher oil resistance. Small amounts of other monomers may also be used in conjunction with the two main components. Thus interesting copolymers can result from the use of small amounts of styrene, methyl methacrylate, ethyl acrylate, or vinylidene chloride. However, the particular improvements obtained are often realized at the expense of other properties. For example, 4% styrene with the usual amount of acrylonitrile improves the plasticity of the rubber but decreases its low-temperature flexibility. The introduction of methacrylic acid results in very reactive copolymers which can be vulcanized with metal oxides without additional accelerators. On account of their increased reactivity, these rubbers are difficult to process; their reactivity is very advantageous for latex purposes, however.

The butadiene/acrylonitrile ratio in nitrile rubber does not affect its tensile strength significantly, nor its tear, abrasion, and heat resistance. A number of other important physical properties such as processibility, resilience, low temperature properties, and swelling behavior are strongly affected by the acrylonitrile content in the rubber, however.

The polarity of the rubber increases with increasing acrylonitrile content, giving rise to incompatibility with nonpolar solvents, i.e., aliphatic hydrocarbons and paraffinic oils. This manifests itself in terms of improved swell resistance and decreased solubility. The resilience and low temperature flexibility of the vulcanizates decrease at the same time. As a consequence of the increasing polarity, the compatibility with polar plastics (e.g., PVC, phenoplasts) increases with increasing acrylonitrile content. Finally, the gas permeability of the vulcanizates increases with increasing acrylonitrile content.

The influence of acrylonitrile content on vulcanizate properties can be summarized as follows:

Effect of Acrylonitrile
Content on Nitrile Rubber

Low   Acrylonitrile content   High

————————————— better processibility ————————————→
———————————————higher oil and gasoline resistance—————————→
←————————————————— higher resilience —————————————
←————————————————lower permanent set ————————————
←————————————— better low-temperature flexibility ——————————
————————————————better compatibility with polar plastics ——————————→
←————————————— higher solubility in aromatic solvents ——————————
←————————————————— lower density ——————————————

3. Polymerization Additives

Since the additives used in the polymerization of nitrile rubber and SBR are identical, they will not be discussed in detail here (cf. Chapter 4.A).

The requirements as to the purity of the *water* (deionized, oxygen-free) are the same as in SBR production.

As in the copolymerization of styrene and butadiene, anionic, cationic and nonionic *emulsifiers* can be used. Anionic types are preferred. Cationic emulsifiers are used for special latices. Nonionic agents are only auxiliary or secondary emulsifiers. The above-mentioned dibutyl sodium naphthalene sulfonate (cf. recipe in Table II) is preferentially employed in combination with, e.g., Mersolate and/or a nonionic emulsifier.

The *initiators* or *polymerization catalysts* are essentially the same as used in SBR production (cf. Chapter 4.A). While catalysts of relatively low activity (e.g., persulfates without activators) were used at comparatively high polymerization temperatures in the early days of nitrile rubber manufacture, redox catalysts are primarily employed today (cold rubber). Examples of such redox systems are, e.g.,

Hydrogen peroxide (20%)	0.35 wt parts
Iron sulfate heptahydrate	0.02 wt parts
tert-Dodecyl mercaptan	0.5 wt parts

or

Potassium persulfate	0.4 wt parts
Iron sulfate heptahydrate	0.0005 wt parts
Triethanol amine	0.4 wt parts

tert-Dodecyl mercaptan is also a modifier.

Since acrylonitrile has a higher polymerization rate than styrene, smaller amounts of initiator, less active systems, or lower temperatures than are used in the SBR process can be employed in nitrile rubber manufacture. The type and amount of initiator and particularly the polymerization temperature are important factors in determining the physical properties, especially the processibility of the polymer.

Like in the case of SBR, nitrile rubbers prepared at low temperatures (cold polymers) have a more regular molecular structure than those obtained at higher temperatures. Consequently, the processibility of the product is strongly affected by the polymerization temperature. At the same Mooney viscosity, cold polymerized nitrile rubbers exhibit better processibility than hot polymerized ones. Furthermore, cold polymerized nitrile rubbers have a lesser cyclization tendency (cyclization = thermal crosslinking) which is advantageous when high processing temperatures are used (internal mixer). On the other hand, due to their somewhat branched structure, hot polymerized nitrile rubbers show less cold flow in green mixes (e.g., open cure of hoses) at the same Mooney viscosity. For the same recipes, the mechanical properties of nitrile rubber vulcanizates are hardly affected by the polymerization temperature.

tert-Dodecyl mercaptan is one of the preferred modern modifiers but it is used in larger amounts than are shown for diisopropyl xanthogen disulfide in Table II. The actual amount used depends on the desired softness of the polymer and ranges between 0.5 and 1.0 parts by wt. However, a small amount of diisopropyl xanthogen disulfide is occasionally used in combination with dodecyl mercaptan to retard the polymerization initiation of acrylonitrile somewhat. Smaller amounts of modifiers can be used for butadiene-acrylonitrile copolymerizations than for butadiene–styrene systems, since, due to its lower molecular weight, acrylonitrile exerts a stronger modifying effect than styrene at equal weight concentrations. The type and quantity of modifier, the conversion level and possibly incipient crosslinking are factors which largely determine the mechanical properties of the polymers. This is shown in Table III.

Polymers obtained without modifiers contain predominantly branched or crosslinked macromolecules and are as a consequence difficult to process. Unmodified or only slightly modified rubbers of the type prepared in Germany before and during World War II have essentially disappeared from the market, therefore. Despite the superior mechanical properties of little modified polymers, rubber manufacturers have nowadays generally adopted the original American method of using modifiers for directly producing a copolymer of a Mooney viscosity which allows direct processing. This means that one is sacrificing some advantages in mechanical properties in order to eliminate separate plasticization.

As with other elastomers, the Mooney viscosity of nitrile rubber

TABLE III

Effect of Various Mercaptans on Plasticity and Tensile
Strength of Butadiene-Acrylonitrile Copolymers

Mercaptan	Conversion % in 16 hr at 28°C	Will. plast. at 80°C, 10 kg load	Tensile strength, psi[a]	Elongation at break, %	Modulus at 300% elong., psi
n-Decyl	66	86.0	2650	775	495
n-Decyl	62	74.0	3150	735	500
n-Dodecyl	65	64–0	3940	705	755
n-Tetradecyl	60	160–67	4200	620	980
n-Hexadecyl	63	214–116	3900	420	2200

[a] Recipe (in parts): rubber, 100; EPC black, 50; dibenzothiazolyl disulfide accelerator.

strongly influences its processing characteristics. Banding on the mill during compounding, the rate of filler and plasticizer uptake, heat buildup during compounding as well as extrusion and calenderability are more favorable for the soft types than for the harder ones. On the other hand, higher viscosity rubbers show less air trapping than the low viscosity types and give a lower percentage of rejects in some cases, therefore. Harder rubbers can also be extended with larger amounts of fillers and plasticizers (cheap mixes). The effect of Mooney viscosity on mechanical properties is relatively slight. Because of their higher tensile strengths, higher viscosity types give better resilience than the corresponding soft rubbers. However, hard and soft rubbers containing the same amount of acrylonitrile exhibit practically identical swelling properties and low temperature flexibility independent of the Mooney viscosity.

Effect of the Viscosity in Nitrile Rubber Types

Low	Mooney viscosity	High
← better processibility		
← higher banding rate on the mill		
← shorter compounding time		
← lower heat buildup during compounding		
← better extrudability		
← better calendering		
better dimensional stability during open cure →		
higher loadings with fillers and plasticizers →		
less trapped air →		
somewhat better mechanical properties →		

The processibility of rubbers produced with *controlled precrosslinking* is excellent in contrast to that of crosslinked products obtained in polymerizations without modifiers. Controlled precrosslinking can be achieved for example, by using one part by weight of divinylbenzene in the polymerization charge. For this reason, precrosslinked nitrile rubbers having a high gel content are sometimes prepared for blending with other types of nitrile rubber to improve the extrudability, the calendering properties as well as the dimensional stability during open cure. Under these conditions the crosslinked polymer acts like a factice because of its spherical structure. However, due to the changes in the molecular structure, precrosslinking strongly affects the polymer properties. The plasticity of the rubber is not only strongly decreased but the vulcanizate properties are also affected. The tensile strength and elongation at break strongly decrease, the stiffness and hardness increase, the rebound resilience and tear resistance as well as the dynamic glass transition temperature deteriorate, whereas by contrast the compression set and swelling properties improve. Because of their relatively low tensile strength and elongation at break, crosslinked polymers are usually not used alone but in combination with noncrosslinked rubbers.

The *antifreeze agents* used at polymerization temperatures below 0°C are the same as those used in the SBR process.

To avoid polymer discoloration, N-phenyl-β-naphthylamine is not used anymore as a *terminator* and *stabilizer* despite its otherwise excellent properties. Frequently one part by weight of hydroquinone or sodium hyposulfide is used as terminator, either alone or in combination with small amounts of sodium dimethyl dithiocarbamate. However, since these terminators, unlike N-phenyl-β-naphthylamine, do not simultaneously act as stabilizers, additional stabilizers must be employed, in which case nondiscoloring types are obviously preferred. *tert*-Butyl-p-cresol as well as some of its derivatives and trinonyl phenyl phosphite are primarily used today.

In regard to *coagulants*, the situation is the same as for SBR production.

C. Commercial Nitrile Rubber Manufacture

1. Apparatus

The copolymerization of butadiene and acrylonitrile is very similar to SBR manufacture. Like SBR, nitrile rubber can be manufactured commercially by two processes, i.e., continuously or in batch primarily depending upon the level of production.

The continuous polymerization can be carried out according to different methods. In the one most frequently used, the polymerization proceeds in a series of consecutive autoclaves, in so-called polymerization lines. The batch process on the other hand is carried out in individual reactors. The size of the autoclave is determined by the size of the desired charge and is normally up to 20 m³. The autoclave is usually made of stainless, chrome-nickel or glass-lined steel.

Measuring pumps are usually employed for the continuous addition of the individual ingredients of the polymerization charges. These pumps introduce the individual components, i.e., monomers, aqueous solutions of emulsifiers, activators and modifiers, and water in the required constant proportions into the first reactor. The charge is continuously transferred after a certain residence time from the first reactor to the next, and so on.

Stirring devices are usually necessary for emulsion polymerization.

These provide thorough mixing and sufficient heat exchange with the cooling walls or cooling baffles.

Finally, depending on the polymerization temperature, heating and cooling units are required. The individual solutions are brought to the required temperature prior to the introduction into the reactor. For the preparation of hot rubber at slightly elevated temperatures, additional heating devices are usually unnecessary, since the exothermic reaction produces sufficient continuous heat. The extent of the heat transfer, which depends on the intensity of the stirring and/or cooling if employed, determines the final operating temperature. For polymerizations at lower temperatures, the cooling medium is chosen according to the required degree of cooling. Cold rubber charges are often cooled with brine.

2. The Polymerization Process

In normal emulsion polymerization of butadiene and acrylonitrile, preemulsifying of the charge, e.g., with a homogenizer, is often superfluous. The monomers can be introduced directly into the charge containing the emulsifier.

The introduction sequence of the individual ingredients is important. For nitrile rubber production according to the recipe in Table II, the following introduction sequence is recommended.

That portion of the water which is not used to dissolve the other components is first charged to the reactor. The solution of the emulsifier which has been mixed previously with the sodium hydroxide solution in an emulsifying vessel is then added. Next, the sodium pyrophosphate solution is introduced. Subsequently, the acrylonitrile, the modifiers, and finally the butadiene are added. The modifier is often dissolved in the butadiene and preemulsified in the soap solution. As the final ingredient, the peroxide activator is introduced, whereupon the vessel is closed tightly, and the polymerization is commenced with gentle stirring under pressure.

Depending on the particular activator used, it is necessary to exclude atmospheric oxygen to a varying degree, since its presence can lead to disturbances. Air can be replaced by simply evaporating a small excess of butadiene. After a certain induction period during which residual amounts of oxygen are consumed, the polymerization commences. On the other hand, when redox initiators are used, the last traces of atmospheric oxygen must be painstakingly removed.

Frequently, not all of the above-mentioned components are added at the start of the polymerization and the introduction of ingredients occurs at different time intervals. This is the case especially when the substances are easily decomposed or consumed or altered during the polymerization.

For example, the catalyst might be added gradually when the conversion threatens to cease due to insufficient catalyst addition, or when the catalyst, e.g., hydrogen peroxide, is easily destroyed.

The emulsifier is usually introduced gradually during the manufacture of concentrated latices to avoid unduly high latex viscosities or coagulation at higher conversions. Slow addition of emulsifier, e.g., sodium alkyl benzyl sulfonate, is advisable in the production of finely dispersed materials.

As a rule, modifiers are introduced gradually since these materials are rapidly consumed during the polymerization. Gradual addition of modifier results in a relatively constant chain length and a more uniform product, since such addition allows the modifier to act throughout the reaction.

In the manufacture of precrosslinked nitrile rubber types, divinylbenzene is generally added at a conversion level of 10–35%.

The charge must be sufficiently buffered in nitrile rubber manufacture to prevent the pH from becoming too alkaline, since, in the more strongly alkaline pH ranges, acrylonitrile may be partially consumed by an addition reaction with water or may cyanoethylate the amines in the system.

After the monomer introduction, the vessel is pressurized. Operation under pressure increases the polymerization rate significantly. The pressure change during the reaction can be used to monitor the process, since the gaseous butadiene is consumed in the course of the polymerization. After the disappearance of the monomer phase (about 50–60% conversion) the pressure drop becomes particularly noticeable.

As mentioned above, the solutions are adjusted to the polymerization temperature prior to their introduction into the reactor. The temperature of nitrile rubber manufacture is normally between 5 and 30°C and depends on the particular catalyst system selected. Higher temperatures are sometimes used. This is for instance the case when the initiation is to be accelerated at the beginning of the polymerization without any risk of a deterioration of the polymer. An increase in the polymerization temperature during the reaction is sometimes advisable, for example, when the reaction slows down or when the last amounts of monomer are to be utilized. Butadiene-acrylonitrile copolymerization can also be carried out at a gradually increasing temperature. However, it should be kept in mind that the vapor pressure of the monomers increases with increasing temperature which might necessitate the use of stronger autoclaves. On the other hand, structurally more regular polymers having improved processing characteristics are obtained at lower temperatures.

High polymerization rates are desirable for economic reasons. The recipes are usually so adjusted that the conversion is at least 12% per hour. A polymerization time in excess of 48 hr is considered uneconomical.

The yield is also a factor influencing the economics of the manufacturing process. The maximum yield obtainable is limited by the processibility and the mechanical properties of the polymer. At higher acrylonitrile and modifier content, higher nitrile rubber conversions can be achieved. With certain latices the polymerization can be carried out to almost 100% conversion.

After the desired conversion level has been reached, the terminator is introduced, preferentially in a separate vessel, whereupon the unreacted monomers are recovered by steam distillation in two consecutive towers. Sometimes the monomers can be recovered prior to the termination, e.g., when the terminators are relatively volatile and would leave together with the monomers and cause their subsequent polymerization to be inhibited. Finally, the stabilizer is added. The removal of monomers and small amounts of dimer possibly present is important, since monomeric acrylonitrile is extremely toxic. Furthermore, residual monomers and dimers confer a strong odor to the rubber. Since nitrile rubber polymerization is usually terminated at between 70 and 80% conversion, the recovery of the monomers is also an important economic problem.

It is extremely difficult to produce a completely odorless nitrile rubber. The odor, which may also stem from modifiers or fatty acids, can be reduced by steam treatment.

3. Coagulation, Drying, and Latex Concentration

The coagulation of a latex, which almost always requires the addition of a coagulant, is a rather complex colloid chemical process. It encompasses at least two main stages, i.e., flocking and coalescence. Flocking is achieved by removing the electrical charge from the particles through bringing them to their isoelectric point and by changing the hydration state of the emulsifier layer. This is sometimes a reversible process, e.g., through recharging with electrolytes. The flocculated latex particles coalesce in the second stage after which precipitation is irreversible.

Differences in the coagulation may occur depending upon whether the emulsifier remains in the polymer or is substantially removed. On account of the swelling properties, one usually tries to wash out the emulsifier as completely as possible from nitrile rubber.

Nitrile latices are customarily precipitated with electrolytes, mostly with sodium chloride solutions. It is important to use the results of preliminary experiments for determining the precipitation conditions so that the electrolytes employed can be washed out carefully again. The possibility of washing out the catalyst and emulsifier should also be determined. The sodium chloride precipitation of latices obtained with soaps can be effectively promoted with acids which convert the soaps to free fatty acids. The amount of acid added determines the size of the crumb. When an excess of acid is used, clumping occurs readily, however. Frequently the latex is first creamed with a concentrated sodium chloride solution and subsequently coagulated by the addition of dilute acid.

After the latex has been coagulated and the crumbs thoroughly washed, the product must be dried. Since nitrile rubber has a relatively great tendency for cyclization it is quite sensitive during drying. Thus the drying process must be carefully controlled so as not to change the processing and solution properties of the polymer. If nitrile rubber is insufficiently stabilized and is carelessly treated thermally or mechanically during the drying process, side chains may be split off by a radical mechanism causing the cyclization tendency of the rubber to be increased. The polymer may also be crosslinked or degraded if it contains small amounts of soluble iron compounds in the presence of phenolic stabilizing agents. The latter effect, which is normally undesirable and should be avoided, is technically utilized in the so-called thermal degradation process. Usually, however, one tries to keep the drying temperature so low that the polymer properties are changed as little as possible. The product should not sinter during drying and should have a porous surface so that difficultly accessible areas in the interior of the crumb can also be satisfactorily dried. To prevent the particles from sticking either together or to other surfaces, powders such as talcum, magnesium stearate, etc., are added. A relatively low temperature, at least not higher than 50°C, is usually employed at the beginning of the drying. Subsequently, the temperature is raised to 70°C or even higher.

In commercial plants, the material is often dried in rotary or belt driers. Special screw-conveyor driers or drying extruders may also be employed.

In the production of crumbs or powders care must be taken that after the coagulation the individual particles do not clump together during drying. This can be accomplished among others by employing a spraying device. A fundamental requirement for this is the correct selection of emulsifier.

When subsequent concentration of the latex is desired, this can be accomplished only if the latex consists of sufficiently large particles. The particle size can be increased by partially coagulating the latex, i.e., by introducing an amount of coagulant insufficient for complete coagulation. Thus salts of inorganic monovalent cations and organic amines, or organic acids (acetic acid, formic acid, lactic acid) can be used. Not all types of latices are suitable for this kind of particle size increase; only those prepared with anionic emulsifiers, i.e., soaps, resinous soaps, alkyl sulfonates or sulfates, can be used. Latices containing long chain amines and quaternary ammonium salts are unsuitable.

Another possibility to increase the particle size is by freezing the latex to a temperature not too much below its freezing point and then thawing it. A number of latices can be subjected to this treatment without coagulation and can be concentrated subsequently.

It is hardly possible to obtain a high nitrile rubber latex concentration by direct polymerization. Thus when higher latex concentrations are required, postpolymerization concentration becomes necessary. Nitrile latices usually coagulate at a concentration of 25 to 30%.

Among the concentration processes developed for natural rubber latices, only a few may be considered for nitrile rubber. The particle size of nitrile rubber latices is usually much smaller than that of natural latex. Thus, centrifugation, electrodecantation, and filtration cannot be employed at all or give unsatisfactory results with nitrile latices. Commercially only two processes, i.e., creaming and evaporation are used to concentrate synthetic latices.

In the creaming process, which is analogous to the creaming of milk, certain high molecular weight colloids which are soluble in or highly swollen by water are added to the latex. These can be synthetic or natural products. The best process is to add the latex gradually to the concentrated, aqueous thickening agent free of swollen particles and to homogenize after each addition. Toward the end, the latex can then be added in one shot.

Another widely used process for the preparation of concentrated nitrile latices is evaporation. This process may yield latex concentrations as high as 50–55%. Although the stability of the latex produced in this manner is normally satisfactory, steps must be taken to avoid undesirable skin and foam formation, since all the emulsifier remains in the end product.

Two commercial processes are available for concentration by evaporation, the Revertex process and the vaccum film process.

IV. MOLECULAR STRUCTURE, IDENTIFICATION, AND DETERMINATION METHODS FOR NITRILE RUBBER (2,3,10,11)

A. Molecular Structure

A large variety of structures can occur in the synthesis of nitrile rubber depending on the manner in which the monomer units enter the growing chain. As will be shown, this variation in enchainment greatly influences the physical properties. The structure of the polymer is largely determined by the type of initiator used and by the polymerization mechanism. The most important factors are the monomer ratio, the polymerization type (emulsion, solution, bulk), the initiator system (radical, ionic), the additives used (e.g., modifiers, etc.), and the temperature profile. Besides growing by adding monomers in a statistical or nonstatistical manner, the polymer radical may react with double bonds or C—H bonds either in its own or in another chain. In the former case, cyclic polymers are obtained, whereas in the latter case, branched species will be formed. Finally, it should be kept in mind that various chemicals are introduced during processing and vulcanization which may affect the chemical and physical properties of the polymer, i.e., by splitting off side chains, by causing degradation or cyclization at higher drying or processing temperatures, by giving rise to isomerization of cis-1,4 units into trans-1,4 units during vulcanization, etc.

Since nitrile rubber is presently prepared almost exclusively in aqueous emulsion with radical initiators, it does not have a strictly linear structure but is more or less branched or crosslinked. Electron microscopy of nitrile rubber reveals a spherical morphology. The degree of branching is lower with the cold rubber than with the hot rubber. The molecular weight distribution is rather broad. Cold and hot nitrile rubbers are noticeably different in this respect, the cold rubber usually having a narrower distribution.

Branching, crosslinking, molecular weight distribution, and average molecular weight are strongly affected by the conversion. For example, the cis-1,4 to trans-1,4 ratio is 12.4–77.6 in a nitrile rubber containing 28% acrylonitrile prepared at about 28°C. Evidently the predominant enchainment is *trans*-1,4.   The proportion of *trans*-1,4 structure increases strongly at the expense of the *cis*-1,4 content with decreasing polymerization temperatures.

Since the molecular weight of acrylonitrile is lower than that of styrene, nitrile rubber contains more monovinyl units per unit weight than SBR when the weight ratio between the monomers is the same in both polymers.   Consequently, the number of double bonds in nitrile rubber is lower than in SBR at the same monomer weight ratio.   This difference influences the properties of these polymers.   It also causes the relative degree of branching to be lower in nitrile rubber than in SBR.

B. Identification of Nitrile Rubber

The presence of nitrile rubber is occasionally suggested by the high density of the crude rubber (depending on the type, 0.94–1.105) which sets it distinctly apart from other elastomers.

Rubber purified by extraction is used for qualitative analysis.   It may contain fillers and factice even after such extraction, however.

Nitrile rubber burns with a sooty flame giving off unpleasant odors. On heating the polymer in a test tube, pyrolysis occurs with formation of a light or slightly discolored, initially acidic distillate which turns alkaline upon further heating at the same time as an unpleasant odor appears (dangerous when inhaled since the pyrolysis vapors contain hydrogen cyanide or cyanogen).   Since nitrile rubber contains 5 to 10% nitrogen, it or its blends can be identified in a preliminary manner by the Lassaigne nitrogen test.

Definite identification of nitrile rubber can be accomplished through pyrolysis according to H. P. Burchfield.   This method is based on a known color reaction between gaseous pyrolysis products or high molecular weight compounds and *p*-dimethylaminobenzaldehyde.   The results are unambigous for most elastomers, provided the sample is not a mixture of several elastomers.   During the pyrolysis of nitrile rubber, various nitrogen containing compounds are split off from the acrylonitrile and give an orange color with *p*-dimethylaminobenzaldehyde.   The orange color turns into burgundy on heating with methanol.   A certain amount of experience is necessary for this analysis, since the color reactions of various elastomers do not differ much in some cases (see Table IV). Under certain conditions, nitrogen-containing additives which yield alkaline degradation products on pyrolysis can yield an orange red color with *p*-dimethylaminobenzaldehyde similar to that caused by the degradation products of acrylonitrile.

TABLE IV
Color Formation in Burchfield Test

Material	Pyrolyzate	After heating with methanol
Control	—	Pale yellow
Poly(vinyl chloride)	Yellow	Yellow
Chlorobutadiene rubber	Yellowish green	Dirty green
Chlorosulfonated polyethylene	Yellow	Yellow
Silicone rubber	Yellow	Yellow
Polyurethane elastomers	Yellow	Yellow
Nitrile rubber	Orange red	Burgundy
Natural rubber (NR)	Reddish brown	Reddish violet
Butyl rubber	Yellow	Blue violet
Styrene butadiene rubber (SBR)	Yellow green	Green
Mixtures of		
3 parts of NR, 1 part SBR	Reddish brown	Dark violet
1 part NR, 1 part SBR	Yellowish green	Greenish blue
1 part NR, 3 parts SBR	Yellowish green	Greenish blue

A modification of this test is the so-called Chloroprene–Nitrile Test. A strip of spot test paper is impregnated with copper acetate and metanil yellow and is dipped in the corresponding solution of benzidine hydrochloride and hydroquinone. Subsequently, the test paper is contacted with the pyrolysis products of the sample. In the presence of nitrile rubber, a green color develops. Chloroprene rubber yields a red color, while other elastomers as controls predominantly yield brownish colors.

Another spot test paper, impregnated with p-dimethylaminobenzaldehyde and hydroquinone, can be used for identification of the pyrolysis vapors of various rubbers. Nitrile rubber yields a red color, polychloroprene and styrene-butadiene rubber a bluish green color, natural rubber a dark blue color, and butyl rubber a violet color.

Ostromow's method is a particularly reliable identification method for nitrile rubber. In the presence of nitrile rubber, hydrogen cyanide is evolved on pyrolysis. The hydrogen cyanide is led into a dilute sodium hydroxide solution which is subsequently treated with a dilute iron(II) chloride solution. After heating and acidification with hydrogen chloride a blue color or blue precipitate (Prussian blue) is obtained in the presence of acrylonitrile.

TABLE V
Degradation Times of Various Rubber Types
(Vulcanizates) in Acid Mixture

	Time until the start of degradation, sec	
Type of rubber	40°C	70°C
Thiokol	0–35	Immediately
Natural rubber	20–70	0–5
Polychloroprene	40–60	5–8
Styrene-butadiene rubber	630–1200	10–25
Nitrile rubber	650–1800	30–80
Butyl rubber	∼5000	120–140

According to Parker, nitrile rubber can be identified by determining the degradation time in a nitric acid–sulfuric acid mixture. Table V shows data obtained by Parker. Carbonate fillers interfere with this method; natural rubber, thioplastics and polychloroprene cannot be distinguished by this acid test. Differentiation is simpler with other elastomers.

To identify nitrile rubber in unknown vulcanizates, Parker recommends swelling in various solvents. The logarithm of the swelling ratio in benzene and in aniline is plotted against the logarithm of the swelling ratio in gasoline and benzene (Eq. 15).

$$\log \frac{\text{benzene swelling}}{\text{aniline swelling}} = f\left(\log \frac{\text{gasoline swelling}}{\text{benzene swelling}}\right) \tag{15}$$

The effects of fillers and the degree of vulcanization are eliminated by this treatment. The values fall in defined areas in the coordinate system as shown in Fig. 6. The values for nitrile rubber are +1.1 to 1.7 on the abscissa and 0 to −0.1 on the ordinate.

V. COMPOUNDING AND PROPERTIES OF VULCANIZATES (1–3,12,13)

A. Compounding

1. Vulcanization Systems

Nitrile rubber is usually vulcanized with sulfur, accelerators, and zinc oxide. The amount of sulfur used with nitrile rubber is somewhat lower

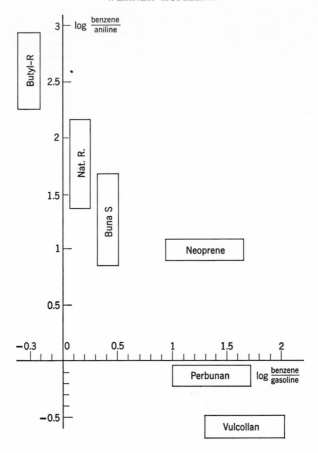

Fig. 6. Swelling analysis by Parker's method.

and the amount of accelerator somewhat higher than that used with natural rubber. Accelerators of the benzothiazyl type, which give safe processing conditions, have been found particularly useful. They can be activated with basic accelerators, thiuram disulfides or dithiocarbamates. Sulfenamide accelerators are often activated with thiuram monosulfide.

With large amounts of sulfur, nitrile rubber can be vulcanized to a hard rubber which, in addition to its oil resistance, has an excellent heat distortion resistance.

Sulfur-free vulcanizates can be prepared with organic peroxides, and particularly with tetramethyl thiuram disulfide. Such vulcanizates have excellent hot air resistance.

2. Antioxidants

Nitrile rubber contains practically always an efficient stabilizer to protect the material against oxidation during storage. Stabilization is particularly necessary when the finished products are exposed to high heat or strong dynamic influences.

To improve the hot air resistance of thiuram or peroxide vulcanizates, particularly aldol-α-napthylamine and 2-mercaptobenzimidazole and their combinations have proven satisfactory.

To enhance hot air resistance of conventional sulfur vulcanizates, usual antioxidants such as the discoloring antioxidants are used. To increase dynamic fatigue resistance, antioxidants of the p-phenylene-diamine class are employed among others. Higher concentrations of these compounds also provide protection against ozone attack.

3. Fillers

In contrast to the situation with similar natural rubber or poly-chloroprene vulcanizates, relatively low mechanical property values are obtained with unfilled nitrile rubber or with inert fillers alone. However, with reinforcing blacks or light-colored reinforcing fillers the tensile strength of nitrile rubber approaches the tensile level of natural rubber vulcanizates. The use of light-colored reinforcing fillers is advantageous from the point of view of aging. As with other rubbers, large amounts of inert blacks and inert light fillers can be used to achieve the same hardness (cost saving). Processibility of the mixture is also improved. As anticipated, the swelling of vulcanizates in gasoline and oils decreases with increasing amounts of fillers (corresponding to the decreasing rubber content).

4. Plasticizers

The viscosity, tackiness, and extrudability of nitrile rubber mixtures as well as the hardness, resilience, low temperature flexibility, and swell properties of vulcanizates can be regulated with various types of plasti-cizers. To improve calenderability and extrudability and partly also the building tack of mixtures, processing plasticizers, e.g., colophonium, coresine, coumarone resin, xylene formaldehyde resins, wool fat, non-swelling and liquid factices, and aromatic oils are used. All plasticizers influence the tensile strength of nitrile rubber vulcanizates. This influence is particularly noticeable with synthetic plasticizers of the ester and ether types. This group of plasticizers improves rebound resilience and low temperature flexibility, however, and can be considered typical in nitrile rubber processing.

Inexpensive mineral oil plasticizers of the types employed extensively with natural rubber, styrene-butadiene rubber, polybutadiene, and similar elastomers have only a limited compatibility with nitrile rubber (swell resistance of NBR toward mineral oils). Essentially only aromatic mineral oils can be used and in limited amounts. Mineral oils do not improve the resilience and low temperature properties of nitrile rubber.

Special antistatic plasticizers increase the electrical conductivity of nitrile rubber articles.

B. Blends of Nitrile Rubber with Other High Polymers

1. Blends with Others Elastomers

The compatibility of nitrile rubber with other elastomers is primarily determined by the polarity of the elastomer. Thus, nitrile rubber and natural rubber or polybutadiene have only a very limited compatibility. However, for special purposes small amounts of natural rubber can sometimes be admixed (for example, to improve building tack).

Nitrile rubber and styrene-butadiene rubber are compatible in all proportions without a significant sacrifice in mechanical properties. This compatibility is exploited economically when only moderate swell resistance is required or when large amounts of plasticizer (low temperature flexibility) are to be used and a negative swell (shrinking) would occur due to plasticizer extraction if nitrile rubber were used alone.

Nitrile rubber and polychloroprene can be blended and such blends improve the ozone and weathering resistance of nitrile rubber. Effective ozone protection can also be achieved by blending with poly(vinyl chloride).

The swell characteristics of these elastomer blends are determined by the swell of the individual components and by the blend ratios.

2. Blends with Poly(vinyl Chloride)

Poly(vinyl chloride) can be blended in all proportions with nitrile rubber. Compatibility with nitrile rubber starts at 23% acrylonitrile content and increases with increasing acrylonitrile content.

Unvulcanized blends with preponderant amounts of poly(vinyl chloride) are plastics in which the nitrile rubber functions as a nonextractable polymeric plasticizer. Blends with preponderant amounts of nitrile rubber are too soft for plastics applications; however, they are vulcanizable and can be employed as elastomers. Poly(vinyl chloride) blends give a variety of both advanatages and disadvantages. Thus, processibility, tensile strength, ozone and weathering resistance, swell resistance, and flame resistance are improved while resilience, low temperature flexibility, and permanent set are impaired.

3. Blends with Phenol-Formaldehyde Resins

Finely dispersed not completely hardened phenol-formaldehyde resins are well compatible with nitrile rubber. The compatibility increases with increasing acrylonitrile content. The softening temperature of such a resin should not be above 100°C, since the phenol-formaldehyde resin must be blended above this temperature to yield a homogeneous dispersion. After the nitrile rubber has been completely blended with the phenol-formaldehyde resin, the other components are added at normal mixing temperatures. Since phenol-formaldehyde resins are thermoplastics, processibility (for example, extrusion, calendering) and sealability of nitrile rubber mixtures will be improved by the addition of such a resin. Adhesion to metals will also be increased.

The tensile strength, tear resistance, abrasion resistance, and swell resistance of nitrile rubber vulcanizates are considerably improved by the addition of phenol-formaldehyde resins. The hardness of vulcanizates is strongly increased so that hard-rubberlike vulcanizates can be obtained. Disadvantages of these blends are decreased resilience and increased permanent set, particularly in the range of high deformation forces and temperatures.

Nitrile rubber/phenol-formaldehyde resin blends are used among others for bushings, ladies' heel points, brake linings, and gaskets.

C. Processing of Nitrile Rubber

The processing technology of nitrile rubber closely resembles that of natural rubber or styrene-butadiene rubber.

D. Properties of Vulcanizates

As mentioned above, nitrile rubber vulcanizates are characterized by excellent oil and gasoline resistance combined with good aging characteristics and high abrasion resistance. They show only little fatigue under dynamic stress.

1. Stress–Strain Behavior

In contrast to vulcanizates of natural rubber and polychloroprene, nitrile rubber vulcanizates which contain no reinforcing fillers exhibit only modest tensile properties. However, outstanding strength characteristics can be obtained through the use of reinforcing fillers. The maximum tensile strength obtainable is dependent upon the Shore hardness of the vulcanizate. The optimum tensile strength occurs at a hardness of about 70 Shore A. The maximum tensile strength obtainable is about 280 kg/cm² (4000 psi).

2. Elastic Properties and Low Temperature Behavior

The resilience and the low temperature behavior of nitrile rubber vulcanizates are less favorable than those of natural rubber or styrene-butadiene rubber. These properties are particularly influenced by the acrylonitrile content, the composition of the mixtures and the particular choice of ester or ether plasticizers.

With suitable compositions, the permanent set of nitrile rubber vulcanizates is very low.

3. High Temperature Behavior

Independent of acrylonitrile content, nitrile rubber vulcanizates exhibit good hot air resistance and optimum compositions can be used for extended periods at temperatures up to 100°C. Even higher temperatures can be tolerated in oil. Suitably compounded nitrile rubber vulcanizates have good resistance toward hot steam.

4. Swell Resistance

Nitrile rubber vulcanizates are highly resistant toward nonpolar solvents, e.g., gasoline, mineral oils, and fats. Swell resistance is of course affected by the swelling medium, the temperature and the composition.

Certain additives used in modified industrial lubricating oils may have a strongly hardening effect on nitrile rubber vulcanizates. Furthermore, these vulcanizates are not resistant toward aromatic hydrocarbons (e.g., benzene, toluene) and polar solvents (e.g., chlorinated hydrocarbons, esters, or ketones).

5. Abrasion Resistance

Nitrile rubber vulcanizates which contain no fillers exhibit relatively high abrasion rates, but such materials are of no technical interest also for other reasons. By contrast, vulcanizates with reinforcing fillers exhibit about 20–30% less abrasion than corresponding natural rubber vulcanizates.

The abrasion properties of nitrile rubber are also clearly better than those of SBR.

6. Gas Permeability

The gas permeability of nitrile rubber vulcanizates is very low and it decreases with increasing acrylonitrile content. The gas permeability of vulcanizates having a high acrylonitrile content is of the same order of magnitude as that of butyl rubber (cf. Chapter 5.A).

E. Applications of Nitrile Rubber

Nitrile rubber vulcanizates are used primarily when excellent swell resistance, combined with good resistance toward aging, heat, and wear, and a low permanent set are indicated. Typical applications are: molded goods of all kinds, gaskets, bushings, membranes, buffers, shock absorbers, shoe heels, soles, roller covers, rubber stamps, printing cloth, V belts, conveyor belts, rubberized cloth, cable sheaths (particularly NBR/PVC), hoses, rubber footwear, gloves, brake and clutch linings, punch plates, closed-cell sponge, and adhesives.

VI. ACRYLONITRILE MODIFIED NATURAL RUBBER (1)

The fact that acrylonitrile confers good oil and gasoline resistance to copolymers containing this monomer prompted a study to learn if the swell resistance of natural rubber could be improved by grafting acrylonitrile onto it.

Studies carried out mainly in England and France have shown that acrylonitrile can be easily attached to natural rubber. Two methods can be used: (1) a natural rubber solution containing acrylonitrile and a peroxide, e.g., benzoyl peroxide, can be heated, or (2) acrylonitrile can be added to a rubber latex and polymerized under usual conditions.

Indeed, acrylonitrile-modified natural rubber shows increased swell resistance in proportion to its acrylonitrile content. This is indicated in Table VI where the volume swell of acrylonitrile-modified natural rubber vulcanizates is given for various acrylonitrile contents. According to these data, the swell of an acrylonitrile-modified natural rubber containing about 50% acrylonitrile is equivalent to that of a nitrile rubber having a high acrylonitrile content.

TABLE VI
Swelling of Acrylonitrile-Modified
Natural Rubber Vulcanizates

Acrylonitrile in rubber, %	Volume swell of vulcanizates	
	In benzene, %	In kerosene, %
0	346	190
20	221	134
56.5	110	50

From an application point of view, acrylonitrile-modified natural rubber stands between natural rubber and nitrile rubber. It can be crosslinked with the same agents as used with other rubbers and the compounding does not present any particular problems.

Acrylonitrile-modified natural rubber has not gained commercial importance, however.

VII. VINYLPYRIDINE RUBBER (4,13–15)

As already mentioned, vinylpyridine was also studied as a possible monovinyl component for copolymerization with butadiene during the early development of synthetic rubber but vinylpyridine rubber as such did not become commercially significant. However, it is used in latex form to improve the adhesion between rubber and fabric in textile impregnation, particularly in combination with resorcinol-formaldehyde resins.

Vinylpyridine rubber is prepared in a manner analogous to that used for nitrile rubber and it can also be vulcanized with sulfur and accelerators. Its vulcanizates exhibit a certain degree of oil resistance. An interesting unorthodox vulcanization process which will be discussed in some detail below can be used with this rubber.

Elastomers containing vinylpyridine units can be crosslinked with compounds containing active chlorine. The crosslinking reaction proceeds via a heteropolar quaternization of the nitrogen. Consequently, these compounds containing active chlorine are called quaternizing agents.

A reactive chlorine containing compound, e.g., benzyl chloride, benzal chloride, benzotrichloride bischloroacetate, and octylchloroacetate, is added as the only crosslinking agent to the rubber mix on the mill. Since some of these substances irritate the mucous membranes, efficient exhaust systems must be used. They also attack and cause corrosion of the mill rolls and the molds. Other disadvantages associated with the use of these materials are that they make the stock adhere strongly to the molds and that they require a very long vulcanization period, e.g., 5 hr at 155°C (312°F). Due to these and other economic factors, this crosslinking process has not yet found industrial application.

As a consequence of the nitrogen quaternization, the vulcanizates become polar and acquire good swell resistance toward gasoline, oil, and many other solvents. The swell resistance increases with increasing amounts of the chlorine compound, but the activity of the quaternizing agent plays a decisive role in this connection, too. For example, 40% swell is obtained in isooctane with the moderately active octyl chloroacetate (4%), whereas the swell obtained with the same amount of bischloroacetate, benzal chloride, and benzotrichloride is 10, 10, and 5%, respectively. Thus, the good swell resistance of nitrile rubber can be reached with active vulcanization systems.

The degree of vulcanization (tensile value) is influenced by the various quaternizing agents in the same manner as the resistance to swell. The best tensile values are obtained with benzotrichloride, which is the physiologically most aggressive compound among the above-mentioned quaternizing agents.

Quaternization is explained on the basis of the formation of a heteropolar bond. The fundamental reactions can be visualized as shown in Eqs. 16–19.

$$(16)$$

$$(17)$$

It is known that 2- and 4-alkyl-pyridines undergo aldol condensations readily (Eq. 18).

$$
\begin{array}{l}
\underset{N}{\diagdown}\!\!=\!\!\underset{(\delta^-)}{\overset{H}{\underset{-CHR}{\mid}}} + (\delta^+)\,\diagup^{C=O} \rightarrow \underset{N}{\diagdown}\!\!=\!\!\underset{\diagdown C-OH}{\overset{-CHR}{\underset{\mid}{}}}
\end{array}
\tag{18}
$$

According to Saville and Moore, two vinylpyridine groups can react similarly with each other giving crosslinked rubber (Eq. 19).

Thus, bifunctional chlorine-containing compounds are not absolutely necessary for obtaining a network, although crosslink formation is even more readily explainable with such components.

References

Instead of individual literature references a list of surveys is provided.

1. W. L. Semon, in *Synthetic Rubber*, G. S. Whitby, Ed., Wiley, New York, 1954, pp. 794–837.
2. W. Hofmann, "Nitrile Rubber," *Rubber Chem. Technol.*, **36,** No. 5 (1963), published in **37,** No. 2 (1964).
3. W. Hofmann, *Nitrilkautschuk*, Verlag Berliner Union, Stuttgart, 1965.

4. P. Schneider, in *Kautschuk-Handbuch*, S. Boström, Ed., Berlag Berliner Union, Stuttgart, Vol. 1, 1959, p. 237.
5. C. E. Morrell, in *Synthetic Rubber*, G. S. Whitby, Ed., Wiley, New York, 1954, p. 56; W. J. Toussaint and J. L. Marsh, *ibid*, p. 86; M. L. Fisher, *ibid*, p. 105.
6. H. Logemann, in *Methoden der organischen Chemie*, (*Houben-Weyl*), E. Müller, Ed., XIV/1, G. Thieme, Stuttgart, 1961.
7. G. Kolb and J. Peter, in *Chemische Technologie IV*, K. Winnacker and L. Küchler, Eds., C. Hauser, Munich, 1960.
8. L. Küchler, *Polymerisationskinetik*, Springer, Berlin, 1951.
9. W. Hofman, in *Kautschuk-Handbuch*, S. Boström, Ed., Verlag Berliner Union, Stuttgart, Vol. 1, 1959, p. 330.
10. M. Hoffmann, in *Methoden der organischen Chemie*, (*Houben-Weyl*), E. Müller Ed., XIV/2, G. Thieme, Stuttgart, 1963.
11. P. Schneider, in *Methoden der organischen Chemie*, (*Houben-Weyl*), E. Müller, Ed., XIV/2, G. Thieme, Stuttgart, 1963.
12. W. Hofmann, in *Kautschuk-Handbuch*, S. Boström, Ed., Verlag Berliner Union, Stuttgart, Vol. 2, 1960, p. 172.
13. W. Hofmann, *Vulkanisation and Vulkanisationshilfsmittel*, Verlag Berliner Union, Stuttgart, 1965, p. 336.
14. S. Boström, Ed., *Kautschuk-Handbuch*, Verlag Berliner Union, Stuttgart, Vol. II, 1960, p. 166.
15. W. B. Reynolds et al., *Proc. Rubber Technol. Conf. 3rd, London, 1954,* 226.

CHAPTER 4

ELASTOMERS BY RADICAL AND REDOX MECHANISMS

C. Neoprene

Chester A. Hargreaves, II

*Elastomer Chemicals Department, E. I. du Pont de Nemours & Co.,
Wilmington, Delaware*

Contents

I. INTRODUCTION

Neoprene is the generic name for polymers of 2-chloro-1,3-butadiene (chloroprene). Polychloroprene was the first commercially successful

227

synthetic elastomer, being introduced by E. I. du Pont de Nemours and Company, Inc. in 1932 under the now abandoned tradename DuPrene. The polymer structure consists predominantly of *trans*-2-chloro-2-butenylidene-1,4 units in which the electronegative chloro function both deactivates the double bond and increases polymer cohesive energy density. These effects are reflected in the unique combination of commercial assets exhibited by neoprene: oil, oxidation, and general chemical resistance.

The Elastomers Department of E. I. du Pont de Nemours and Company, Inc. is the major producer of neoprene, offering more than 30 types of dry and latex polymers (1,2) from manufacturing locations in the United States, Great Britain, and Japan. Other manufacturers of polychloroprene elastomers are listed below:

Company	Location	Trade name
Farbenfabriken Bayer AG	West Germany	Perbunan C
—	USSR	Sovprene, Nairit
Distugil, S. A. (in 1966) (3)	France	Butachlor
Denki Kagaku Kogyo KK	Japan	Denka Polychloroprene
Petro-Tex Chemical Corp. (in 1968) (4)	Texas, USA	—

II. CHLOROPRENE (2-CHLORO-1,3-BUTADIENE) (5)

Chloroprene was discovered by Carothers and Collins (6,7) in 1930 during their classical studies on acetylene. It is a colorless, highly reactive liquid with a characteristic ethereal odor. It is only slightly soluble in water but is miscible with common organic solvents. Chloroprene is both toxic and flammable and forms explosive mixtures with air in the range of 4–20 vol % concentration. Continued exposure to air containing in excess of 25 ppm chloroprene produces toxic symptoms.

A. Properties

The free radical reactivity of the conjugated butadienyl structure is markedly enhanced by the electron donating chlorine atom. The electronic interaction of the chlorine atom with the conjugated double bonds is shown by shifts of the ultraviolet spectral bands, relative to butadiene, to longer wavelengths. Infrared and Raman spectral data also reveal the molecule to be transoid and planar (5).

Chloroprene polymerizes spontaneously at room temperature and undergoes free radical addition reactions with a variety of inorganic and organic compounds at rates generally much greater than those for comparable reactions with butadiene and isoprene. Chloroprene combines with oxygen in a light catalyzed reaction to form peroxides which are believed to have the following polymeric structure (8):

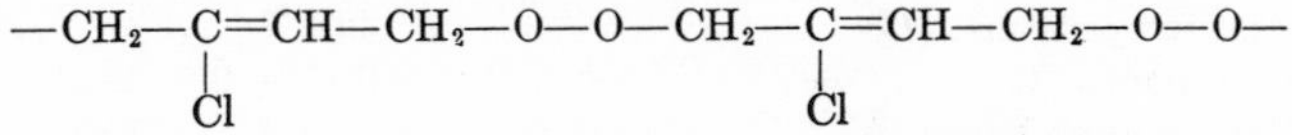

$$-CH_2-\underset{\underset{Cl}{|}}{C}=CH-CH_2-O-O-CH_2-\underset{\underset{Cl}{|}}{C}=CH-CH_2-O-O-$$

Accordingly, chloroprene should be stored in an oxygen-free atmosphere and at low temperatures ($< -15°C$), preferably in the presence of suitable antioxidants such as phenyl-2-naphthylamine, phenothiazine, or hydroquinone.

Chloroprene also undergoes typical addition reactions by an ionic mechanism but, in these instances, the electronegative chlorine atom has a deactivating influence. Similarly, in Diels-Alder reactions, the reactivity of chloroprene is reduced compared to that of butadiene or isoprene. The activation energy for dimerization is approximately 22 kcal/mole (9). The vinylcyclohexene and octadiene dimer structures illustrated here have been isolated and identified (7,8,10–13).

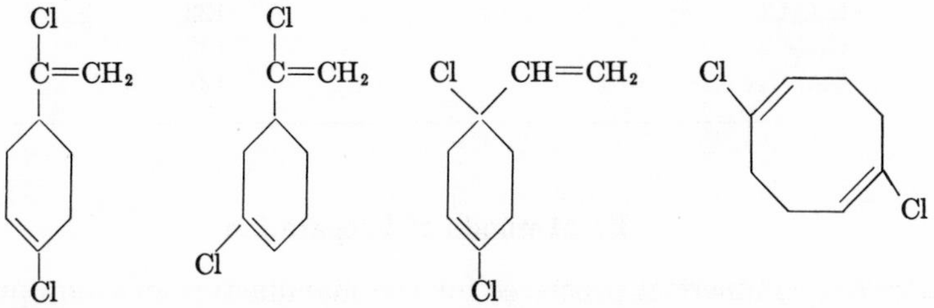

The octadiene dimer results from head-to-head dimerization and is predominantly in the boat configuration with *cis-cis* double bond configuration.

The chlorine substituent in chloroprene is of low activity similar to a vinylic chlorine atom. The Grignard reagent from chloroprene, 2-(1,3-butadienyl)magnesium chloride, has been prepared, however, by reaction of the isomeric 4-chloro-1,2-butadiene with magnesium (14) (reaction 1). Selected physical properties of chloroprene are listed in Table I.

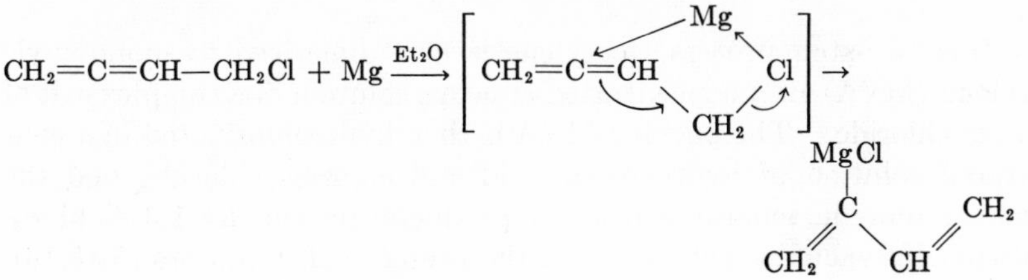

TABLE I
Physical Properties of Chloroprene (5)

Critical temperature, °C	261.7
Melting point, °C	-130 ± 2
Boiling point at 1 atm, °C	59.4
Viscosity at 25°C, cp	0.394
Refractive index, n_D^{20}	1.4583
Density	
d_4^{20}	0.9585
d_{20}^{20}	0.9583
Vapor pressure–temperature relationship mm Hg: °K	$Log\ p = (1545.3/T) + 7.527$
Dipole moment (in benzene), esu	1.42×10^{-18}
Dielectric constant, 27°C	4.9
Intramolecular distances (planar), Å	
r_{C-C}	1.36
r_{C-C}	1.46 ± 0.04
r_{C-Cl}	1.70 ± 0.02
Intramolecular angles (planar), deg	
C_1C_2Cl	122
$C_1C_2C_3$	122
$C_2C_3C_4$	127 ± 3

B. Methods of Preparation

Present commercial processes for the manufacture of chloroprene are based on either acetylene or butadiene, but a number of other routes have been disclosed in the literature.

1. Acetylene Process

$$2HC{\equiv}CH \xrightarrow[80°C]{Cu_2Cl_2,NH_4Cl} CH_2{=}CH{-}C{\equiv}CH$$

$$CH_2{=}CH{-}C{\equiv}CH + HCl \xrightarrow[20-50°C]{Cu_2Cl_2,NH_4Cl} CH_2{=}\underset{\underset{Cl}{|}}{C}{-}CH{=}CH_2$$

In this two-step process, acetylene is first dimerized to monovinylacetylene (MVA) in a concentrated aqueous solution of a complex salt of cuprous chloride. The purified MVA is then hydrochlorinated in a concentrated solution of hydrochloric acid and cuprous chloride, and the dried chloroprene separated from by-products (principally 1,3-dichloro-2-butene) by vacuum distillation in the presence of inhibitors (5,15,16).

Early studies (17,18) indicated that the hydrochlorination of MVA proceeded by 1,4 addition to yield the allenic isomer, 4-chloro-1,2-

$$\overset{\delta+}{CH_2}{=}CH{-}C{\equiv}\overset{\delta-}{CH} + HCl \rightarrow Cl{-}CH_2{-}CH{=}C{=}CH_2 \xrightarrow{Cu_2Cl_2}$$

$$CH_2{=}\underset{\underset{Cl}{|}}{C}{-}CH{=}CH_2 \quad (2)$$

butadiene (reaction 2), which is isomerized to chloroprene under the influence of cuprous chloride. Dologopol'skii and Trenke (19) have recently suggested that the activating influence of cuprous chloride results from complex formation with MVA. In the complex, the polarizing influence of the Cu atom on the acetylene bond leads to direct 1,2 addition of HCl (reaction 3).

$$H^+[CH_2{=}CH{-}\overset{\delta+}{C}{\equiv}\overset{\delta-}{CH} \cdots HCuCl_3]^- \rightleftharpoons$$

$$H^+[CH_2{=}CH{-}\underset{\underset{Cl}{|}}{\overset{-}{C}}{=}CH \cdots HCuCl_2] \rightarrow CH_2{=}CH{-}\underset{\underset{Cl}{|}}{C}{=}CH_2 + H^+ + CuCl_2^- \quad (3)$$

2. Butadiene Process

The three essential steps to produce chloroprene from butadiene (20–22) are chlorination, isomerization, and caustic dehydrochlorination as shown in equations 4–6.

$$CH_2{=}CH{-}CH{=}CH_2 + Cl_2 \rightarrow$$

$$\underset{\underset{Cl}{|}}{CH_2}{-}\underset{\underset{Cl}{|}}{CH}{-}CH{=}CH_2 + ClCH_2{-}CH{=}CH{-}CH_2Cl \quad (4)$$

$$60\% \qquad\qquad 40\% \text{ (cis and trans)}$$

$$Cl{-}CH_2{-}CH{=}CH{-}CH_2Cl \xrightarrow[Cu]{Cu_2Cl_2} \underset{\underset{Cl}{|}}{CH_2}{-}\underset{\underset{Cl}{|}}{CH}{-}CH{=}CH_2 \quad (5)$$

$$ClCH_2{-}\underset{\underset{Cl}{|}}{CH}{-}CH{=}CH_2 + Na^+OH^- \rightarrow CH_2{=}\underset{\underset{Cl}{|}}{C}{-}CH{=}CH_2 + Na^+Cl^- + H_2O$$

$$(6)$$

A variety of preferred process conditions have been disclosed (23).

3. Alternate Processes

Other routes for the preparation of chloroprene involve dehydrochlorination of products obtained by chlorination of butane and butenes (24); thermal dehydrochlorination of 1,3-dichloro-2-butene (25), oxidative dehydrogenation of 2-chloro-2-butene (26), dehydrochlorination of 1,2,3-trichloro-3-butene with metals (27), and dehydrochlorination of the 3,4-dichloride of butadiene sulfone (28).

III. POLYMERIZATION (29)

A. Free Radical Initiation

Chloroprene polymerizes readily by a free radical mechanism; and, as a consequence, the majority of experimental polymerization studies and commercial polymerization practices are concerned with free radical initiated emulsion systems. The behavior of chloroprene in emulsion polymerization is unusual in several aspects. The locus of initiation is not restricted to the soap micelles or derived latex particles, but initiation can also occur in the dissolved monomer or at the monomer–water interface (30). Further, chloroprene does not follow the classical kinetics of emulsion polymerization as developed by Smith and Ewart (31) since the propagation rate per latex particle is not independent of type and amount of initiator (this subject is discussed in Chapter 4.A). The extreme radical reactivity of chloroprene is believed to be primarily responsible for these anomalous aspects of chloroprene emulsion polymerization (30).

The heat of polymerization is 16.2 ± 0.3 kcal/mole of monomer (170 cal/g) at $334.5°K$ (32).

1. Emulsifying Agents

Anionic surfactants of the rosin acid, fatty acid, or alkyl sulfonate type are usually employed in the emulsion polymerization of chloroprene (30,33–35). Cationic soaps such as cetyl pyridinium bromide or betaine surfactants can be used in the stabilization of neutral or acidic emulsions of chloroprene (36,37). Polymerization may also be effected using protective colloids as emulsifiers (38).

2. Initiators and Stabilizing Agents

A wide variety of redox systems using common inorganic and organic peroxy compounds can serve as initiators in chloroprene polymerization (39,40). The sensitivity of chloroprene is such, however, that it must be stabilized with antioxidants against uncontrolled polymerization from adventitious contaminants such as trace oxygen (41). Since the common initiators are not very efficient in overriding the stabilizer action, a number of specialized initiators have been devised for this purpose. Representative classes include acyclic azo compounds (42), nitrogen sulfide compounds such as heptasulfur imide (43), trialkyl boron compounds with cocatalysts (44), borazanes (45)

$$[(R_3B)_n(N{\overset{\displaystyle R_1}{\underset{\displaystyle R_{111}}{\diagup}}}R_{11})]$$

and formamidine sulfinic acid (46).

Free radical inhibitors are used as stabilizers to stop the polymerization at the desired monomer conversion: e.g., thiurams (47), oxalate anions (48), aromatic amines and phenolic compounds (49), and N,N-dialkylhydroxyl amines (50).

3. Molecular Weight Control

Molecular weight control of polychloroprene can be accomplished by standard chain transfer reactions with such modifying agents as alkyl mercaptans, diethyl xanthogen disulfides, dialkylthiuram disulfides, and iodoform (51–54) or by postpolymerization reactions on chloroprenesulfur copolymers.

Primary alkyl mercaptans, from C_5 to C_{12}, are consumed in chain transfer reactions with chloroprene with the same rate constant as that for chloroprene consumption ($C_T = 1.0$) (55). Earlier studies using dodecylmercaptan tagged with radioactive sulfur had shown the required presence of RS groups in the polymer backbone (56). The rate of mercaptan disappearance is independent of chain length (C_5 to C_{12}) and is only slightly affected by polymerization temperature and polymerization rate. These observations, coupled with the fact that these mercaptans have only limited solubility in aqueous base, permit the conclusion that the chain transfer reaction is a function of the inherent reactivity of the mercaptan at the polymerization site and is not diffusion or solubility controlled. The observation that the chain transfer constant for tert-dodecyl mercaptan is only one-half that of the corresponding primary mercaptan is compatible with this conclusion (57).

Chain transfer constants for other agents have been reported (54): diisopropylxanthogen disulfide 0.99, diethylthiuram disulfide 1.60, tetradecyl mercaptan 0.64.

Regardless of modifier activity, the plot of molecular weight as a function of conversion shows a sharp upturn at conversion above 70%. This phenomenon is not unexpected in view of the increasing probability of branching and crosslinking reactions as the polymer/monomer ratio increases. These reactions presumably occur between growing polymer chains ($P_n \cdot$) and previously formed polymer (P_m) followed by further reactions with monomer (M).

$$\sim\!P_n\cdot \quad + \overset{\{}{\underset{\{}{P}}_m \rightarrow \sim\!\overset{\{}{\underset{\{}{P}}}_{n+m}\cdot$$

$$\sim\!\overset{\{}{\underset{\{}{P}}}_{n+m}\cdot + oM \rightarrow \sim\!\overset{\{}{\underset{\{}{P}}}_{n+m+o}\!\sim$$

The resulting branched, high molecular weight species (P_{n+m+o}) are especially vulnerable to network formation through subsequent reactions of a similar nature which need occur only rarely to result in an insoluble product.

The so-called sulfur modified neoprenes (the "G" types) are prepared by polymerization of chloroprene in the presence of small amounts of elemental sulfur with subsequent treatment of the latex with thiuram disulfides or other thiophilic agents (54,58–61). This postpolymerization treatment is trivially termed "plasticization" or peptization. Studies with radioactive sulfur (56) demonstrated that sulfur was acting as a comonomer rather than a chain transfer agent (modifier) to produce a polymer

$$-[CH_2-\underset{\underset{Cl}{|}}{C}=CH-CH_2]_n-S_x-[CH_2-\underset{\underset{Cl}{|}}{C}=CH-CH_2]_n-S_x-$$

where x is 2–6 and n is, on the average, 80–110x. The copolymer prepared at conversions above about 10% is a gel prior to treatment with the thiuram disulfide. Since the peptization process is effective only when sulfur is present during the polymerization, there is clear implication of involvement of the thiuram disulfide with the polysulfide part of the polymer backbone (62). Klebanskii and co-workers (63) have confirmed and extended these findings and suggested the following free radical mechanism for molecular weight control of chloroprene-sulfur copolymers:

$$-(CH_2-\underset{\underset{Cl}{|}}{C}=CH-CH_2)_n-S-S-S-S-(CH_2-\underset{\underset{Cl}{|}}{C}=CH-CH_2)_m-S-S- \xrightarrow[\substack{or \\ peroxides}]{\Delta}$$

$$-(CH_2-\underset{\underset{Cl}{|}}{C}=CH-CH_2)_n-S\cdot \quad etc. \quad (1)$$

$$-(CH_2-\underset{\underset{Cl}{|}}{C}=CH-CH_2)_n-S\cdot + (R)_2N\underset{\underset{S}{\parallel}}{C}-S-S-\underset{\underset{S}{\parallel}}{C}-N(R_2) \rightarrow$$

$$-(CH_2-\underset{\underset{Cl}{|}}{C}=CH-CH_2)_n-S-S-\underset{\underset{S}{\parallel}}{C}-N(R)_2 + (R)_2N\underset{\underset{S}{\parallel}}{C}-S\cdot \quad (2)$$

The molecular weight distribution curve of sulfur modified poly-chloroprene (Neoprene GN), based on osmotic pressure measurements, shows a pronounced maximum at 100,000 but extends to molecular weights over 1 million indicating the presence of branched molecules (64). Calibration of the intrinsic viscosity–molecular weight relationship by osmometry affords the following relationship $[\eta] = 1.4 \times 10^{-4} \, M^{0.73}$. The "W" neoprenes which are prepared in the absence of sulfur exhibit a narrower molecular weight distribution with the maximum shifted to the 200,000 mol wt region. The intrinsic viscosity–molecular weight relationship is $[\eta] = 1.55 \times 10^{-4} \, M^{0.71}$ (66).

4. Polymer Structure

The microstructure of polychloroprene prepared by free radical polymerization processes consists primarily of a linear sequence of trans-2-chloro-2-butenylidene-1,4 units (1) with minor contributions from structures obtained by cis-1,4 polymerization (2), 1,2 polymerization (3), and 3,4 polymerization (4).

The contribution of structures 2–4 increases with increasing poly-merization temperature from a total of about 5% at a polymerization temperature of −40°C to about 30% at polymerization temperatures approaching 100°C (66). An improved high resolution infrared spectro-photometric method has been developed for the determination of poly-chloroprene isomers (67).

More extensive polymer structure determination, using high resolution nuclear magnetic resonance spectroscopy, has permitted the quantitative identification of sequence isomers derived from head-to-tail (**5**), head-to-head (**6**), and tail-to-tail (**7**) monomer addition (68).

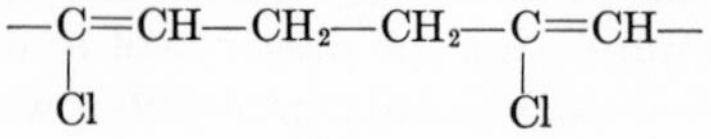

$$-C=CH-CH_2-CH_2-C=CH-$$
$$\ \ \ |\qquad\qquad\qquad\quad |$$
$$\ \ \ Cl\qquad\qquad\qquad\quad Cl$$

(**5**)

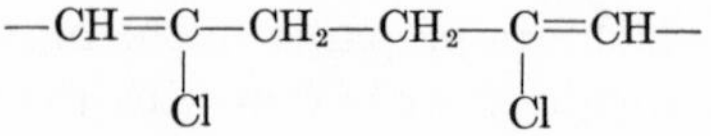

$$-CH=C-CH_2-CH_2-C=CH-$$
$$\qquad\quad |\qquad\qquad\qquad |$$
$$\qquad\quad Cl\qquad\qquad\qquad Cl$$

(**6**)

$$-C=CH-CH_2-CH_2-CH=C-$$
$$\ \ \ |\qquad\qquad\qquad\qquad\quad |$$
$$\ \ \ Cl\qquad\qquad\qquad\qquad\quad Cl$$

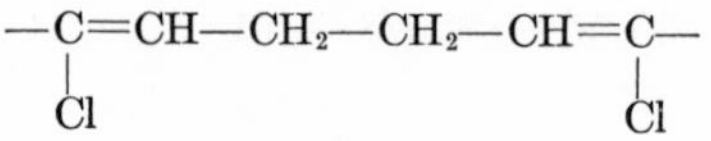

(**7**)

Structures **6** and **7** each account for 10–15% of polychloroprene sequence distribution.

The elastomeric properties of polychloroprene are largely controlled by polymer microstructure—especially such properties as rate of crystallization and degree of crystallinity (69). Accordingly, microstructure control and information on the effect of specific microstructures on polymer properties are of major technological importance. In contrast to the experiences with polybutadiene and polyisoprene, there have been no reports on the use of catalysts, per se, to control polychloroprene microstructure. Recently an elegant indirect route (reactions 3–5) has been devised for the preparation of *cis*-1,4 polychloroprene (70). The

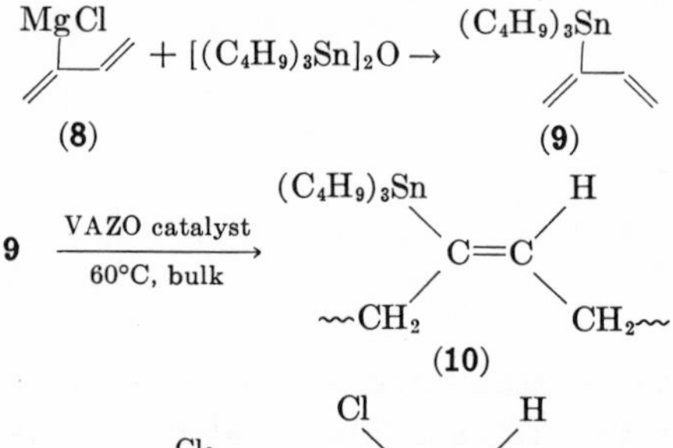

$$\underset{(\mathbf{8})}{\overset{MgCl}{\diagdown\diagup}} + [(C_4H_9)_3Sn]_2O \rightarrow \underset{(\mathbf{9})}{\overset{(C_4H_9)_3Sn}{\diagdown\diagup}} \tag{3}$$

$$\mathbf{9} \xrightarrow[60°C,\ bulk]{VAZO\ catalyst} \underset{(\mathbf{10})}{\overset{(C_4H_9)_3Sn}{\underset{\sim\!CH_2}{}}C=C\overset{H}{\underset{CH_2\sim}{}}} \tag{4}$$

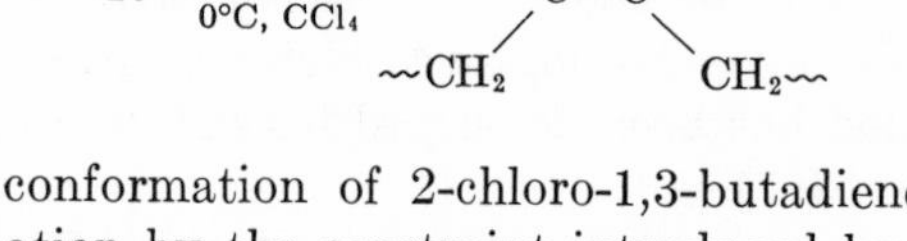

$$\mathbf{10} \xrightarrow[0°C,\ CCl_4]{Cl_2} \overset{Cl}{\underset{\sim\!CH_2}{}}C=C\overset{H}{\underset{CH_2\sim}{}} \tag{5}$$

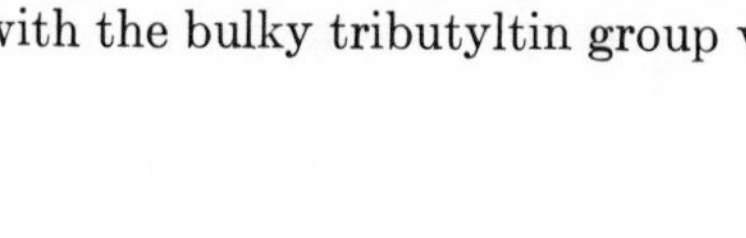

normal S-*trans* conformation of 2-chloro-1,3-butadiene was converted to S-*cis* conformation by the constraint introduced by the replacement of the chloro group with the bulky tributyltin group via the intermediate

Grignard compound. Bulk polymerization of 2-(tri-n-butyltin)-1,3-butadiene (9), followed by chlorinolysis of the resulting cis-1,4-poly-2-(tri-n-butyltin)-1,3-butadiene (10) yielded essentially all cis-1,4 polychloroprene as evidenced by ozonolysis and spectroscopic studies. Isomer sequence for this polymer has been estimated as 50–60% head-to-tail and 25–20% head-to-head and tail-to-tail.

Compared with a 95% $trans$ polymer, the cis polymer exhibits higher liquid density (1.283 vs. 1.243) and a higher glass transition temperature (-20 vs. $-45°C$). Melt transition temperatures for the isomeric polymers are in the range 70–80°C.

The early studies of Maynard and Mochel on the effect of polymerization temperature on polychloroprene structure and properties (66,69) clearly demonstrated that crystallinity, rate of crystallization, and melt transition temperature were inverse functions of polymerization temperature. This inverse relationship obviously results from increased introduction of structural irregularities (cis-1,4, 1,2, and 3,4 monomer addition) at higher polymerization temperatures. The degree of crystallization has also been shown to decrease on polymer aging (71). As is the case for natural rubber, the melt transition temperature for polychloroprene is a direct function of the temperature at which the crystallites are formed. Again, as with natural rubber, polychloroprene crystal structures orient and increase in size under stress (72,73). More detailed studies by Kargin and co-workers (74) have shown that the spherulitic structures which form with time in polychloroprene films lead to a considerable strengthening of the elastomer. Further, the fact that the crystalline structuration does not result in a loss of elastic properties implies that the elasticity of crystallizing elastomers is a function of both the elasticity of individual molecules and of supramolecules (dendrites, spherulites, and faceted crystals).

The kinetics of polychloroprene crystallization have been studied by differential thermal analysis (75,76). The heat of fusion of the crystalline phase of polychloroprene is approximately 23 cal./g; the activation energy for crystallization has been reported as 24.8 kcal/mole (77). Variations in molecular weight and degree of crosslinking have only slight effects on polychloroprene crystallization rate. Vulcanizates crystallize most rapidly at $-12°C$; the thaw temperature is about 15°C higher than the crystallization temperature (78).

Crystallinity values for polychloroprene have been calculated from known values of liquid polymer density ($d_{25°C} = 1.23$) and crystalline phase density ($d_{25°C} = 1.35$). Polymer prepared at $-40°C$ is approximately 38% crystalline $(T_m \sim 73°C)$; polymer prepared at $+40°C$ is about 12% crystalline $(T_m \sim 45°C)$. Crystalline polychloroprene exhibits an x-ray diffraction pattern identity period of 4.7 Å along the *trans*-2-chloro-2-butenylidene-1,4 structure. Wide angle x-ray diffraction studies of crosslinked high *trans* content polychloroprene reveal that the chain axis aligns along the fiber axis, and the chain axis distribution is much sharper than the distribution of amorphous statistical segments which exist before crystallization. This phenomenon has been interpreted to permit determination of the critical size of crystallization nuclei (79). The crystal morphology of polychloroprene has not been completely identified (80,81).

5. Copolymerization

Copolymers of chloroprene with diverse monomers have been widely reported. In practice, however, significant modification of polychloroprene properties by copolymerization is difficult because of the extreme radical reactivity of chloroprene relative to other monomers as evidenced by the representative monomer reactivity ratios (82) given in Table II.

A Q value of 8.07 and an e value of 0.46 have been determined for chloroprene by copolymerization with butadiene (95). These values

TABLE II
Monomer Reactivity Ratios
(M_1 = 2-chloro-1,3-butadiene)

r_1	M_2	r_2	Ref.
3.4 ± 0.07	Butadiene	0.059 ± 0.014	—
3.65 ± 0.11	Isoprene	0.133 ± 0.025	83
0.355 ± 0.055	2,3-Dichloro-1,3-butadiene	2.15 ± 0.25	84
5.47	Hexachlorobutadiene	0.10	85
6.07 ± 0.53	Acrylonitrile	0.01 ± 0.01	86, 87
5–8	Styrene	0.00–0.05	—
3.6	Methyl isopropenyl ketone	0.1	88
2.68	Methacrylic acid	0.15	88
3.9 ± 0.25	Methyl methacrylate	0.18 ± 0.06	89
6.12 ± 0.2	Methyl methacrylate	0.08 ± 0.007	90
50	Vinyl acetate	0.01	91, 92
5.19 ± 0.03	2-Vinylpyridine	0.06 ± 0.01	93
2.10 ± 0.13	2-Vinylquinoline	0.38 ± 0.03	93
0.016	Vinylidene cyanide	0.0048	94

reflect both the extreme tendency of chloroprene to react with radicals and its negligible tendency to alternate with comonomers containing either electron attracting or electron withdrawing functions. In copolymerization with vinylidene cyanide, in which both r_1 and r_2 are close to zero, alternation of monomer sequences has been reported (94).

Infrared spectra have confirmed that the polychloroprene sequences in copolymers contain isomeric units representing all possible modes of monomer addition in the same ratio as in the homopolymer prepared at the same polymerization temperature (88).

A standard procedure to prepare reasonably homogeneous chloroprene copolymers in spite of the generally disparate monomer reactivity ratios is to program the addition of the more reactive chloroprene to the polymerizing emulsion so as to maintain the desired monomer ratio according to the equations relating copolymer composition and monomer ratio in the charge (82,87,96) (cf. Chapter 4.B, pp. 192ff.)

6. Commercial Manufacture

The polymerization process for the preparation of the general purpose polychloroprene, Neoprene GN, is representative of commercial manufacture of both dry polymers and polychloroprene latexes (97).

A solution of sulfur and rosin in chloroprene is emulsified with an aqueous solution of caustic soda and the sodium salt of naphthalenesulfonic acid–formaldehyde condensation product. The sodium rosin emulsifier is formed *in situ;* the condensation product is used to stabilize the latex when it is subsequently acidified for polymer isolation. Emulsification is carried out by recirculating the chloroprene–water mixture through a centrifugal pump to give particles about 3 μ in diameter. The emulsion is pumped to a jacketed, glass-lined kettle equipped with a glass coated agitator. An aqueous solution of potassium persulfate is used to initiate and maintain the polymerization, and the polymerization temperature is kept at 40 $\pm$ 1°C by circulating -15°C brine through the kettle jacket and by varying agitator speed.

The monomer conversion is followed by measuring the specific gravity of the emulsion (98). The polymerization is stopped at 91% conversion (s.g. = 1.069) by adding an aqueous emulsion of a xylene solution of tetraethyl thiuram disulfide. The emulsion is cooled to 20°C and aged at this temperature for about 8 hr to solubilize the polymer (plasticization) by the action of the thiophilic thiuram disulfide on the polysulfide segments. The thiuram also serves to stabilize the finished dry product.

After aging, the alkaline latex is acidified to pH 5.5–5.8 with 10% acetic acid. This arrests the plasticizing action of the thiuram disulfide, precipitates the rosin which is retained by the polymer, and prepares the latex for isolation of the polymer (99).

Neoprene is isolated from latex by the continuous coagulation of a polymer film on a freeze drum followed by washing and drying. The dry polymer is formed into ropelike sections and bagged. The success of this process depends upon having a latex that is completely coagulated within a few seconds at −10 to −15°C and that gives a film strong enough to withstand the stresses imposed upon it during washing and drying.

The acidified latex containing the rosin and the acid stable dispersing agent is fed through porcelain pipes to a pan in which a stainless steel freeze roll, 9 ft in diameter, rotates partly immersed in the latex at a peripheral speed of about 36 ft/min. The roll is cooled to about −15°C by circulating brine. The polymer in the frozen film of the latex deposited on the roll is coagulated as the drum makes part of a revolution. The film is stripped from the roll by a stationary knife and placed on a continuous woven stainless steel belt where it is thawed and washed. Water is sprayed onto the film and then forced through the film as a result of reduced pressure applied underneath. The washed film is fed to squeeze rolls where pressure of about 15 psi reduces the water content to 25–30% of the dry weight. The film is dried in a current of air at about 120°C as it is carried continuously through a multipass dryer by an endless conveyor of fabric covered aluminum girts. The conveyor travels about 6–8 ft/min slower than the squeeze rolls, since the film shrinks during the drying. The last two sections of the dryer, held at 50°C, comprise a cooling compartment. The dried film is discharged over a driven stripper roll and fed over a water cooled roll to a roper from which the rope is conveyed to the cutter and bagger.

The W type neoprenes are made and isolated by a similar process but the recipe contains no sulfur or thiuram disulfide.

The polymerization process as described is also used in the preparation of stable dispersions suitable for latex processing.

B. Other Polymerization Systems

1. Ionic Polymerization

The polymerization of chloroprene with a variety of ionic initiators has been reported, but information is generally lacking on the effect of initiator type on polymer structure.   Maynard and Mochel (100) have shown that cationic initiators such as $AlCl_3$ and BF_3 yield unstable polychloroprenes in which structures derived from 1,2 monomer addition processes are absent.   Copolymerization of isobutylene and chloroprene with BF_3 is reported to be slow and to afford only low molecular weight products (101).   Overberger and Kamath have determined cationic monomer reactivity ratios as a function of solvent polarity for the copolymerization of chloroprene and styrene (102).   A cationic mechanism has been postulated for the polymerization of chloroprene with complex catalysts composed of alkyl or arylmagnesium bromide or $Al(Et)_3$ with excess $TiCl_4$ (103,104).   The insoluble powdery product is believed to be a ladder cyclopolymer consisting of a sequence of fused cyclohexane rings.   Initially, the polymerization is presumed to proceed by a 3,4 isotactic or syndiotactic addition.   Intramolecular cyclization can occur by a reversal of polymerization direction (shown below for 3,4 isotactic case) or by copolymerization of a monomer unit or growing chain with pendant double bonds.

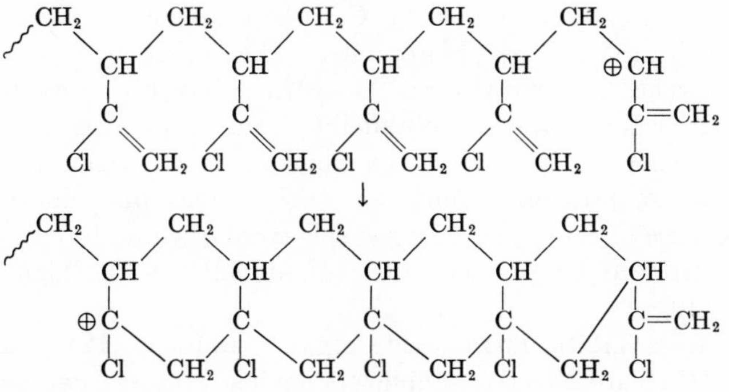

The polymerization of chloroprene with Ziegler-type catalysts has been disclosed (105,106) and claims have been made for the cis-1,4 addition polymerization of chloroprene using a mixture of $AlCl_3/CoCl_2$ with or without the addition of aluminum alkyl sesquichloride (107).

Polymerization of chloroprene in anionic systems has received only limited attention.   The patent literature discloses the use of lithium (108,109), lithium alloys (110), other metal alloys (111), and monovalent magnesium halides (112) as chloroprene polymerization catalysts. $R_3Al–RLi$ catalysts are reported to yield a black polychloroprene containing conjugated double bonds derived by elimination of active chlorine as HCl (113). In the polymerization of chloroprene in hydrocarbon media with butyl lithium, termination reactions predominate and preclude attainment of high conversion. The basic polymer unit arises from *trans*-1,4 monomer addition (114). The catalyst system Bu₂Mg/BuMgI is active with chloroprene at 40–60°C; the product is only partly soluble and has not been characterized (115).

2. Thermal Polymerization

Information has been developed recently which suggests that the thermal polymerization of inhibited chloroprene proceeds by dimer addition reactions, and little or no monomer is involved in the propagation process (116).

3. ω-Polychloroprene

An unusual shiny, granular, insoluble, highly crosslinked polychloroprene was originally observed by Carothers and co-workers (117) on ultraviolet irradiation of chloroprene. This product was termed ω-polychloroprene; the trivial term "popcorn polymer" is now applied to this polymeric form which is obtainable from all polymerizable dienes. In the presence of monomer, the popcorn polymerization continues without apparent external initiation (118). This phenomenon is due to the presence of polymeric free radicals (seeds) whose termination rate is grossly reduced because of restricted mobility in the high viscosity medium (119).

Nitric oxide (120), nitric oxide–organic adducts (121), and nitrosoamines (122) are effective stabilizers against popcorn polymerization.

IV. VULCANIZATION

The vulcanization chemistry of neoprene differs from that of the other diene polymers. The double bonds of neoprene are sufficiently deactivated by the electronegative chlorine atom so that direct vulcanization with sulfur is limited. Rather, the major crosslinking site is the tertiary allylic chloride structure generated by 1,2 addition polymerization. The average allylic chloride content of homopolychloroprene (Neoprene W) is approximately 1.5% of the theoretically available chloride or

1 active chloride per 67 chloroprene units.   For neoprene of 200,000 av mol wt, this value corresponds to 34 crosslinks per molecule.

Metal oxides are necessary but not sufficient vulcanization ingredients. Their role involves an acid (HCl) accepting function as well as other mechanistically undefined actions as vulcanization promoters (ZnO) and retarders (MgO).   It has been hypothesized, but never unequivocally demonstrated, that the metal oxides can generate ether crosslinks directly (reaction 1).

$$
2\text{~C~} \begin{matrix} \text{Cl} \\ | \\ | \\ \text{CH=CH}_2 \end{matrix} \quad + M^{2+}O^{2-} \rightarrow \begin{matrix} \text{~C~} \\ \| \\ \text{CH} \\ | \\ \text{CH}_2 \\ | \\ \text{O} + M^{2+}2\text{Cl}^- \\ | \\ \text{CH}_2 \\ | \\ \text{CH} \\ \| \\ \text{~C~} \end{matrix}
\tag{1}
$$

Practically, a combination of ZnO and MgO is used in most neoprene vulcanization systems (1).

Two mechanisms are believed operative for neoprene vulcanization. Difunctional curatives such as diamines and bisphenols can be bis-alkylated by the polymer active chlorides, probably with simultaneous 1,3 allylic shifts, to yield stable crosslinks (123) (reaction 2).   The by-product

$$
2\text{~C~} \begin{matrix} \text{Cl} \\ | \\ | \\ \text{CH=CH}_2 \end{matrix} \quad + H_2NRNH_2 \rightarrow \begin{matrix} \text{~C~} \\ \| \\ \text{CH} \\ | \\ \text{CH}_2 \\ | \\ \text{NH} \\ | \\ \text{R} + 2\text{HCl (as MgCl}_2 \text{ or ZnCl}_2) \\ | \\ \text{NH} \\ | \\ \text{CH}_2 \\ | \\ \text{CH} \\ \| \\ \text{~C~} \end{matrix}
\tag{2}
$$

$ZnCl_2$, a strong Lewis acid, can serve to accelerate the alkylation reaction. These difunctional curing agents do not cure neoprene which has been pretreated (with piperidine or aniline) to remove allylic chlorides.

An alternative mechanism has been proposed to explain the vulcanization of neoprene with substituted thioureas such as NA-22 (ethylene thiourea) (124). In this instance, the active chlorine atoms participate through an isothiuronium intermediate.

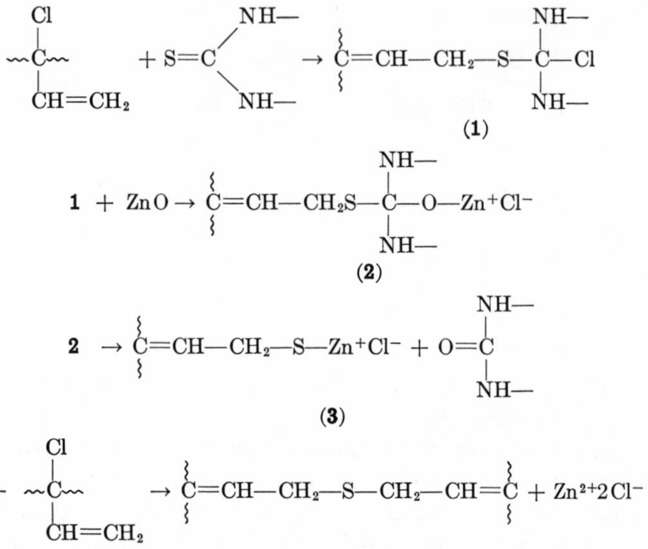

(1)

(2)

(3)

The vulcanization of sulfur-chloroprene copolymers differs from that of chloroprene homopolymers in that rearrangement of the backbone polysulfide segmers is involved as well as the accepted interactions with allylic chloride structures (125). Free radical structures such as RS· and ·SS· presumably arise from interaction of the polysulfide segmers with the thiuram disulfide present in the polymer. These radicals react with each other or with the polymer backbone to yield sulfidic crosslinks. ZnO has also been shown to react with the polysulfide units, reducing the number of sulfur atoms present per unit but increasing the number of polymer polysulfide units.

Martel and Smith have studied the vulcanization of polychloroprenes with N,N'-dinitroso-p-phenylene-bis(hydroxylamine) salts and propose the following bisalkylation crosslinking reactions (126):

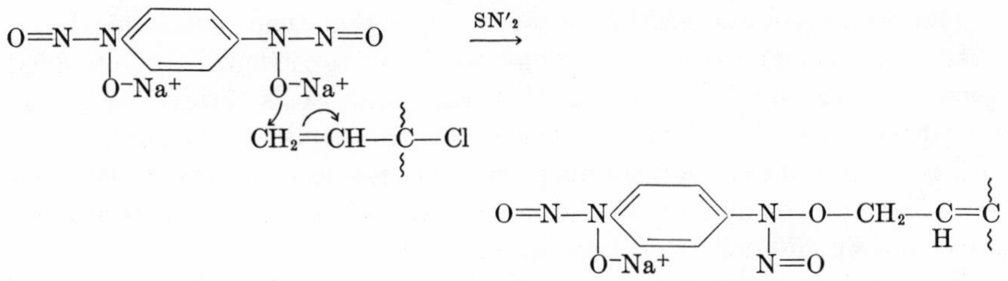

$$O=N-N\underset{\underset{O^-Na^+}{|}}{\diagdown}-N-N=O \quad \xrightarrow{\text{SN}'_2}$$

$$\underset{\underset{\overset{|}{CH_2=CH-C-Cl}}{\overset{|}{\curvearrowright}}}{\overset{O^-Na^+}{}}$$

$$O=N-N\underset{\underset{O^-Na^+}{|}}{\diagdown}\underset{\underset{N=O}{|}}{\diagup}-N-O-CH_2-C=C{\overset{H}{}}$$

(β-form)

(4)

$$4 \;+\; CH_2=CH-\overset{\}{C}-Cl \rightarrow \sim C\sim$$

$$\overset{\|}{CH}$$

$$\overset{|}{CH_2}$$

$$\overset{|}{O}$$

$$\overset{|}{N-N=O}$$

$$N-N=O$$

$$\overset{|}{O}$$

$$\overset{|}{CH_2}$$

$$\overset{\|}{CH}$$

$$\sim C\sim$$

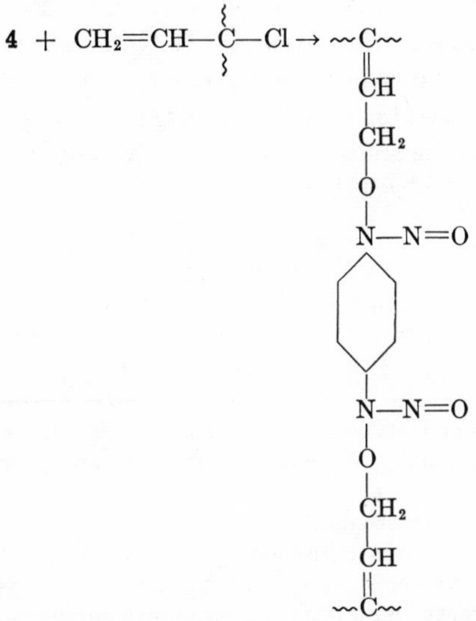

Vulcanization of chloroprene rubbers with epoxy resins (127), with dicumyl peroxide (128), with thiosemicarbazide accelerators (129), and with various agents adsorbed in molecular sieves (130) has been reported.

Crystallization processes in polychloroprene as a function of vulcanization conditions have been studied by electron microscopy (131).

V. NEOPRENE TYPES AND USES

Neoprene maintains a strong market position in diverse elastomer applications because of its exceptional combination of useful properties in both dry and latex forms. Neoprene property assets include good processing characteristics, high extendability with fillers and oils, versatile control of vulcanization rate and state; and remarkable resistance to deterioration by oils, solvents, heat and flame, weather, oxygen, ozone, and a variety of other chemicals.

The major volume use for neoprene is in the varied mechanical and automotive goods market, including hose, belts, tire sidewalls, and molded goods. Wire and cable jackets, highway joint seals, bridge pads, soil pipe gaskets, roof and maintenance coatings, caulks and sealants, and adhesives are other important neoprene market areas. Latex applications include foams for mattresses and railroad journal box lubricators, binders for cellulosic and asbestos fibers, and adhesives.

Neoprene types can be considered in five categories: two general purpose categories which include the sulfur copolymers (G types) and nonsulfur modified polymers (W types); the adhesive grades; other specialty dry types; and the latexes. Within each category, numerous types are available in which specific properties or combinations of properties are optimized for specific end use applications. Type distributions and associated property advantages within the above categories are summarized in Tables III–VII.

TABLE III
General Purpose G Types

Type	Special property
GN	Basic polymer of G types which are known for general neoprene property assets and especially for peptizability, tack and green strength, and processing safety
GNA	Improved raw polymer stability
GRT	Improved crystallization resistance
GT	Superior raw polymer stability, improved control of peptizability, improved processing safety, crystallization resistance equivalent to GRT

TABLE IV
General Purpose W Types

Type	Special property
W	Basic polymer of W types which possess general neoprene property assets and exhibit raw polymer stability superior to G types
WM-1	Reduced bulk viscosity, improved processing characteristics
WHV	Increased bulk viscosity, greater extendability with fillers, oils
WX, WRT	Increased crystallization resistance
WD	Crystallization resistance equivalent to WRT, bulk viscosity comparable to WHV
WB	Superior processing characteristics

TABLE V
Adhesive Grades

Type	Special property
AC	High crystallinity
AD	High crystallinity, improved stability
HC	Highest crystallinity with associated increased melt transition temperature, low bulk viscosity, balata-like properties
GG	High crystallinity, peptizability
AF	Low crystallinity, room temperature vulcanizability with metal oxides, superior use strength at elevated temperatures

TABLE VI
Specialty Types

Type	Special property
KNR	High peptizability, affords low viscosity, high solids content solutions
FB	Very low viscosity, fluid processing characteristics
FC	Lowest bulk viscosity, high crystallinity, fluid processing characteristics
S	Ultra-high viscosity
ILA	Superior oil resistance, specific adhesion to poly(vinyl chloride)

TABLE VII
Latex Types

Type	Special property
571	Basic latex type exhibiting typical neoprene property assets, high tensile strength
572	Increased crystallization rate
750	Excellent flexibility and wet gel extensibility
650	Analog of Type 750 with higher solids content
842A	Increased vulcanization rate, versatile combination of use properties
601A	Analog of Type 842A with higher solids content
400	Very high crystallization rate, increased ozone resistance, outstanding tensile properties
450	Excellent oil resistance, excellent processing characteristics
460	Excellent stability, noncrystallizing, very high conformability to substrate
950	Cationic soap system
735	Low viscosity base polymer
736	Low viscosity base polymer
635	Higher solids version of Type 735
60	High hot wet strength, high set resistance in hot oil

248 CHESTER A. HARGREAVES

The standard principles of compounding and processing dry and latex elastomers generally apply to the neoprenes, but a wealth of information specifically relevant to neoprene has been developed. The reader is referred to general references 1 and 2, to recent publications on the new types and uses for neoprene (132–138) and to the numerous product and application bulletins available from E. I. du Pont de Nemours and Co., Inc.; Elastomer Chemicals Department, Wilmington, Delaware.

References

1. R. M. Murray and D. C. Thompson, *The Neoprenes*, Elastomer Chemicals Department, E. I. du Pont de Nemours & Co., Inc., Wilmington, Delaware (1963).
2. John C. Carl, *Neoprene Latex*, Elastomer Chemicals Department, E. I. du Pont de Nemours & Co., Inc., Wilmington, Delaware (1963).
3. *Rubber Plastics Age*, **46**, 256 (1965).
4. *Chemical Week*, p. 71, Jan. 8, 1966; p. 19, Jan. 15, 1966.
5. P. S. Bauchwitz, "Chloroprene," in *Kirk-Othmer Encyclopedia of Chemical Technology*, 2nd ed., Vol. 5, A. Standen Ed., Interscience, New York, 1964, p. 215-229.
6. W. H. Carothers and A. M. Collins, U.S. Pat. 1,950,431 (1934) (to Du Pont).
7. W. H. Carothers, I. Williams, A. M. Collins, and James E. Kirby, *J. Am. Chem. Soc.*, **53**, 4203 (1931).
8. W. Kern, H. Jockusch, and A. Wolfram, *Makromol Chem.*, **3**, 223–226 (1949); *ibid.*, **4**, 213–221 (1950).
9. N. C. Billingham, P. A. Leeming, R. S. Lehrle, and J. C. Robb, *J. Polymer Sci.*, to be published.
10. A. C. Cope and W. R. Schmitz, *J. Polymer Sci.*, **72**, 3056 (1950).
11. J. D. Roberts, *J. Polymer Sci.*, **72**, 3300 (1950).
12. R. E. Foster and R. S. Schreiber, *J. Polymer Sci.*, **70**, 2303 (1948).
13. I. N. Nazarov and A. I. Kuznetsova, *J. Gen. Chem. USSR* (*Eng. Transl.*), **30**, 143 (1960).
14. C. A. Aufdermarsh, Jr., *J. Org. Chem.*, **29**, 1994 (1964).
15. J. A. Nieuwland and R. R. Vogt, *The Chemistry of Acetylene*, A. C. S. Monograph No. 99, Reinhold, New York, 1945, p. 160–175.
16. A. S. Carter and F. W. Johnson, U.S. Pat. 2,200,057 (1940) (to Du Pont); A. S. Carter, U.S. Pat. 2,207,784 (1940) (to Du Pont); Knapsack-Griesheim AG, Brit. Pat. 1,024,695 (1966); Knapsack-Griesheim AG, U.S. Pat. 3,048,639 (1962); G. P. Colbert, U.S. Pat. 3,175,012 (1962) (to Du Pont); Denki Kagaku Kogyo KK, Brit. Pat. 1,008,819 (1963).
17. W. H. Carothers and C. J. Berchet, *J. Am. Chem. Soc.*, **55**, 786,2807 (1933).
18. W. H. Carothers, C. J. Berchet, and A. Collins, *J. Am. Chem. Soc.*, **54**, 4066 (1932).
19. I. M. Dologopol'skii and Yu V. Trenke, *Zh. Obshch. Khim.*, **33**, 773 (1963).
20. J. E. Muskat and H. E. Northrup, *J. Am. Chem. Soc.*, **52**, 4043 (1930).
21. A. A. Petrov and N. P. Sopov, *J. Gen. Chem.* (*USSR*) (*Eng. Transl.*) **17**, 1105 (1947).
22. W. H. Carothers, U.S. Pat. 2,038,538 (1936) (to Du Pont).

23. G. W. Hearne and D. S. LaFrance, U.S. Pat. 2,430,016 (1947), 2,446,475 (1948) (to Shell Development Co.); P. A. Jenkins (to Distillers Co. Ltd.) U.S. Pat. 2,942,037 (1960), 2,942,038 (1960); C. W. Capp and H. P. Crocker, U.S. Pat. 2,948,760 (1960) (to Distillers Co. Ltd.); R. F. Stahl and C. Woolf, U.S. Pat. 3,049,572, 3,049,573, (1962) (to Allied Chemical Corp.); F. J. Bellinger, U.S. Pat. 3,015,679 (1962) (to Distillers Co. Ltd.); S. K. Lachowicz U.S. Pat. 3,026,360 (1962) (to Distillers Co. Ltd.).

24. J. H. Blumberg, U.S. Pat. 3,188,357 (1965) (to Allied Chemical Corp.); R. P. Arganbright, U.S. Pat. 3,079,445 (1963) (to Monsanto Chemical Co.).

25. H. P. Crocker, Brit. Pat. 825,609 (1959) (to Distillers Co., Ltd.). A. L. Klebanshii, K. K. Chevychalova, and A. P. Belenkaya, *J. Appl. Chem. USSR (Eng. Transl.)*, **9**, 1955 (1936); *Chem. Abstr.*, **31**, 2580 (1937).

26. R. P. Arganbright, U.S. Pat. 3,149,171 (1964) (to Monsanto Chemical Company).

27. H. Weiden, K. Kaiser, and P. Komischke, Can. Pat. 692,463 (1964) (to Knapsack Griesheim A. G.).

28. H. L. Johnson and A. P. Stuart, U.S. Pat. 2,922,826 (1960) (to Sun Oil Company).

29. C. A. Hargreaves, II and D. C. Thompson, "2-Chlorobutadiene Polymers," in *Encyclopedia of Polymer Science and Technology*, Vol. 3, Wiley, New York, 1965, pp. 705–730.

30. M. Morton, J. A. Cala, and M. W. Altier, *J. Polymer Sci.*, **19**, 547 (1956) and references cited therein.

31. W. V. Smith and R. H. Ewart, *J. Chem. Phys.*, **16**, 592 (1948).

32. S. Ekegren, O. Öhrn, K. Granath, and P. O. Kinell, *Acta Chem. Scand.*, **4**, 126 (1950), *Chem. Abstr.*, **44**, 8758 (1950); Unpublished data, E. I. du Pont de Nemours and Co., Inc., Wilmington, Delaware.

33. A. M. Collins, U.S. Pat., 1,967,861 (1934) (to Du Pont).

34. D. E. Andersen and R. G. Arnold, *Ind. Eng. Chem.*, **45**, 2727 (1953).

35. D. Rosahl and H. Esser, U.S. Pat. 3,074,899 (1963) (to Farbenfabriken Bayer).

36. B. Dales and F. B. Downing, U.S. Pat. 2,138,226 (1938) (to Du Pont).

37. H. W. Walker and F. N. Wilder, U.S. Pat. 2,263,322 (1941) (to Du Pont).

38. A. M. Collins, U.S. Pat. 2,010,012 (1935) (to Du Pont).

39. A. M. Collins, U.S. Pat. 1,967,861 (1934) (to Du Pont); P. O. Bare, U.S. Pat. 2,426,854 (1947) (to Du Pont).

40. E. J. Lorand, U.S. Pat. 2,569,480 (1951) (to Hercules Powder Co.).

41. W. H. Carothers, A. M. Collins, and J. E. Kirby, U.S. Pat. 1,950,438 (1934) (to Du Pont).

42. J. T. Maynard, U.S. Pat. 2,707,180 (1955) (to Du Pont).

43. Farbenfabriken Bayer A. G., Brit. Pat. 854,979 (1964).

44. A. R. Heinz, H. Haberland, and D. Rosahl, Fr. Pat. 1,187,771 (1961) (to Farbenfabriken Bayer A. G.).

45. G. Jennes, H. Sutter, and K. Nutzel, U.S. Patent 3,236,823 (1966) (to Farbenfabriken Bayer A. G.).

46. A. R. Heinz, D. Rosahl, and W. Graulich, U.S. Pat. 3,013,000 (1961) (to Farbenfabriken Bayer A. G.).

47. H. W. Walker, U.S. Pat. 2,259,122 (1941) (to Du Pont).

48. A. M. Hutchinson, Can. Pat. 661,997 (1963) (to Distillers Co., Ltd.).

49. J. R. Goertz, U.S. Pat. 2,576,009 (1951) (to Du Pont).

50. F. P. Demme, Ger. Pat. 1,138,944 (1962) (to Pennsalt Chemical Co.).
51. H. W. Walker, U.S. Pat. 2,259,122 (1941) (to Du Pont).
52. J. R. Vincent, U.S. Pat. 2,463,225 (1949) (to Du Pont).
53. W. E. Mochel, U.S. Pat. 2,567,117 (1951) (to Du Pont).
54. A. L. Klebanskii, N. Ya. Tsukerman, V. N. Kartsev, A. L. Labutin, Yu. V. Trenke, L. P. Mal'shina, N. A. Borovikova, G. G. Karelina, and Yu. P. Rozhkov, *Kauchuk i Rezina*, **20**, 1 (1961); *Chem. Abstr.*, **56**, 1574 (1962).
55. J. W. McFarland and R. Pariser, *J. Appl. Polymer Sci.*, **7**, 675 (1963).
56. W. E. Mochel and J. H. Peterson, *J. Am. Chem. Soc.*, **71**, 1426 (1949).
57. M. Morton and I. Piirma, *J. Polymer Sci.*, **19**, 563 (1956).
58. W. H. Carothers and J. E. Kirby, U.S. Pat. 1,950,439 (1934) (to Du Pont).
59. M. A. Youker, U.S. Pat. 2,234,215 (1941) (to Du Pont).
60. A. M. Collins, U.S. Pat. 2,264,173 (1941) (to Du Pont).
61. A. A. Sparks and R. C. Moore, Brit. Pat. 1,019,917; 1,019,918 (1966) (to Distillers Co., Ltd.).
62. W. E. Mochel, *J. Polymer Sci.*, **8**, 583 (1952).
63. A. L. Klebanskii, N. Ja. Zukerman, and L. P. Fomina, *J. Polymer Sci.*, **30**, 363 (1958): *Rubber Chem. Technol.*, **32**, 588 (1959).
64. W. E. Mochel, J. B. Nichols, and C. J. Mighton, *J. Am. Chem. Soc.*, **70**, 2185 (1948).
65. W. E. Mochel and J. B. Nichols, *Ind. Eng. Chem.*, **43**, 154 (1951).
66. J. T. Maynard and W. E. Mochel, *J. Polymer Sci.*, **13**, 227 (1954).
67. R. C. Ferguson, *Anal. Chem.*, **36**, 2204 (1964).
68. R. C. Ferguson, *J. Polymer Sci.*, **A2**, 4735 (1964).
69. J. T. Maynard and W. E. Mochel, *J. Polymer Sci.*, **13**, 235 (1954).
70. C. A. Aufdermarsh and R. Pariser, *J. Polymer Sci.*, **A2**, 4727 (1964).
71. I. Koessler and L. Soob, *J. Polymer Sci.*, **54**, 77 (1961).
72. N. M. Kocharyan et. al., *Dokl. Akad. Nauk. Arm. SSR*, **38**, 149 (1964).
73. J. B. Campbell, *Science*, **141**, 329 (1963).
74. V. A. Kargin, T. I. Sogolova, and T. K. Shaposhnikova, *Vysokomolekul. Soedin*, **6**, 1022 (1964); *Chem. Abstr.*, **61**, 7206 (1964).
75. B. Ya Teitelbaum and N. P. Anoshina, *Vysokomolekul. Soedin.*, **7**, 978 (1965).
76. B. Ya Teitelbaum, T. A. Yagfarova, N. P. Anoshina, and V. A. Naumov, *Dokl. USSR, Phys. Chem. Sec. (Eng. Transl.)*, **150**, 463 (1963).
77. M. Hanok and I. N. Cooperman, *Proc. Intern. Rubber Conf. Preprints Papers*, Washington, D.C., **1959**, p. 582.
78. R. M. Murray and J. D. Detenber, *Rubber Chem. Technol.*, **34**, 668 (1961).
79. W. R. Krigbaum and R. J. Roe, *J. Polymer Sci.*, **A2**, 4391 (1964).
80. W. R. Krigbaum et al., *Polymer*, **7**, 61 (1966).
81. A. N. Gent, *J. Polymer Sci.*, **3**, 3787 (1965).
82. *Copolymerization*, (High Polymers, Vol. XVIII), G. E. Ham, Ed., Interscience, New York, 1964.
83. W. H. Carothers, A. M. Collins, and J. E. Kirby, U.S. Pat. 2,066,329 (1937) (to Du Pont).
84. N. G. Karapetyan, I. S. Boshnyakov, and A. S. Margaryan, *Vysokomolekul. Soedin.*, **7**, 1993 (1965); *Chem. Abstr.*, **64**, 6864 (1966).
85. A. L. Klebanskii and O. A. Timofeev, *J. Polymer Sci.*, **52**, 23 (1961).
86. F. C. Wagner, U.S. Pat. 2,395,649 (1946) (to Du Pont).

87. Farbenfabriken Bayer A. G., Brit. Pat. 858,444 (1961).
88. G. S. Wich and N. Brodoway, *J. Polymer Sci.*, **A1**, 2163 (1963).
89. V. I. Eliseeva, N. G. Karapetyan, I. S. Boshnyakov, and A. S. Margaryan, *Lakokrasochyne Materialy i ikh Primenenie*, **3**, 15 (1965); *Chem. Abstr.*, **63**, 8495 (1965).
90. K. W. Doak and D. L. Dineen, *J. Am. Chem. Soc.*, **73**, 1084 (1951).
91. USSR Patent 136,041 (1961); *Chem. Abstr.*, **55**, 19297 (1961).
92. K. K. Georgieff, K. G. Blaikie, and R. C. White, *J. Appl. Polymer Sci.*, **8**, 889 (1964).
93. M. M. Koton, *Chem. Abstr.*, **55**, 16546 (1961).
94. H. Gilbert et al., *J. Am. Chem. Soc.*, **78**, 1169 (1956).
95. R. Simha and L. A. Wall, *J. Res. Natl. Bur. Std.*, **41**, 521 (1948).
96. R. Kobayashi, Can. Pat. 694,809 (1964) (to Denki Kabaken Kogyo KK).
97. A. M. Collins, U.S. Pat. 2,264,173 (1941) (to du Pont).
98. R. S. Barrows and G. W. Scott, *Ind. Eng. Chem.*, **40**, 2193 (1948).
99. M. A. Youker, *Chem. Eng. Progr.*, **43**, 391 (1947).
100. J. T. Maynard and W. E. Mochel, *J. Polymer Sci.*, **13**, 251 (1954).
101. V. I. Anosov and A. A. Korotkov, *Vysokomolekul. Soedin.*, **2**, 354 (1960); *Chem. Abstr.*, **54**, 20302 (1960).
102. C. G. Overberger and V. G. Kamath, *J. Am. Chem. Soc.*, **85**, 446 (1963).
103. N. G. Gaylord, I. Koessler, M. Stolka, and J. Vodehnal, *J. Am. Chem. Soc.*, **85**, 641 (1963).
104. N. G. Gaylord, I. Koessler, M. Stolka, and J. Vodehnal, *J. Polymer Sci.*, **A2**, 3969 (1964).
105. F. P. Gintz, Brit. Pat. 1,018,142 (1966) (to Distillers Co., Ltd.).
106. I. M. Robinson and R. C. Schreyer, U.S. Pat. 3,118,864 (1964) (to du Pont).
107. Shell Int. Res., Brit. Pats. 906,054, 906,055 (1962).
108. H. L. Jackson and K. L. Seligman, U.S. Pat. 3,004,012 (1961) (to du Pont).
109. H. L. Jackson, U.S. Pat. 3,004,011 (1961) (to du Pont).
110. H. L. Jackson, U.S. Pats. 2,908,672; 2,908,673 (1959) (to du Pont).
111. Farbenfabriken Bayer A. G., Brit. Pat. 920,472 (1963).
112. R. Stroh and K. Nützel, U.S. Pat. 3,048,571 (1962) (to Farbenfabriken Bayer A. G.).
113. B. A. Dolgoplosk and E. I. Tinyakova, *Dokl. Akad. Nauk SSSR*, **146**, 362 (1962); *Chem. Abstr.*, **58**, 2507 (1963).
114. B. L. Erusalimskei, I. G. Krasnoselskaya, and V. V. Mazurek, *Rubber Chem. Technol.*, **38**, 991 (1965).
115. B. A. Dolgoplosk, B. L. Erussalimskii, E. N. Kropatsheva, and E. L. Tinyakova, *J. Polymer Sci.*, **58**, 1333 (1962).
116. P. A. Leeming, R. S. Lehrle et al., *Nature*, **207**, 403 (1965).
117. W. H. Carothers, I. Williams, A. M. Collins, and J. E. Kirby, *J. Am. Chem. Soc.*, **53**, 4203 (1931).
118. A. K. Banbrook, R. S. Lehrle, and J. C. Robb, *J. Polymer Sci.*, **C4**, 1161 (1964).
119. G. H. Miller, D. Chock, and E. P. Chock, *J. Polymer Sci.*, **A3**, 3353 (1965).
120. F. J. Bellringer, Can. Pat. 672,457 (1963), (to Distillers Co., Ltd.).
121. G. P. Colbert, U.S. Pat. 3,175,012 (1965) (to du Pont).
122. W. Vogt, H. Weiden, K. Gehrmann, and K. Sennewald, Ger. Pat. 1,148,230 (1963) (to Knapsack Griesheim A. G.).

123. P. Kovacic, *Ind. Eng. Chem.*, **47**, 1090 (1955), and references therein.
124. R. Pariser, *Kunststoffe*, **50**, 623 (1960).
125. N. D. Zakharov, N. A. Bogdanovick, Z. D. Tyuremnova, and V. S. Glavina, *Vysokomolekul. Soedin.*, **5**, 910 (1963) (English Summary); *Chem. Abstr.*, **59**, 7737 (1963).
126. R. F. Martel and D. E. Smith, *Rubber Chem. Technol.*, **34**, 658 (1961).
127. N. D. Zakharov and G. A. Maiorov, *Soviet Rubber Technol. (English Transl.)*, **22**, 11 (1963).
128. L. D. Loan and J. Scanlon, *Rubber Plastics Age*, **44**, 1315 (1963).
129. F. Lober, U.S. Pat. 3,022,275 (1962) (to Farbenfabriken Bayer A. G.).
130. F. M. O'Connor, Can. Pats. 701,362 and 701,368 (1965) (to Union Carbide Corp.).
131. V. G. Kalashnikova, M. V. Kazhdan, Z. Ya. Berestneva, and V. A. Kargin, *Dokl. USSR, Phys. Chem. Sect. (English Transl.)*, **158**, 920 (1964); *Chem. Abstr.* **62**, 4189 (1965).
132. G. B. Oks and J. G. Carpenter, *Rubber Plastics Age*, **44**, 1211 (1963).
133. A. C. Stevenson, *Chem. Age (India)*, **15**, 901 (1964).
134. W. R. Abell and C. H. Gelbert, *Tappi*, **48**, 97A (1965).
135. R. E. Eckert, *J. Appl. Polymer Sci.*, **7**, 1715 (1963).
136. C. H. Gelbert, *Tappi*, **46**, 172A (1963).
137. T. P. Yin and R. Pariser, *J. Appl. Polymer Sci.*, **8**, 2427 (1964).
138. T. P. Yin and R. Pariser, *J. Appl. Polymer Sci.*, **7**, 667 (1963).

CHAPTER 4

ELASTOMERS BY
RADICAL AND REDOX MECHANISMS

D. Acrylic Elastomers

H. A. Tucker

The B. F. Goodrich Company Research Center, Brecksville, Ohio

and A. H. Jorgensen

*The B. F. Goodrich Chemical Company, Development Center,
Avon Lake, Ohio*

Contents

I. INTRODUCTION

The acrylate elastomers are included among the specialty rubbers (1). The specialty rubbers are polymers which find use in products other than tires and tubes and have some unusually favorable combination of properties such as high thermal stability, oxygen and ozone resistance, and resistance to swell in solvents. They include the butadiene-acrylo-

253

nitrile copolymers, neoprene, the silicone rubbers, the fluorinated elastomers, and ethylene-propylene copolymers and terpolymers. In price and properties the acrylates fit in between the nitriles and the silicones. The major components in commercial acrylate elastomers are ethyl and butyl acrylates.

A number of general reviews on polyacrylates and more specific ones on acrylate elastomers have been written during the last decade or so (2–7). For the sake of completeness we will be duplicating some of this material, but we hope to bring the field up to date in this chapter and to include some topics which have not been covered elsewhere.

The main impetus for the commercial development of an acrylate elastomer came from the automobile industry because of the need for a material for seals that would withstand, at high temperatures, oil containing sulfur-bearing additives. Under the more severe conditions that were being imposed, the unsaturated butadiene-acrylonitrile copolymers used heretofore were inadequate and a saturated polar elastomer was needed. About 1948, the B. F. Goodrich Chemical Company answered this need by commercializing a development of the U.S. Department of Agriculture's Eastern Regional Laboratories and placed on the market a copolymer of ethyl acrylate and 2-chloroethyl vinyl ether (8), PA-21, which was later given the designation Hycar 4021. Since then, both producers and acrylate polymers have increased in number.

II. MONOMER SYNTHESIS

The first process for preparing acrylate esters started with the addition of hydrogen cyanide to ethylene oxide (9). The hydracrylonitrile thus formed is then simultaneously dehydrated and hydrolyzed to acrylic acid which may be converted to any desired ester. When acrylonitrile became available, by the addition of hydrogen cyanide to acetylene, hydrolysis of it provided a better route to acrylic acid. A continuous process based on this route has been described (10).

A still better route to acrylates would be the direct conversion of acetylene to acrylic esters without going through acrylonitrile. Reppe did just that by adding carbon monoxide and ethanol to acetylene in the presence of a nickel catalyst (11). Some recent publications show that there is still interest in this process (12,13).

One of the newer routes to acrylates makes a departure from acetylene as a raw material and uses β-propiolactone, which in turn is made from ketene and formaldehyde (14). The lactone is polymerized and then depolymerized to give acrylic acid, reportedly in exceptionally pure form

(15).   A continuous process that circumvents isolation of β-propiolactone is also described (16).

A recent patent discloses the synthesis of acrylic acid by the oxidation of propylene going through acrolein in a two-stage vapor phase process (17).   This suggests a very attractive route for making acrylates in the future from a cheap petroleum product.

The acrylates are a remarkably versatile class of monomers because of the great variety of esters that can be made.   They can be made by direct esterification of the acid with alcohols, of course, but ester exchange is perhaps an even more generally useful synthetic method because of the ready availability of ethyl acrylate.   Titanate catalysts have been found very useful for this reaction (18) and where the exchange is with a high boiling alcohol, the reaction may be driven to completion by distilling off the ethyl acrylate-ethanol azeotrope.   The presence of a polymerization inhibitor is necessary during the synthesis since ethyl acrylate is a very reactive monomer.

III. POLYMERIZATION

Acrylate esters can be polymerized in a large variety of ways.   Initiation may be through a free radical mechanism and the free radical may be formed from the thermal decomposition of an azo or peroxide catalyst. The initiating free radical may be formed also using a redox system consisting of a hydroperoxide and ferrous ion, or of a boron alkyl and hydrogen peroxide.   Furthermore, an organometallic compound or ultraviolet radiation may initiate polymerization of acrylates.   The polymerization system may be bulk, solution, aqueous emulsion, or suspension.

A. In Bulk

Because of the high heats of polymerization of acrylates (19) and their rapid rates (20), polymerization in bulk is difficult to control.   Also, because of chain transfer at the α-hydrogen of the polymer chain, bulk polymerization often leads to branched and crosslinked polymers even at low conversions (21).   These difficulties as well as problems in handling the product, make bulk polymerization impractical for acrylates.

B. In Suspension

Riddle (2) considers suspension polymerization to be a number of bulk polymerizations carried out simultaneously with water as a heat transfer medium.   With the large surface area of the many droplets and the high heat capacity of the water, the problem of heat buildup encountered in bulk polymerization is solved. A monomer soluble catalyst such as benzoyl peroxide is used, and the monomer–initiator mixture is suspended as droplets in water by agitating in the presence of a protective colloid such as poly(vinyl alcohol). The polymer forms as beads, which can be recovered from suspension by filtration. Suspension polymerization is most easily used when the resultant polymer is relatively nontacky, so that the beads do not coalesce into large agglomerates.

C. In Solution

Polymerization in solution, although not commercially important for acrylate elastomers at present, is a convenient and practical method. High molecular weight polymers can be obtained using a free radical initiator if a solvent with a low chain transfer constant, such as *tert*-butanol is used (22). A study of chain transfer to solvent for methyl acrylate has been made recently (23). As might be expected, it was found that the chain transfer constants of various solvents with methyl acrylate are higher than the corresponding values for less reactive monomers such as styrene and methyl methacrylate. Other solvents besides *tert*-butanol which have particularly low values for chain transfer in polymerizing methyl acrylate are cyclohexane, benzene, chloroform, and acetone. Secondary butanol, isopropyl benzene, and methyl ethyl ketone have particularly high values.

Polymerization in a one-phase solution has some advantage over an emulsion system if more than one monomer is involved and if one of the monomers has an appreciable solubility in water. In such a case the solution system gives a polymer that is nearly homogeneous in composition (22). Molecular weights and rates of polymerization in solution are lower than for emulsion systems; however, the molecular weight can be raised by polymerizing at a lower temperature. A convenient initiator for use in organic solvents at low temperature is a trialkyl boron plus oxygen or hydrogen peroxide. This catalyst has been shown to generate free radicals (24). A study of the copolymerization of ethyl acrylate and allyl chloride was made recently using this initiation system (25).

D. In Emulsion

By far the most important polymerization medium for acrylate elastomers is an aqueous emulsion. A recent review of this type of polymerization has been written by Van der Hoff (26). The chief advantages of such a system are that high molecular weight polymer is produced rapidly and easily and that emulsions have a much lower viscosity at high concentrations of polymer than solutions. A recent paper studying the emulsion polymerization of methyl acrylate shows that the reason for the high molecular weights produced is the same as that found for SBR polymerizations; namely, the rate constant for the termination step is reduced substantially compared to the rate constants for initiation and propagation (27).

Acrylate monomers, in contrast to methacrylates, are readily hydrolyzed (28) and therefore control of the pH during polymerization in aqueous emulsion is important if formation of acrylic acid from the ester is to be avoided. Mast and Fisher (29) have shown that under basic conditions hydrolysis is fairly rapid even at room temperature. This means, then, that soaps, such as potassium oleate, commonly used in emulsion polymerization, are not suited to the preparation of polyacrylate esters. Instead, the salts of long chain sulfonic acids are used, since they operate well under neutral or acid conditions, being derived from a strong acid.

Polymerization of acrylates has been studied using 2,2′-azoisobutyronitrile initiator, and the rates obtained with various esters are reported by Luskin and Myers (6). Azo initiators have the advantage over the peroxides in that they generate exclusively tertiary carbon free radicals, which are quite inactive in comparison to oxygen free radicals in removing hydrogen from a substrate (30). Thus, they should reduce branching when used to initiate acrylate polymerization. It has been claimed that the rate of decomposition of azo initiators is not affected by the solvent used (31), but this has been contested more recently (32).

In spite of some advantages that azo initiators have, peroxides and especially water soluble peroxides such as potassium persulfate are probably most commonly used for polymerizing acrylates industrially.

E. Stereoregular

In recent years acrylates have also been polymerized by anionic initiators. For example, Furakawa has used strontium tetraethyl zincate to polymerize n-butyl, isobutyl, sec-butyl, and tert-butyl acrylate. All polymers obtained were crystalline except the normal butyl derivative (33). However, even though this poly(butyl acrylate) did not crystallize, it was subsequently shown that it was stereoregular both by its infrared spectrum and by the crystalline x-ray diagram obtained from analysis of polyacrylic acid derived from it (34).

A detailed account of the preparation and properties of isotactic poly(isopropyl acrylate) has been published recently (35). As an initiator, an ether solution of phenyl magnesium bromide is used to produce a crystalline polymer of melting point 162°C. In contrast, poly(isopropyl acrylate) prepared by free radical polymerization is rubbery at room temperature. A recent patent claims that high conversions and high molecular weights are obtained with the Grignard catalyst if the acrylic ester is first mixed with a metal halide rather than directly with phenyl magnesium bromide or diphenyl magnesium in combination with a metal halide (36).

Ueno and Schuerch (37) have shown that high molecular weight isotactic poly(isopropyl acrylate) can be isomerized with sodium isopropoxide without hydrolyzing the polymer or reducing its molecular weight, thus converting the crystalline polymer to a rubbery one.

Besides the zinc and the Grignard catalysts already mentioned, a Ziegler-Natta catalyst consisting of butyl lithium and titanium tetrachloride has been used to polymerize tert-butyl acrylate (38). Here again, a crystalline polymer was obtained.

In most of the cases cited above stereospecificity has been claimed because crystalline polyacrylates were obtained from esters of bulky, branched chain alcohols. However, Yoshino and coworkers (39) have shown that such bulky acrylate esters are not necessary for obtaining high stereospecificity. They polymerized α,β-dideuteromethyl acrylate with lithium aluminum hydride and showed by NMR analysis that a highly isotactic polymer was obtained. NMR analysis of a polymer obtained by free radical initiation, on the other hand, showed very low stereospecificity. This indicates that it is possible to obtain homopolymers of highly regular structure that are nevertheless amorphous and rubbery.

Ultraviolet irradiation in solution at very low temperatures also results in stereospecific polymerization of isopropyl acrylate. A detailed description of the preparation and properties of syndiotactic poly(isopropyl acrylate) prepared in this way has been published (40). In contrast to the isotactic form, this polymer is rubbery and only partially crystalline.

Since it has been shown that stereoregular polyacrylates can be obtained that are rubbery and amorphous or only partially crystalline, it is quite possible that this method for producing stereoregular polymers will become important in the technology of acrylate elastomers. Presently one can only speculate on the advantages to be obtained from these new forms of polyacrylates. Perhaps the more regular structure will result in an improvement in gum strength in polymers which typically have very low strength when unreinforced.

IV. COPOLYMERIZATION

Copolymerization is a very important consideration in the preparation of acrylate elastomers because it is the most practical means, first, of introducing a site for crosslinking and, second, of modifying the properties of the elastomer to make it fit a variety of specialty uses. Nevertheless, no attempt will be made to cover this subject in great detail, because an excellent and recent review has been written by Luskin (5) and a previous review is available by Riddle (2). Rather, in subsequent sections, we will discuss mainly the use of copolymerization in producing acrylate elastomers with certain desired properties.

The Q and e parameters of Alfrey characterize the acrylate monomers. A few values along with the value for methyl methacrylate, are given in Table I. Copolymerization reactivity ratios r_1 and r_2 can be computed from them using Eqs. 1 and 2 (41):

$$e_2 = e_1 \pm (- \ln r_1/r_2)^{\frac{1}{2}} \tag{1}$$

$$Q_2 = (Q_1/r_1)^{-e_1/e_1 - e_2} \tag{2}$$

The low Q value for acrylates indicates that a highly reactive free radical with low stability is obtained from this type of monomer. The Q for methyl methacrylate is higher than for any of the acrylate esters, reflecting the more stable free radical produced by this monomer. The positive values of e indicate an electron deficient double bond. As would be expected, the e value of methyl methacrylate is lower showing that it is less electron deficient than the other derivatives in Table I.

A few selected values for copolymerization reactivity ratios for acrylates are given in Table II. Additional reactivity ratios for acrylic monomers with a variety of other monomers have been published by Young (42).

Since a copolymer of butyl acrylate and acrylonitrile has commercial importance, it is of interest to point out how well-matched these monomers

TABLE I
Q and e Values of Acrylic Monomers (6)

Monomer	Q	e
Methyl acrylate	0.43	0.73
Ethyl acrylate	0.34	0.58
Butyl acrylate	0.43	0.53
Methyl methacrylate	0.74	0.40

TABLE II

Reactivity Ratios for n-Butyl and Ethyl Acrylates With
Various Comonomers (5)

Monomer$_1$	r_1	M_2	r_2
n-Butyl acrylate	1.005 ± .005	Acrylonitrile	1.003 ± .012
	.08 ± .02	Butadiene	0.99 ± .07
	.15 ± .04	Styrene	0.48 ± .04
	4.4	Vinyl chloride	0.07
Ethyl acrylate	0.67 ± .02	Acrylonitrile	1.17
	5.7	Sodium acrylate	1.5
	0.20	Styrene	0.80

are in copolymerization.   In less favorable cases where the reactivity ratios are widely different, a product of heterogeneous composition will ordinarily be obtained in a batch charge.   In order to avoid this, the more reactive monomer may be proportioned during the polymerization or the polymerization may be stopped at a low conversion.

The reactivity ratios cited, of course, are valid only for free radical initiated copolymerizations.   For polymerizations initiated by ionic or coordinated ionic catalysts, the choice of monomer pairs that will copolymerize is much more limited.   Very little has been published so far on reactivity ratios in these systems.

V. VULCANIZATION

Polyacrylates, although saturated, do contain reactive sites (the α-hydrogen and the ester group) and use has been made of these for cross-linking the homopolymer of poly(ethyl acrylate).   For example, Hansen (43) has reported the use of diamines which presumably form a crosslink by aminolysis of the ester group and Seeger (44) has an early paper describing a peroxide cure where the attack very likely is at the α-hydrogen.   Attack at the α-hydrogen is proposed as the most likely mechanism for curing with some alkaline reagents such as sodium meta-silicate described by Semegen (45).

More recently, Mendelsohn (46) has reported that copolymers of butyl acrylate and styrene and other combinations of monomers can be cured with compounds containing benzoyl peroxide, dicumyl peroxide, and ethylene dimethacrylate.   Vulcanizates are weak, however, and aging is not as good as can be obtained with more conventional cures of commercial poly(ethyl acrylates) such as Hycar 4021 (47).   Work performed at the B. F. Goodrich Research Center in 1961 gave results similar to those of Mendelsohn. Our data showed, too, that vulcanizates of high strength cannot be obtained by peroxide cures even if an additional reactive site for free radical attack is built in through copolymerization with p-isopropylbenzyl acrylate (48).

In general, the more fruitful means of making vulcanizable acrylate elastomers has been to introduce a reactive site for crosslinking by copolymerization. As was mentioned earlier, one of the first comonomers used commercially for that purpose was 2-chloroethyl vinyl ether, a component of Hycar 4021. A review of the curing of that polymer has been published by Fram (7). But it might be worth adding that 2-mercaptoimidazoline (DuPont Accelerator Na-22), used in neoprene compounding technology, has been employed recently in Hycar 4021 recipes along with an acid acceptor such as lead oxide, to give faster cures than afforded by triethylenetetramine (49).

The early literature is replete with descriptions of cure systems involving other comonomers used to supply a crosslinking site, but which have not been commercialized. Two examples use butadiene (50) and acrylic acid (51). With the first monomer as a component of a polyacrylate a more or less conventional sulfur cure becomes possible. However, the introduction of unsaturation also makes the polymer more susceptible to oxidative degradation and to embrittlement in contact with oils containing sulfur-bearing additives. Therefore, this type of cure system is not advantageous for most purposes. With a carboxyl function in the polymer a cure is possible with a divalent metal ion such as zinc. Vulcanizates of this sort, however, show high compression set and, therefore, are not suited for many uses.

Work has continued beyond these early examples of vulcanization systems with the goals of obtaining more stable vulcanizates and faster curing systems. Use of β-hydroxyethyl methacrylate as a comonomer with ethyl acrylate to produce a heat resistant vulcanizate, when compounded with calcium hydroxide and ethylene glycol, has been reported (52). We tried this system in our own laboratory and found it gave an excellent vulcanizing system.

The use of still another reactive monomer, vinyl chloroacetate, has been described by the American Cyanamid Company in recent patents (53). The increased reactivity of the chlorine in this monomer allows the use of ammonia, in the form of ammonium benzoate, as the crosslinking agent rather than the more usual diamines.

Another reactive functional group that has been exploited for vulcanization of acrylate elastomers is the 1,2-epoxide. Simms reports that copolymers of ethyl acrylate with glycidyl methacrylate can be cured with amines or diacids (54). We have shown that ammonium benzoate or sulfur plus a fatty acid soap can also serve as crosslinking agents for this type of reactive site (55). The sulfur-soap reagent will crosslink at a site provided by vinyl chloroacetate as well (56).

The use of various derivatives of acrylamide and methacrylamide for the purpose of introducing a reactive site into a polymer were described by Müller, Dinges, and Graulich (57). The following formulas illustrate the types of monomers investigated:

$$CH_2{=}CH\overset{O}{\overset{\|}{C}}NHCH_2NR \tag{3}$$

$$CH_2{=}CH\overset{O}{\overset{\|}{C}}NHCH_2OR \tag{4}$$

$$CH_2{=}CH\overset{O}{\overset{\|}{C}}NHCH_2NH\overset{O}{\overset{\|}{C}}R \tag{5}$$

They are listed in the order of decreasing reactivity. In the absence of acid, functional groups introduced by these monomers are stable; under acid catalysis they form a methylene-bis-amide crosslink. Müller and co-workers present evidence for a two-step mechanism for crosslinking and suggest that some free amide forms first and then the methylene-bis-amide crosslink forms through interaction of the amide and the original amide derivative. Further support for this view is given by Arbuzova and coworkers (58) who have shown that a mixed methylene-bis-amide is formed by the interaction of an N-methylol amide with an amide of another acid. Müller demonstrated the crosslinking of a copolymer of ethyl acrylate and N-methoxymethyl methacrylamide through swell measurements. The simpler acrylamide derivative, N-hydroxymethyl acrylamide, has also been used to afford a cure site to polyacrylates (59). However, the alkoxymethyl acrylamide has the advantage of lower reactivity thus preventing premature crosslinking of its copolymers. Also, the water insolubility of the higher alkoxymethyl compounds is of great advantage in emulsion copolymerizations.

An American Cyanamid brochure (60) describing the interaction of an amide with their product Cymel 300 (hexakis methoxymethyl melamine) suggests still another cure system somewhat similar to the one discussed above.

Such a variety of curing systems indicate the tremendous versatility of acrylate elastomer copolymers in this respect. Of course, not all of these will be of commercial value. In practical systems some of the characteristics demanded of a curing system are: rapid but controlled curing, low corrosivity to molds, low toxicity, good-aging vulcanizates, and reasonable cost. Those systems which satisfy these demands best will be exploited in the newer acrylate elastomers.

VI. GLASS TRANSITION TEMPERATURES AND OIL RESISTANCE

We have discussed some of the minor monomer components of an acrylate elastomer used to supply the cure sites. Now we will discuss the major monomer components and how they affect the balance between low temperature flexibility and resistance to swell in aromatic oils. Generally, these two properties are opposed to each other and the best that can be done is to choose a composition that will give an optimum compromise.

Before getting into a discussion of the balance of these two properties, it is of interest to discuss, separately, the glass transition temperature, and its relationship to composition. The glass transition temperature, or second order transition temperature, is the temperature at which a polymer changes from the rubbery to the glassy state. Determination of this temperature may be carried out by measuring the change in any one of a number of physical properties of a polymer with temperature. If the physical property is plotted against temperature, a change in the slope of the resulting curve marks the transition. (This is in contrast to a discontinuity in the curve which marks a crystalline melting point.) Commonly, change in volume is measured and Loshaek has described a dilatometer suited for use with elastomers (61). Refractive index has been used as a measure also and glass transition temperatures for a number of acrylate homopolymers have been reported from such measurements (62,63). Measurements of these types usually are rather difficult and time consuming. Where highly exacting results are not demanded, a brittle point (64) or Gehman freeze point (65) is determined to approximate the glass transition temperatures of elastomers.

Usually, but not always, these various measurements give values that are not too far apart. Hughes and Brown (66) have shown good agreement between dilatometrically determined glass transitions and those determined by torsional modulus measurements for a number of polyacrylates. (The torsional modulus method is similar to the one involved in the determination of the Gehman freeze point.) These workers also

TABLE III
Glass Transition Temperatures of Acrylic Homopolymers
(References in parentheses)

Polyacrylate ester	Glass transition temperature, °C
Methyl	8 (68)
Ethyl	−24 (68)
Isopropyl	−3 to −6 (68)
n-Butyl	−54 (66)
sec-Butyl	−22 (68)

show the importance of obtaining a homogeneous sample in cases where a glass transition temperature truly characteristic of a copolymer is desired. T_g values for a few acrylate homopolymers are listed in Table III.

A more extensive list of glass transition temperatures for polyacrylates has been published recently. In this publication the interesting generalization is made that the glass transition temperature becomes lower for polyacrylates containing bulky groups in a side chain, as the bulky group is moved farther from the backbone (67).

One of the earliest attempts to establish a mathematical relationship between the glass transition temperature of a copolymer and composition was published by Gordon and Taylor (69). The equation may be given in the following form:

$$(T_g - T_{gA})C_A + K(T_g - T_{gB})C_B = 0 \tag{6}$$

where T_g is the glass temperature of the copolymer of components A and B, T_{gA} and T_{gB} are the glass temperatures of the homopolymers of A and B respectively, C_A and C_B are the weight fractions of components A and B. K is a constant which is a function of the ratio of the differences in the thermal coefficients of expansion of the rubbery and glassy states of the components A and B. The form of the equation was verified by Wood (70) for ten different copolymers including those of ethyl and butyl acrylates. However, the constant K actually found was significantly smaller than calculated from the thermal coefficients of expansion of the homopolymers.

On the other hand, Illers (71) showed that the Gordon-Taylor equation held in only eight out of eighteen copolymer systems investigated. He points out that the curves relating the glass transition temperature to copolymer composition can have a wide variety of shapes: curves with minima, maxima, and linearity, as well as positive and negative departures from linearity have been found. He shows also the following types of curves for acrylate copolymers:

Styrene-methyl acrylate: linear
Styrene-ethyl acrylate: slightly convex
Styrene-butyl acrylate: S-shaped with inflection near 60 wt % acrylate
Acrylonitrile-methyl acrylate: linear but evaluated only between 50 and 100% acrylate
Vinylidene chloride-methyl acrylate: maximum in T_g at 50 °C and 50% acrylate
Methyl methacrylate-methyl acrylate: slightly concave
Ethyl methacrylate-ter t-butyl acrylate: deeply convex

Copolymers of ethylene and ethyl acrylate and ethylene and butyl acrylate have been shown to give a minimum in the T_g composition plot (72).

The above discussion indicates that it will be difficult to find a single mathematical relationship that will describe how T_g varies with composition for copolymers with complete reliability. However, T. G Fox suggests that the following simple equation is often useful for making an estimate (73):

$$1/T_g = (W_a/T_{ga}) + (W_b/T_{gb}) \tag{7}$$

where T_g is the glass transition temperature of the copolymer, T_{ga} and T_{gb} are the glass temperatures of the homopolymers of components a and b, and W_a and W_b are the weight fractions of a and b. All temperatures are expressed in °K.

It is of interest to compare T_g for acrylates of different tacticities. Shetter (68) reports that in almost all cases there is little difference in T_g's among acrylate polymers prepared by three different procedures and which presumably have different degrees of tacticity: (1) ambient temperature free radical polymerization, (2) low temperature free radical polymerization, and (3) anionic polymerization in nonpolar medium. The T_g's were determined dilatometrically. However, there was one exception. Poly(isopropyl acrylate) did show an overall difference in T_g of 22°C in a comparison of three polymer samples prepared by the three different methods. The polyacrylates stand in contrast to the polymethacrylates which showed large differences in T_g between their syndiotactic and isotactic forms.

Having discussed low temperature flexibility alone, in terms of the glass transition temperature, we are now ready to describe the balance of this property with swell in aromatic oil (such as ASTM #3 oil) for a number of polymers. Perhaps the most lucid way to compare polymers with regard to this balance is to plot their swell as ordinate against their Gehman freeze point as abscissa. If the ordinate is on a logarithmic scale and the abscissa on a linear scale, then a very nearly linear curve with a negative slope results for many families of copolymers of varying composition. For example, copolymers of butyl acrylate with various amounts of ethyl acrylate and a minor amount of some reactive monomer for a cure site can be prepared, vulcanized, and these properties measured and plotted. Such a plot then gives a line with a fairly low slope, which means that the Gehman freeze point for the copolymer increases considerably for a moderate decrease in oil swell as the ethyl acrylate content of the polymer is raised. The question next arises whether other copolymers exist which give a more favorable relationship as indicated by a steeper slope of their linear plot. In our laboratories we prepared a family of curves representing various copolymers of butyl acrylate as shown in Fig. 1. The plot indicates that two copolymers are worse than the copolymer with ethyl acrylate. These are copolymers with acrylonitrile and with N,N-dimethyl acrylamide. Also indicated are two copolymers which are better. These are copolymers with 2-cyanoethyl and 1,1,4-trihydroperfluorobutyl acrylate, respectively.

Others have made similar comparisons of acrylate copolymers. Bovey and co-workers (74), for example, have compared copolymers of butadiene with acrylonitrile and with 1,1-dihydroperfluorobutyl acrylate in a plot of their swell in benzene, on a linear scale, against the T_{10} temperature obtained from the Gehman test measurement. Curvilinear plots are obtained with these coordinates, but they show clearly that the copolymer of the fluorinated acrylate has the more favorable balance of properties. That is, at a given T_{10} temperature it has the lowest swell in benzene.

Bovey and Abere (75), in a subsequent paper, go on to show the effectiveness of introducing an ether linkage in the fluorinated alcohol part of the acrylate ester for producing a polymer with a still better balance of low temperature flexibility and swell in aromatic solvents. Bovey suggests that the increased flexibility of the carbon–oxygen bond compared to a carbon–carbon bond is responsible for imparting better flexibility to the polymer. In the same paper, another interesting correlation between structure and property balance is pointed out for some unfluorinated polyacrylates. A comparison is made between the polymers of two esters having ether linkages at different positions in their

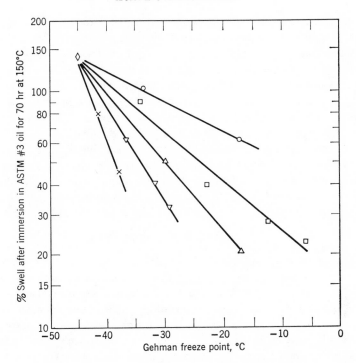

Fig. 1. Balance of oil resistance and Gehman freeze point for various copolymers of *n*-butyl acrylate. ◇, *n*-BA only; ○, *n*-BA/*N*,*N*-dimethyl acrylamide; □, *n*-BA/acrylonitrile; △, *n*-BA/ethyl acrylate; ▽, *n*-BA/2-cyanoethyl acrylate; ✕, *n*-BA/1,1,4-trihydroperfluorobutyl acrylate.

side chains, but equal numbers of carbon atoms. The polymer having three carbon atoms rather than two between the ester and ether structures has the lower T_g although the volume swell for both polymers is the same.

Still another study of the consequences of structure changes on the balance of T_g and the swell in solvents was made recently by McCurdy and Prager (76). They measured T_g by refractive index and measured swells in a solvent mixture of 30 vol % of toluene in isooctane. Polymers of acrylate esters having equal numbers of atoms in their side chains, but with one CH_2 group substituted by oxygen and by sulfur were compared. Both the T_g and swell dropped sharply in replacing $—CH_2—$ by $—O—$ or $—S—$. Another paper was published (77) describing the same balance of properties for polyacrylates having both a sulfide linkage and a terminal nitrile in their side chains. However, no great advantage in achieving a better balance of properties resulted from introducing the nitrile function.

In our own laboratories (48) we have shown that the effectiveness of the nitrile group is increased in giving a good balance of low swell and good low temperature flexibility, if it is moved away from the polymer backbone, and especially if it is attached to a flexible linkage such as an ether. This was demonstrated by comparing the swell in ASTM #3 oil and the Gehman freeze points for vulcanized copolymers consisting mainly of butyl acrylate and acrylonitrile, butyl acrylate and 2-cyanoethyl acrylate, and butyl acrylate and 2-(2-cyanoethoxy) ethyl acrylate. Making the comparison at the same Gehman freeze point, the 2-(2-cyanoethoxy) ethyl acrylate copolymer has a considerably lower swell.

VII. AGING OF ACRYLATES

Carbon reinforced vulcanizates of acrylate elastomers age very well in air even at temperatures as high as 180°C. For this reason, Steele and Jacobs expressed surprise that appreciable degradation occurred for the raw polymer at 80 to 120°C in their studies of the oxidation of poly(ethyl acrylate) films (78). The breakdown of the polymer was followed by intrinsic viscosity measurements, absorption of oxygen, and titration for acidity. Very little degradation occurred on heating in vacuum; however, if benzoyl peroxide was added, breakdown was rapid. In air the oxidative reaction was reported to be completely inhibited by the presence of 1% of hydroquinone. A mechanism involving autocatalytic oxidation was proposed for the reaction of oxygen with the polymer.

A more recent study of oxidative degradation of poly(ethyl acrylate) conflicts with the work of Steele in that only slow changes in the polymer were found even at 160°C (79). In this case, infrared analysis was used to follow the polymer breakdown and no acid formation was found under conditions where Steele found 38%.

Work has been published by Tobolsky and coworkers (80) which is more illuminating, perhaps, because they compare the oxidation of poly(ethyl acrylate) with other saturated polymers such as polypropylene and a copolymer of ethylene and propylene. In this study the degradation is followed by oxygen absorption in the temperature range 85 to 110°C using benzoyl peroxide as an initiator. Poly(ethyl acrylate) had the lowest composite rate constant for the oxygen absorption while ethylene-propylene copolymer was intermediate. This was explained as a consequence of the electron-withdrawing effect of the carbethoxy group.

Since peroxy radicals are electrophilic they attack electron-poor sites less readily. Poly(ethyl acrylate) also had the lowest activation energy for the oxidative process and this was ascribed to the resonance stability that the carbethoxy group affords a free radical formed at the tertiary carbon atoms of the polymer.

For most purposes acrylate elastomers are used in carbon black reinforced recipes and the carbon black surely contributes to the oxidative stability of these polymers just as it has been shown to do in polyethylene (81). Recently, even the type of black has been shown to influence the oxidative stability of polyacrylate compounds (82). This same paper also claims that an antioxidant, such as phenyl-β-naphthylamine is effective in improving the stability of a polyethyl acrylate vulcanizate.

TABLE IV
Commercially Available Acrylic Elastomers
(References in parentheses)

Polymer	Manufacturer	Base polymer[a]	Cure site	Method of manufacture
Chemigum AC (83)	Goodyear	EA	2-Chloroethyl vinyl ether	Emulsion polymerized
Cyanacril R	American Cyanamid	EA	Vinyl chloroacetate[b]	—
Hycar 2121X38	B. F. Goodrich Chemical Co.	n-BA and VCN	—	Emulsion polymerized
Hycar 4021	B. F. Goodrich Chemical Co.	EA	2-Chloroethyl vinyl ether	Emulsion polymerized
Krynac 880 (84)	Polymer Corporation, Ltd.	EA	2-Chloroethyl vinyl ether	Emulsion polymerized
Krynac 881 (84)	Polymer Corporation, Ltd.	EA	—	Emulsion polymerized
Paracril 9500 (85)	U.S. Rubber	EA	2-Chloroethyl vinyl ether	Emulsion polymerized
Paracril 9411 (85)	U.S. Rubber	n-BA and VCN	—	Emulsion polymerized
Thiacril 36 & 44 (86)	Thiokol	EA	2-Chloroethyl vinyl ether[a]	—
Thiacril 55 (86)	Thiokol	EA and n-BA	2-Chloroethyl vinyl ether[a]	—

[a] EA, ethyl acrylate; n-Ba, n-butyl acrylate; VCN, acrylonitrile.
[b] Assumed from literature information and from behavior of product.

This is one of the few instances where the effectiveness of an antioxidant has been demonstrated for a polyacrylate vulcanizate. It may be that the stress-relaxation measurements used in this study are more sensitive in distinguishing differences among vulcanizates than the usual comparative tensile, elongation, and modulus measurements.

VIII. CONCLUSION

The outlook for the growth of polyacrylate elastomers in volume and importance seems to be optimistic (49), judging from the entry of new producers into competition with the oldest producer, B. F. Goodrich. Furthermore, each of the various producers is manufacturing a multiplicity of acrylate polymers to make available acrylates with better low temperature flexibility and faster curing rates and thus to make these elastomers more widely applicable.

A description of the commercial acrylate elastomers is given in Table IV. A few other acrylic elastomers are available in semi-commercial quantities, such as Cyanacril LT3, Hycars 2121X58 and 4031, and Thiacril 76.

References

1. "Specialty Elastomers," *Materials in Design Engineering*, **62**, No. 2, 119 (1965).
2. E. H. Riddle, *Monomeric Acrylic Esters*, Reinhold, New York, 1954.
3. C. H. Fisher, G. S. Whitby, and E. M. Beavers, in *Synthetic Rubber*, G. S. Whitby, Ed., Wiley, New York, 1954, p. 892.
4. G. A. Daum, in *Introduction to Rubber Technology*, M. Morton, Ed., Reinhold, New York, 1959, p. 285.
5. Leo S. Luskin, in *Copolymerization*, George E. Ham, Ed., Interscience, New York, 1964, p. 653.
6. Leo S. Luskin and Robert J. Meyers, in *Encyclopedia of Polymer Science and Technology*, Vol. 1, N. G. Gaylord and N. M. Bikales, Eds., Interscience, New York, 1964, p. 246.
7. Paul Fram, in *Encyclopedia of Polymer Science and Technology*, Vol. 1, N. G. Gaylord and N. M. Bikales, Eds., Interscience, New York, 1964, p. 226.
8. Harry L. Fisher, *J. Chem. Educ.* **37**, 369 (1960).
9. W. Bauer, U.S. Pat. 1,829,208 (Oct. 27, 1931) (to Rohm and Haas A. G.).
10. *Chemical and Engineering News*, **43**, 62 (1965).
11. Walter Reppe, *Ann. Chem.*, **582**, 1 (1953).
12. T. H. Toepel, *Chim. Ind. (Paris)*, **91** (2); *Chem. Abstr.*, **60**, 12638e (1964).
13. S. K. Bhattacharyya and A. K. Sen, *Ind. Eng. Chem., Proc. Des. Develop.*, **3** (2) 169 (1964).
14. F. E. Kung, U.S. Pat. 2,356,459 (Aug. 22, 1944) (to B. F. Goodrich Co.).
15. N. C. Wearsch and A. T. de Paola, U.S. Pat. 3,002,017 (Sept. 26, 1961) (to B. F. Goodrich Co.); J. E. Jansen and W. L. Beears, U.S. Pat. 2,568,635 (Sept. 18, 1951) (to B. F. Goodrich Co.); A. B. Japs, U.S. Pat. 2,568,636 (Sept. 18, 1951) (to B. F. Goodrich Co.).

16. G. J. Fisher, U.S. Pat. 2,844,622 (July 22, 1958) (to Celanese Corp. of America).
17. Brit. Pat. 996,898 (Nov. 11, 1964) (to Distillers Company, Ltd.).
18. U.S. Pat. 2,848,753 (Aug. 26, 1958) (to General Aniline and Film Corp.).
19. K. G. McCurdy and K. J. Laidler, *Can. J. Chem.*, **42** (4), 818 (1964).
20. Ref. 6, p. 264.
21. T. G. Fox and S. Gratch, *Ann. N.Y. Acad. Sci.*, **57**, 367 (1953).
22. E. F. Jordan, W. E. Palm, and W. S. Port, *J. Am. Oil Chem. Soc.*, **38**, 231 (1961).
23. I. N. Sen, U. Nandi, and S. R. Palit, *J. Ind. Chem. Soc.*, **40** (9) 729 (1963).
24. N. L. Zutty and F. J. Welch, *J. Polymer Sci.*, **43**, 445 (1960).
25. T. L. Dawson and R. D. Lundberg, *J. Polymer Sci.*, **A3**, 1801 (1965).
26. B. M. E. Van der Hoff, *Advan. Chem. Ser.*, **34**, 6 (1962).
27. D. Hummel, G. Ley, and C. Schneider, *Advan. Chem. Ser.*, **34**, 60 (1962).
28. J. D. R. Thomas and H. B. Watson, *J. Chem. Soc.*, **1956**, 3958.
29. W. C. Mast and C. H. Fisher, *Ind. Eng. Chem.*, **41**, 790 (1949).
30. P. Gray and A. Williams, *Chem. Soc. (London), Spec. Publ.*, **9**, 97 (1957).
31. L. M. Arnett, *J. Am. Chem. Soc.*, **74**, 2027 (1952).
32. R. C. Peterson, J. H. Markgraf, and S. D. Ross, *J. Am. Chem. Soc.*, **83**, 3819 (1961).
33. J. Furukawa et. al., *Makromol. Chem.*, **42**, 165 (1960).
34. A. Kawasaki, J. Furukawa, T. Tsurata, G. Wasai, and T. Makimoto, *Makromol. Chem.*, **49**, 76 (1961); T. Makimoto et. al., ibid, **50**, 116 (1961).
35. W. E. Goode, R. P. Fellmann, and F. H. Owens, *Macromol. Syn.*, **1**, 25 (1963).
36. U.S. Pat. 3,100,761 (1963) (to Rohm and Haas Company).
37. A. Ueno and C. Schuerch, *Polymer Letters*, **3**, 53 (1965).
38. E. A. H. Hopkins and M. L. Miller, *Polymer*, **4**, 75 (1963).
39. T. Yoshino, J. Komiyama, and M. Shinomiya, *J. Am. Chem. Soc.*, **86**, 4481 (1964).
40. C. F. Ryan and J. J. Gormley, *Macromol. Syn.*, **1**, 30 (1963).
41. T. Alfrey, J. Bohrer, and H. Mark, *Copolymerization* (High Polymer Ser., Vol. VIII), Interscience, New York, 1952, p. 80.
42. L. J. Young, *J. Polymer Sci.*, **54**, 411 (1961).
43. J. E. Hansen, W. E. Palm, and T. J. Dietz, "The Amine Vulcanization of Ethyl Acrylate," *U.S. Bur. Agr. Ind. Chem.*, **AIC-205**, Sept., 1948.
44. N. V. Seeger et. al., *Ind. Eng. Chem.*, **45**, 2538 (1953).
45. S. T. Semegen and J. H. Wakelin, *Rubber Age (NY)*, **71**, 57 (1952).
46. Morris A. Mendelsohn, *Ind. Eng. Chem., Prod. Res. Develop.*, **3** (1), 67 (1964).
47. *Hycar Polyacrylic Rubber*, Manual HM-3, B. F. Goodrich Chemical Co., Cleveland, Ohio, March, 1959.
48. H. A. Tucker, unpublished work, B. F. Goodrich Laboratories, Brecksville, Ohio.
49. *Chem. Eng. News*, **41**, Nov. 18, 27 (1963).
50. W. C. Mast and C. H. Fisher, *India Rubber World*, **119**, 596, 727 (1949).
51. J. E. Hansen, P. E. Meiss, T. J. Dietz, "A Mechanism for the Amine Vulcanization of Acrylic Rubber," paper presented before the 3rd Meeting-in-Miniature, Philadelphia Sect., Am. Chem. Soc., Jan. 20, 1949.
52. WADC Quarterly Rept. No. 1, Firestone Tire and Rubber Co., "Compounding and Evaluation of Elastomeric Materials Suitable for Fabrication into Heat Resistant Tires," March 15, 1957, p. 31.
53. Belg. Pats. 647,926 and 647,927 (May 14, 1964) (to American Cyanamid).
54. J. A. Simms, *Am. Chem. Soc. Div. Paint, Plastics, Printing Inks Preprints*, **19** (2), 138 (1959); *Chem. Abstr.*, **57**, 4855g (1962).

55. P. H. Starmer, unpublished report, B. F. Goodrich Development Center, Avon Lake, Ohio.
56. U.S. Pat. 3, 201,373 (Aug. 15, 1965) (to American Cyanamid).
57. E. Müller, K. Dinges, and W. Graulich, *Makromol. Chem.*, **57**, 27 (1962).
58. I. A. Arbuzova, E. N. Rostovskii, A. L. Lis, and A. G. Eliseeva, *J. Gen. Chem. USSR, Eng. Transl.*, **29**, 3902 (1959).
59. Brit. Pat. 605,335 (Oct. 18, 1961) (to B. F. Goodrich).
60. *Cymel* 300, 301, Tech. Brochure CRT 70, American Cyanamid Company, Plastics and Resins Division, Wallingford, Conn.
61. S. Loshaek, *Phys. Rev.*, **86**, 652 (1952).
62. R. H. Wiley and G. M. Brauer, *J. Polymer Sci.*, **3**, 455 (1948).
63. R. H. Wiley and G. M. Brauer, *J. Polymer Sci.*, **3**, 647 (1948).
64. ASTM D746.
65. ASTM D1053-54T.
66. L. J. Hughes and G. L. Brown, *J. Appl. Polymer Sci.*, **5**, 580 (1961).
67. Sonja Krause, James J. Gormley, Nicholas Roman, J. A. Shetter, and Warren A. Watanabe, *J. Polymer Sci.*, **A3**, 3573 (1965).
68. John A. Shetter, *Polymer Letters*, **1**, 209 (1963).
69. M. Gordon and J. S. Taylor, *J. Appl. Chem.*, **2**, 493 (1952); *Rubber Chem. Technol.*, **26**, 323 (1953).
70. Lawrence A. Wood, *J. Polymer Sci.*, **28**, 319 (1958).
71. K. H. Illers, *Kolloid-Z.*, **190**, 16 (1963).
72. F. P. Reding, J. A. Faucher, and R. D. Whitman, *J. Polymer Sci.*, **57**, 483 (1962).
73. T. G. Fox, *Bull. Am. Phys. Soc.*, **1**, 123 (1956); see also Ref. 69.
74. F. A. Bovey and J. F. Abere, G. B. Rathmann, C. L. Sandberg, *J. Polymer Sci.*, **15**, 520 (1955).
75. F. A. Bovey and J. F. Abere, *J. Polymer Sci.*, **15**, 537 (1955).
76. R. M. McCurdy and J. H. Prager, *J. Polymer Sci.*, **A2**, 1185 (1964).
77. J. H. Prager, R. M. McCurdy, and G. B. Rathman, *J. Polymer Sci.*, **A2**, 1941 (1964).
78. Richard Steele and Harvey Jacobs, *J. Appl. Polymer Sci.*, **2**, 86 (1959).
79. R. T. Conley and P. L. Valint, *J. Appl. Polymer Sci.*, **9**, 785 (1965).
80. A. V. Tobolsky, P. M. Norling, N. H. Frick, and H. Yu, *J. Am. Chem. Soc.*, **86**, 3925 (1964).
81. W. L. Hawkins, M. H. Worthington, and F. H. Winslow, *Rubber Age (NY)*, **88**, 279 (1960).
82. Ajaib Singh and L. Weissbein, "Stress Relaxation of Polyacrylate Elastomers," paper presented at the Am. Chem. Soc. Rubber Div. Meeting, Chicago, September, 1964.
83. R. A. Durdin (Goodyear) to A. H. Jorgensen, correspondence, Oct. 14, 1965.
84. E. B. Storey (Polymer Corporation) to A. H. Jorgensen, correspondence, Sept. 30, 1965 and Oct. 21, 1965.
85. R. D. Gilbert (U.S. Rubber Company) to A. H. Jorgensen, correspondence, Oct. 12, 1965.
86. T. A. Dean (Thiokol) to A. H. Jorgensen, correspondence, Sept. 30, 1965.

CHAPTER 4

ELASTOMERS BY RADICAL AND REDOX MECHANISMS

E. Fluorine-Containing Elastomers*

J. R. COOPER

*Elastomer Chemicals Department,
E. I. du Pont de Nemours & Co.,
Wilmington, Delaware*

Contents

I. INTRODUCTION

For the past fifteen years, a considerable amount of research effort has been devoted to the preparation of extremely stable elastomers. This research was stimulated by the critical need, especially in the aircraft and allied fields, for fluid resistant elastomers capable of functioning over a wide range of temperatures. Because of the well-established stability of fluorine-bearing organics, much of this effort was directed at the preparation and evaluation of fluorinated elastomers.

* The following elastomers are registered trademarks of E. I. du Pont de Nemours & Co.: VITON®, TEFLON®, HYPALON®, ADIPRENE®, NORDEL®. FLOUREL® and KEL-F® are registered trademarks of the Minnesota Mining and Manufacturing Co.

273

The first successful polymerization of a completely fluorinated olefin was carried out by R. J. Plunkett (1) in 1938 on tetrafluoroethylene to produce Teflon. Teflon is an extremely tough plastic having serviceable use at both low and elevated temperatures and, in addition, is chemically inert to acids, alkalis, and most solvents. Several additional fluorinated plastics were developed during the 1940's, but there was much doubt as to the possibility of preparing a functionally useful elastomer containing a high percentage of fluorine.

Polyfluoroprene (the homopolymer of 2-fluorobutadiene-1,3) was the first fluorine-containing elastomer to be synthesized (3). The monomer is prepared by the gas phase addition of anhydrous hydrogen fluoride to monovinyl acetylene. Polyfluoroprene has physical properties which are intermediate between those of polychloroprene (neoprene) and polybutadiene synthesized by a redox emulsion polymerization (2).

Many fluorinated elastomers have since been prepared which contain a higher proportion of fluorine in the polymer. These materials generally have excellent heat, oil, and solvent resistance. A comparison of the properties of a commercially available fluoroelastomer with those of other elastomers is shown in Table I and Fig. 1.

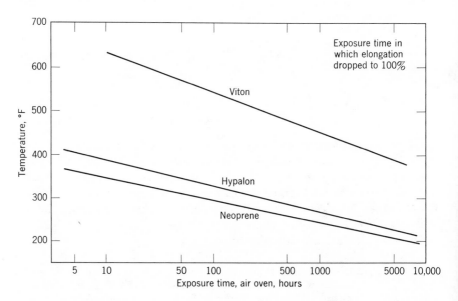

Fig. 1. Service life of Viton (5), Hypalon, and neoprene (7) at elevated temperatures.

TABLE I

Comparative Properties of a Fluoroelastomer with Other Elastomers (6)

Properties	Viton fluoroelastomers	Neoprene (chloroprene)	Hypalon synthetic rubber (chlorosulfonated polyethylene)	Adiprene urethane rubber	Nordel hydrocarbon rubber (ethylene-propylene-diene polymer)	Natural rubber
Tensile strength (psi)						
Pure gum	Over 2000	Over 3000	Over 2500	Over 4000	—	Over 3000
Black loaded stocks	Over 2000	Over 3000	Over 3000		Over 3000	Over 3000
Hardness range (durometer A)	60–95	40–95	40–95	50–99+ (up to 80 durometer D)	40–90	30–90
Specific gravity (base material)	1.85	1.23	1.12–1.28	1.06	0.86	0.93
Vulcanizing properties	Good	Excellent	Excellent	Excellent	Excellent	Excellent
Adhesion to metals	Good to excellent	Excellent	Excellent	Excellent	Excellent	Excellent
Adhesion to fabrics	Good to excellent	Excellent	Good	Excellent	Good to excellent	Excellent
Tear resistance	Fair	Good	Fair	Excellent	Fair	Good
Abrasion resistance	Good	Excellent	Excellent	Outstanding	Good	Excellent
Compression set	Very good	Fair to good	Fair	Good	Excellent	Good
Rebound						
Cold	Good	Very good	Good	Poor at low temp.	Very good	Excellent
Hot	Excellent	Very good	Good	Good at room temp.	Very good	Excellent
Dielectric strength	Good	Good	Excellent	Excellent	Excellent	Excellent
Electrical insulation	Fair to good	Fair to good	Good	Fair	Excellent	Good to excellent
Permeability to gases	Very low	Low	Low to very low	Fair	Fair	Fair
Acid resistance						
Dilute	Excellent	Excellent	Excellent	Fair	Excellent	Fair to good
Concentrated	Excellent	Good	Very good	Poor	Excellent	Fair to good
Solvent resistance						
Aliphatic hydrocarbons	Excellent	Good	Good	Excellent	Poor	Poor
Aromatic hydrocarbons	Excellent	Fair	Fair	Fair to good	Poor	Poor
Oxygenated (ketones, etc.)	Poor	Poor	Poor	Poor	Good	Good
Lacquer solvents	Poor	Poor	Poor	Poor	Poor	Poor
Resistance to:						
Swelling in lubricating oil	Excellent	Good	Good to excellent	Excellent	Poor	Poor
Oil and gasoline	Excellent	Good	Good	Excellent	Poor	Poor
Animal and vegetable oils	Excellent	Good	Good	Excellent	Good	Poor to good
Water absorption	Very good	Good	Very good	Good at room temp. Poor at 212°F	Very good	Very good
Oxidation	Outstanding	Excellent	Excellent	Excellent	Excellent	Good
Ozone	Outstanding	Excellent	Outstanding	Excellent	Outstanding	Fair
Sunlight aging	Very good	Very good	Outstanding	Good	Outstanding	Poor
Heat aging	Outstanding	Excellent	Excellent	Good	Excellent	Good
Flame	Good	Good	Good	Fair	Poor	Poor
Heat	Excellent	Excellent	Excellent	Good	Excellent	Good
Cold	Good	Good	Good	Excellent	Excellent	Excellent

The chemical and heat resistance of the fluorocarbons can be traced to at least two factors (4). The first of these is the stability of the carbon–fluorine bond. This bond is stronger than the carbon–carbon bond and is very resistant to attack. There are no known methods for the controlled replacement of fluorine in a fluorocarbon by other groupings. Only complete degradation can be achieved under extremely drastic conditions. The second factor is steric in nature. The size of the fluorine atom is such that in a highly fluorinated organic the fluorine atoms form a protective shield which prevents chemical attack on the weaker carbon linkages.

Fluoroelastomers have been prepared by condensation polymerization (fluoropolyesters and fluorinated silicones) and by the free radical or redox polymerization of fluoroolefins. Heterofluoropolymers have also been prepared by free radical polymerizations. Only the fluorinated elastomers made by radical or redox polymerizations will be considered in this section.

II. FLUORINATED ELASTOMERS FROM OLEFINS

A. Polymerization

It was at one time believed that the spatial requirements for 3 or 4 fluorine atoms attached to an olefinic double bond were great enough to inhibit polymerization. The discovery in the late 1930's that tetrafluoroethylene could be readily polymerized with free radicals as catalysts (1) paved the way to the polymerization of a large number of fluorinated olefins. The polymers made initially were plastic or resinous materials, and there was much doubt as to the possibility of synthesizing elastomers containing large quantities of fluorine in the molecule. The substitution of fluorine for hydrogen in the olefins tended to produce plastics, presumably because of the higher potential barrier to free rotation in the CF_2 chains which results in a decrease in flexibility of chain molecules. For example, polyperfluorobutadiene is a plastic, whereas polybutadiene is an elastomer. In addition, the vulcanization of a highly fluorinated elastomer seemed unlikely, since no vulcanizable elastomer existed that did not contain hydrogen atoms in the molecule (2). However, extensive investigations beginning in the early 1950's have led to the synthesis of a large number of fluorinated elastomers by the free radical polymerization of fluoroolefins. The majority of these elastomers have been prepared from three general types of olefins:

1. Fluorinated butadienes
2. Fluoroacrylates
3. Vinylidene fluoride with comonomers

Polyfluoroprene, the first fluorine-containing elastomer (3), was synthesized using a polymerization system similar to that employed for the polymerization of chloroprene. In order to improve the chemical and thermal stability of the elastomer, more highly fluorinated butadienes were synthesized and polymerized to form homo- and copolymers. A summary of some of the systems investigated is shown in Table II. In overall properties, the polymers prepared from the fluorinated butadienes did not show significant advantages over neoprene or nitrile rubber.

The esters of acrylic acid with 1,1-dihydroperfluoroalcohols have also been extensively used for the preparation of elastomeric polymers. A general synthesis of the fluoroacrylates is shown below:

$$CH_3(CH_2)_nCOOH \xrightarrow[\text{Simons process}]{\text{HF (ref. 8)}} CF_3(CF_2)_nCOOH \xrightarrow{ROH}$$

$$CF_3(CF_2)_nCOOR \xrightarrow{H_2} CF_3(CF_2)_nCH_2OH \xrightarrow[SO_3]{CH_2=CHCOOH}$$

$$CF_3(CF_2)_nCH_2O\overset{\displaystyle O}{\overset{\|}{C}}-CH=CH_2$$

The fluoroacrylate polymers which possessed the best combination of physical and chemical properties were those prepared from 1,1-dihydroperfluorobutyl acrylate (poly-FBA) and perfluoromethoxy-1,1-dihydroperfluoropropyl acrylate (poly-FMFPA) (2). The fluoroacrylate polymers are most commonly prepared using a water–emulsion persulfate system.

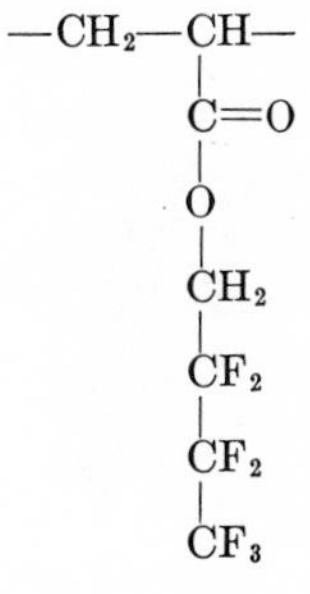

$$
\begin{array}{cc}
-CH_2-CH- & -CH_2-CH- \\
| & | \\
C=O & C=O \\
| & | \\
O & O \\
| & | \\
CH_2 & CH_2 \\
| & | \\
CF_2 & CF_2 \\
| & | \\
CF_2 & CF_2 \\
| & | \\
CF_3 & O \\
 & CF_3 \\
\text{Poly-FBA} & \text{Poly-FMFPA}
\end{array}
$$

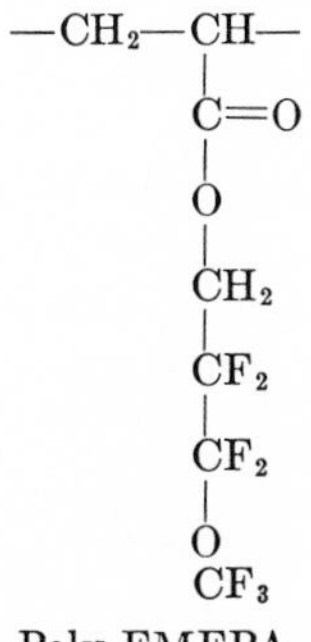

TABLE II
Elastomers from Fluorinated Butadienes
(References in parentheses)

Fluorinated butadiene	Comonomer
$CH_2{=}\overset{\text{F}}{\underset{\text{H}}{C}}{-}C{=}CH_2$	— (3) Styrene (19) 2-Methylstyrene (19) Methyl methacrylate (19) Isoprene (19) Acrylonitrile (19) 2-Chloro-3,3,3-trifluoropropylene (20)
$CH_2{=}\overset{\text{H}}{C}{-}\overset{\text{H}}{C}{=}CF_2$	Vinyl chloride (21) Vinylidene chloride (21) Styrene (22) 2,3,3,3-Tetrafluoropropylene (23) 1,1,2,2-Tetrafluoroethyl vinyl ether (24) Butadiene (25)
$H_2C{=}\overset{\text{H}}{C}{-}\overset{\text{CH}_3}{C}{=}CF_2$	Styrene (26)
$CH_2{=}\overset{\text{CH}_3}{C}{-}\overset{\text{H}}{C}{=}CF_2$	Styrene (26)
$CH_2{=}\overset{\text{H}}{C}{-}\overset{\text{F}}{C}{=}CF_2$	Fluoroprene (27) 1,1,3-Trifluorobutadiene (2) Styrene (22)
$CH_2{=}\overset{\text{F}}{C}{-}\overset{\text{H}}{C}{=}CF_2$	— (2) 1,1,1-Trifluoro-3-trifluoromethyl-2-butene (28) Dichlorodifluoroethylene (29) 1,2,3-Trifluorobutadiene (2) 1,1,2,2-Tetrafluoroethyl vinyl ether (24)
$CH_2{=}\underset{\text{F}}{C}{-}\underset{\text{F}}{C}{=}\underset{\text{F}}{C}{-}H$	— (2)
$CF_3{-}\overset{\text{H}}{C}{=}\underset{\text{H}}{C}{-}\underset{\text{H}}{C}{=}CH_2$	Halogenated olefins (30)

(continued)

TABLE II (*continued*)

Fluorinated butadiene	Comonomer
$H_2C{=}C{-}\overset{\overset{\displaystyle CF_3}{\mid}}{C}{=}CH_2$ $\quad\;\;\mid$ $\quad\;\;H$	Fluoroprene (31) Vinylidene fluoride (32) Trifluoroethylene (32) 1,1,1-Trifluoro-2-trifluoromethyl-2-butene (28) 1,1,2,2-Tetrafluoroethyl vinyl ether (24)
$CF_2{=}CF{-}CF{=}CH_2$	— (33)
$CF_2{=}CF{-}CH{=}CFH$	— (2)
$CF_2{=}CF{-}CH{=}CF_2$	1,1-Dichloro-2,2-difluoroethylene (34)
$CF_2{=}CF{-}\overset{\overset{\displaystyle CH_3}{\mid}}{C}{=}CF_2$	Styrene (22)
$CF_2{=}\overset{\overset{\displaystyle CF_3}{\mid}}{C}{-}CH{=}CH_2$	— (2)
$CF_3{-}\overset{\overset{\displaystyle F}{\mid}}{C}{=}\overset{\overset{\displaystyle F}{\mid}}{C}{-}\overset{\overset{\displaystyle H}{\mid}}{C}{=}CH_2$	— (35)
$CF_3{-}CH{=}CH{-}CH{=}CH{-}CF_3$	— (2)

Of the many fluorinated elastomers that have been investigated, only those prepared by the polymerization of vinylidene fluoride with comonomers have been of commercial significance. The commercially available polymers are summarized in Table III. Vinylidene fluoride readily forms homo- and copolymers. Poly(vinylidene fluoride) is a resinous material showing little or no elasticity. The use of a comonomer containing a bulky group or atom, e.g., hexafluoropropylene

$$\left[\begin{array}{c} F \\ C{=}CF_2 \\ CF_3 \end{array}\right]$$

or chlorotrifluoroethylene

$$\left[\begin{array}{c} F \\ C{=}CF_2 \\ Cl \end{array}\right]$$

breaks up the symmetry of the polymer chain thereby destroying polymer crystallinity and producing elasticity. The vinylidene fluoride based

TABLE III
Commercial Vinylidene Fluoride Elastomers

Trade name	Comonomer(s)
Viton A, A-HV	Hexafluoropropylene
Fluorel	Hexafluoropropylene
Viton B	Hexafluoropropylene Tetrafluoroethylene
Kel-F	Chlorotrifluoroethylene

polymers show outstanding heat and fluid resistance. Vulcanizates of Viton A, a copolymer of vinylidene fluoride and hexafluoropropylene, remain virtually unaffected for extended periods of exposure at temperatures of 250–300°F. Most conventional rubbers embrittle or revert to a plastic state after prolonged exposure at these temperatures. After 28 days exposure at 450°F, Viton A loses only 10% of its tensile strength, shows slight hardening, and retains 50% of its original elongation. A comparison of the heat resistance of Viton A with neoprene and Hypalon is shown in Fig. 1. Both neoprene and Hypalon are considered to have excellent heat resistance characteristics.

Viton A also shows excellent resistance to many oils, lubricants, and solvents, both at room temperature and at elevated temperatures. The only materials that swell Viton severely are highly polar solvents such as low molecular weight esters and ketones (5). Table I shows a comparison of the properties of Viton vulcanizates with those of other elastomers.

The commercial vinylidene fluoride based elastomers which have received widest acceptance also contain hexafluoropropylene as a comonomer (Table III). Adams and Bovey determined that hexafluoropropylene, as well as perfluoro derivatives of 1-butene, 2-butene, isobutene, 1-pentene, and 1-nonene do not homopolymerize but are readily copolymerized with other olefins (16). The difficulty of homopolymerization was attributed to the polarization of the double bond by the electronegative trifluoromethyl group. It would therefore be expected that hexafluoropropylene and similar perfluoroolefins would tend to alternate during copolymerization.

Ferguson (17) investigated the structure of copolymers of vinylidene fluoride-hexafluoropropylene using nuclear magnetic resonance spectroscopy. Polymers containing 60–85 mole % vinylidene fluoride were studied. The gross structure of the copolymer deduced from the spectra was:

$$\left[\underset{0.93}{-\!\!\left(CH_2CF_2CF_2\overset{\displaystyle CF_3}{\underset{|}{C}F}\right)\!\!-}\ \underset{0.07}{-\!\!\left(CH_2CF_2\overset{\displaystyle CF_3}{\underset{|}{C}F}CF_2\right)\!\!-}\right]_{1-n}^{-}$$

$$\left[\underset{0.95}{-\!\!\left(CH_2CF_2CH_2CF_2\right)\!\!-}\ \underset{0.05}{-\!\!\left(CH_2CF_2CF_2CH_2\right)\!\!-}\right]_{n/2}$$

where n is a function of the mole fraction of vinylidene fluoride. It was concluded that (1) little or no homopolymerization of hexafluoropropylene occurs; (2) that little or no chain branching occurs; and (3) that certain modes of addition of the monomers during polymerization are highly preferred. The data also indicated that the four basic repeat units are randomly distributed.

The vinylidene fluoride based elastomers have in most cases been prepared by water-emulsion polymerization at elevated temperatures and pressures, using as a catalyst system a mixture of a persulfate and a bisulfite, though other redox or free radical systems have been successfully employed (9–15). Highly fluorinated surfactants, such as ammonium perfluorooctanoate, are most commonly used since it is important to use a dispersing agent which will not enter into the polymerization by a chain transfer reaction.

Bier, Schaff, and Kahrs (18) studied the redox polymerization of fluorinated olefins using activator systems (persulfate and bisulfite) containing radioactive sulfur. Depending on the activator system employed, the polymer produced was or was not radioactive. The initiators investigated were:

(1) $K_2S_2{}^*O_8 + Fe^{2+}$

(2) $Fe^{3+} + NaHS^*O_3$

(3) $K_2S_2{}^*O_8 + NaHS^*O_3$

Both redox systems employing bisulfite ion (2 and 3) produced polymer having radioactivity. However, polymer made using the persulfate–ferrous ion system was not radioactive. It was concluded that either the persulfate radical reacts with water to produce an $\cdot OH$ radical which initiates the polymerization or that the persulfate radical results in a sulfuric acid ester endgroup which is subsequently hydrolyzed:

$$S_2O_8{}^{2-} + Fe^{2+} \rightarrow SO_4{}^{2-} + \left[\overset{\displaystyle O}{\underset{\displaystyle O}{OS}}\!\!-\!\!O\cdot\right]^{-} + Fe^{3+}$$

$$\left[\overset{\displaystyle O}{\underset{\displaystyle O}{OSO\cdot}}\right]^{-} + CF_2{=}C \rightarrow \left[\overset{\displaystyle O}{\underset{\displaystyle O}{OS}}\!\!-\!\!O\overset{\displaystyle F}{\underset{\displaystyle F}{C}}\!\!-\!\!C\cdot\right]^{-} \xrightarrow{\ H_2O\ } HO\overset{\displaystyle F}{\underset{\displaystyle F}{C}}\!\!-\!\!C\cdot\ + HSO_4{}^{-}$$

or

$$HO\cdot + CF_2\!=\!C \rightarrow HO\!-\!CF_2\!-\!C\!\!\diagdown\!\!^{\diagup}$$

The group $HO\overset{F}{\underset{F}{C}}$— is unstable and readily hydrolyzes to a carboxylic acid.

$$HO\overset{F}{\underset{F}{C}}\!-\!C\!\!\diagdown\!\!^{\diagup} + H_2O \rightarrow HOOC\!-\!C\!\!\diagdown\!\!^{\diagup} + 2HF$$

The bisulfite ion initated system results in a sulfonic acid endgroup. Perfluoroalkane sulfonic acids are stable to hydrolysis (36).

$$HSO_3^- + Fe^{3+} \rightarrow Fe^{2+} + \cdot\overset{H}{\underset{O}{\overset{O}{S}}}O$$

$$\overset{H}{\underset{O}{\overset{O}{O}}}S\cdot + CF_2\!=\!C\!\!\diagdown\!\!^{\diagup} \rightarrow HO\overset{O}{\underset{O}{S}}\!-\!\overset{F}{\underset{F}{C}}\!-\!C\!\!\diagdown\!\!^{\diagup}$$

Therefore, depending on the fluorinated monomers employed, persulfate-bisulfite initiation should result in polymers containing alcohols, sulfonic acids, and carboxylic acids as endgroups.

B. Vulcanization

In order to develop functionally useful properties, most elastomeric materials must be crosslinked or vulcanized. This is also true of the fluorinated elastomers. The polymers prepared from the fluorinated butadienes contain unsaturation and can be vulcanized by the same techniques used for the curing of neoprene stocks (cf. Chapter 4.C). However, the commercial vinylidene fluoride based polymers are completely saturated and the relative chemical inertness of these materials makes them difficult to crosslink. Consequently, a discussion of the vulcanization chemistry of these polymers is in order.

The known methods of crosslinking saturated fluoroelastomers include the use of (1) diamines and derivatives, (2) dithiols and tertiary amines, (3) peroxides, (4) radiation, and (5) thermal cure in the presence of tertiary amines (37). Of these, only the use of the derivatives of diamines has been of commercial importance. The most commonly employed curing agents are shown in Table IV. The saturated fluoroelastomers in commercial practice are vulcanized by first compounding with fillers

TABLE IV
Commercial Vulcanization Agents for Fluoroelastomers

Product name	Name	Formula
Diak No. 1	Hexamethylenediamine carbamate	$H_3{}^+N$—$(CH_2)_6N$ with $CO_2{}^-$ and H
Diak No. 2	Ethylenediamine carbamate	$H_3{}^+N(CH_2)_2$—N with $CO_2{}^-$ and H
Diak No. 3	N,N-Dicinnamylidene-1,6-hexanediamine	$\left[\bigcirc\!\!-C=C-C=N+CH_2+_3 \right]_2$
Diak No. 4	An alicyclic amine salt	—

(carbon black, silica, etc.), metal oxides, and the diamine derivative, subjecting the compounded stocks to a short press cure, then oven curing the press cured articles for several hours at around 200°C.

A rather comprehensive study of the vulcanization of Viton A, a vinylidene fluoride-hexafluoropropylene copolymer, was carried out by Smith and Perkins (37,38). They have proposed that the polymer is crosslinked by a three-stage process as follows: (1) the reaction of a base or radiation to eliminate HF and form double bonds; (2) the reaction of difunctional agents at the double bonds; and (3) the formation of carbon-to-carbon double bonds by mutual reaction of the unsaturated centers.

Step 1. In the first step, hydrogen fluoride is eliminated from the polymer upon treatment with bases or ionizing radiation with the formation of double bonds.

$$-CH_2-CF_2-CH_2-\ \xrightarrow[\text{radiation}]{\text{base}}\ -\underset{H}{C}=CF-CH_2 + HF$$

This process takes place very rapidly and probably occurs during milling when using a diamine curing system.

This reaction has also been postulated by several other workers as the initial one to occur during the crosslinking of saturated fluoroelastomers (39–44).

Bro (44) measured the time required for embrittlement of films of polymers in amines at reflux. He observed that both polyethylene $(-CH_2CH_2-)_n$ and polytetrafluoroethylene $(-CF_2CF_2-)_n$, among others, were stable in refluxing amines; whereas a random copolymer of tetrafluoroethylene and ethylene $[(-CF_2CF_2)_n(CH_2CH_2-)_m]_x$ was unstable. Poly(vinylidene fluoride) $(-CH_2CF_2-)_n$, a random copolymer of tetrafluoroethylene and vinylidene fluoride $[(-CF_2CF_2)_n(CH_2CF_2-)_m]_x$, poly-1,1,2-trifluorobutadiene $(-CF_2CF=CHCH_2-)_n$, and polytrifluoroethylene $(-CHFCF_2-)_n$ were also unstable. The polymers substituted completely with hydrogen or fluorine were inert to amines whereas the structures containing fluorine and hydrogen on adjacent carbon atoms were susceptible to attack by amines by a proposed dehydrofluorination mechanism.

Paciorek (43) and coworkers investigated the reaction of primary, secondary, and tertiary mono- and diamines with the copolymers of (1) vinylidene fluoride and chlorotrifluoroethylene (Kel-F), and (2) vinylidene fluoride and hexafluoropropylene (Viton A). Treatment of the copolymer of chlorotrifluoroethylene and vinylidene fluoride with the amines produced the amine hydrochloride in high yield with very little amine hydrofluoride being formed. It was concluded that the structure

$$\begin{matrix} & \text{F} & \\ -CF_2CH_2&\!\!\!\overset{|}{C}\!\!\!&CCF_2- \\ & \text{Cl} & \end{matrix}$$

was the main unit and that the amine hydrochloride resulted from dehydrohalogenation. Polychlorotrifluoroethylene was unreactive with amines which discounted the possibility of the displacement (Eqs. 1a and 1b).

$$R-NH_2 + \overset{\diagdown}{\underset{\diagup}{\underset{\text{Cl}}{\text{C}}}}\!\!\!\overset{\text{F}}{-}CF_2 \rightarrow \overset{\diagdown}{\underset{\diagup}{\underset{\underset{R}{\overset{|}{N}-H}}{\text{C}}}}\!\!\!\overset{\text{F}}{-}CF_2 + HCl \qquad (1a)$$

$$HCl + RNH_2 \rightarrow RNH_2 \cdot HCl \qquad (1b)$$

From their studies on the vinylidene fluoride-hexafluoropropylene copolymer, it was shown that dehydrofluorination takes place in the presence of amines. It was speculated that the tertiary fluorine atom of the hexafluoropropylene unit was most easily removed (Eq. 2). This

$$-CF_2-CH_2-\underset{\text{F}}{\overset{\text{CF}_3}{\overset{|}{\underset{|}{C}}}}-CF_2- \xrightarrow[\text{RNH}_2]{} -CF_2-\overset{\text{H}}{\text{C}}=\overset{\text{CF}_3}{\text{C}}-CF_2- + RNH_2 \cdot HF \qquad (2)$$

conclusion was also reached by other workers (40). The initially formed double bonds activate the elimination of hydrogen fluoride from neighboring atoms leading to conjugated double bonds (38). This process leads to the formation of the color which is always in evidence after amine treatment of the vinylidene fluoride-hexafluoropropylene copolymer.

Step 2. The second step in the crosslinking reaction involves addition of the curing agent to the site of unsaturation (Eq. 3). Hydrogen fluoride

$$
\begin{array}{c} | \\ \text{CH} \\ \| \\ \text{CF} \\ | \\ \text{CH}_2 \\ | \end{array}
\xrightarrow{\text{RNH}_2}
\left[\begin{array}{c} | \\ \text{CH}_2 \\ | \\ \text{CF—NHR} \\ | \\ \text{CH}_2 \\ | \end{array}\right]
\tag{3}
$$

is eliminated from the intermediate (Eq. 4). For a diamine, the following crosslink would result:

$$
\left[\begin{array}{c} | \\ \text{CH}_2 \\ | \\ \text{CF—NH—R} \\ | \\ \text{CH}_2 \\ | \end{array}\right]
\xrightarrow{-\text{HF}}
\begin{array}{c} | \\ \text{CH} \\ \| \\ \text{C—NHR} \\ | \\ \text{CH}_2 \\ | \end{array}
\rightleftharpoons
\begin{array}{c} | \\ \text{CH}_2 \\ | \\ \text{C=NR} \\ | \\ \text{CH}_2 \\ | \end{array}
\tag{4}
$$

lowing crosslink would result:

$$
\begin{array}{cc} | & | \\ \text{CH}_2 & \text{CH}_2 \\ | & | \\ \text{C=N—R—N=C} \\ | & | \\ \text{CH}_2 & \text{CH}_2 \\ | & | \end{array}
$$

These reactions probably occur during the press cure operation. Other workers have shown that the amine vulcanization of fluoroelastomers results in a C=N crosslink (45).

In the vulcanization of saturated fluoroelastomers such as Viton, it is essential to have an acid acceptor present both to develop the full state of cure and to react with the hydrogen fluoride present since free hydrogen fluoride is deleterious to heat aging properties. Metal oxides (for example, MgO) are generally employed as the acid acceptor. These materials, of course, react with the hydrogen fluoride to form metal fluorides and water. To develop the full state of cure, it is necessary to remove the liberated water and consequently a rather lengthy oven or post cure is required.   The necessity for water removal can be explained by a consideration of the equilibrium reaction (Eq. 5).

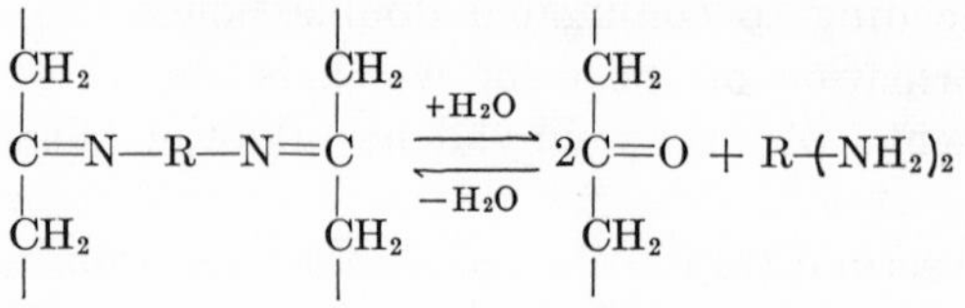

$$
\begin{array}{ccc}
| & | & | \\
CH_2 & CH_2 & CH_2 \\
| & | \quad {\scriptstyle +H_2O} & | \\
C{=}N{-}R{-}N{=}C & \underset{-H_2O}{\overset{}{\rightleftarrows}} & 2C{=}O + R{+}NH_2)_2 \\
| & | & | \\
CH_2 & CH_2 & CH_2 \\
| & | & |
\end{array}
\tag{5}
$$

Step 3.   The final step in the cure involves the thermal formation of additional unsaturation and subsequent crosslinking (38,41,42) (Eq. 6).

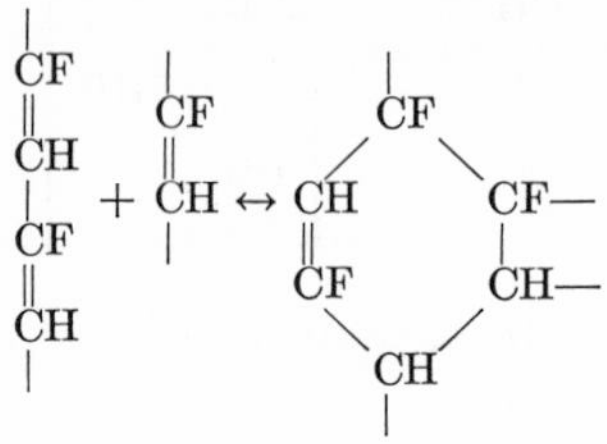

$$
\begin{array}{c}
| \\
CF \\
\| \qquad | \qquad | \\
CH \qquad CF \qquad CF \\
| \quad + \quad \| \quad \leftrightarrow \quad \diagup \quad \diagdown \\
CF \qquad CH \quad CH \qquad CF{-} \\
\| \qquad | \qquad \| \qquad \| \\
CH \qquad \qquad CF \qquad CH{-} \\
| \qquad \qquad \diagdown \quad \diagup \\
\qquad \qquad \qquad CH \\
\qquad \qquad \qquad |
\end{array}
\tag{6}
$$

These reactions, which take place at temperatures around 200°C (42), occur during the ovencure or postcure operation.

III. HETEROFLUOROELASTOMERS

A. Fluorinated Nitrosoelastomers

In 1955, Barr and Haszeldine (46) reported the reaction of trifluoronitrosomethane (CF_3NO) and tetrafluoroethylene (C_2F_4) in the dark at room temperature giving rise to 30–65% of a colorless gas and 35–70% of an almost colorless viscous oil.   The gas was identified as an oxazetidine (**1**).   The oil had the same empirical structure and was shown to be

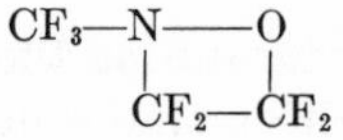

$$
\begin{array}{c}
CF_3{-}N{-}\!\!-\!\!-O \\
| \qquad | \\
CF_2{-}CF_2
\end{array}
$$

(**1**)

a 1:1 copolymer of the monomers having the molecular arrangement shown as structure (**2**).   The copolymer had a molecular weight of about

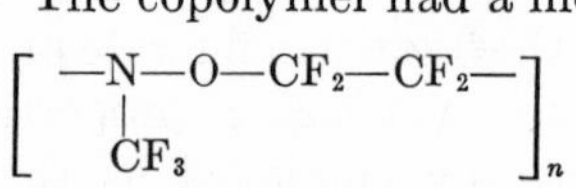

$$
\left[
\begin{array}{c}
{-}N{-}O{-}CF_2{-}CF_2{-} \\
| \\
CF_3
\end{array}
\right]_n
$$

(**2**)

7,000.   It was later shown that elastomeric materials resulted when the reaction conditions were controlled to produce polymers with molecular weights of 50,000 and higher (2,47). The polymerization of several other fluorine-containing nitrosoalkanes and fluorinated olefins also resulted in elastomeric polymers (48–50).

The fluorinated nitrosoalkanes can be prepared by the reaction of a fluorinated alkylbromide or iodide with nitric oxide (NO) in equimolar quantities in the presence of mercury and ultraviolet light. The reaction can be readily followed since the mixture of the alkyl halides and nitric oxide is brown and the fluorinated nitrosoalkanes are deep blue in color.

Fluorinated olefins and nitrosoalkanes have been polymerized in bulk, in solvents, and in aqueous emulsion systems. In general, no catalysts are employed. In a typical bulk polymerization, the monomers are condensed into an evacuated Pyrex tube which is then sealed and maintained at a temperature of 15–20°C for a period of about 2 weeks. The blue color of the nitroso compound slowly fades as the polymerization progresses.

Liquid phase polymerizations can be carried out in fluorinated organic solvents, such as perfluorotributylamine (49), under autogenous pressure at temperatures from −50 to 100°C. Hydrocarbon solvents result in the production of low molecular weight polymers as a consequence of chain transfer reactions.

The ease of polymerization of the fluorinated nitrosoalkanes with the fluoroelfins varies depending on the olefinic material employed. For example, hexafluoropropylene (51) or vinylidene fluoride (52) are difficult to polymerize with trifluoronitrosomethane. Higher temperatures and pressures are required than those used for the polymerization of tetrafluoroethylene with the nitrosoalkanes.

It was originally believed that the polymerization of the fluorolefins with the fluorinated nitrosoalkanes occurred by an ionic mechanism (53) (Eq. 7). This mechanism would explain the formation of a 1:1 copolymer

$$\overset{\delta^+}{N}=\overset{\delta^-}{O} + CF_2=CF_2 \rightarrow \,^+N-O-CF_2-CF_2^- \xrightarrow{C_2F_4, CF_3NO}$$
$$\underset{CF_3}{|} \qquad\qquad\qquad \underset{CF_3}{|}$$

$$\,^+N-O-(CF_2-CF_2-NO)_x-CF_2CF_2^- \quad (7)$$
$$\underset{CF_3}{|} \qquad\qquad\qquad\qquad \underset{CF_3}{|}$$

and the by-product oxazetidine from the intermediate

$$\,^+N-O-CF_2-CF_2^-$$
$$\underset{CF_3}{|}$$

However, an investigation of the influence of catalysts on the polymerization indicated that the reaction occurred by a free radical mechanism.
Exposure of the reaction mixture to ultraviolet radiation for a short
period, or the addition of *tert*-butylperoxide lead to a slightly increased
rate of reaction and a doubling of the polymer:oxazetidine ratio. Reaction in the presence of hydroquinone approximately halved the overall
rate of reaction and decreased the polymer:oxazetidine ratio eightfold.
Furthermore, cationic catalysts, such as $TiCl_4$ and BF_3-etherate which
are known to inhibit anionic reactions did not have any effect upon the
molecular weight of the polymer. A cationic-type mechanism would be
highly unlikely for the perfluoromonomers (49). Kinetic studies also
indicate that the polymerization occurs by a free radical mechanism (2)
(Eq. 8). The initiation step most likely involves homolytic cleavage of

$$R\cdot + CF_3NO \rightarrow R-\underset{\underset{CF_3}{|}}{N}-O\cdot$$

$$\underset{\underset{CF_3}{|}}{R}NO\cdot + CF_2{=}CF_2 \rightarrow R-\underset{\underset{CF_3}{|}}{N}OCF_2-CF_2\cdot \xrightarrow{CF_3NO, C_2F_4}$$

$$R(\underset{\underset{CF_3}{|}}{N}OCF_2-CF_2)_n-\underset{\underset{CF_3}{|}}{N}OCF_2CF_2\cdot \quad (8)$$

the nitrosoalkane (Eq. 9). The nitroso group of a perfluoronitrosoalkane

$$CF_3NO \rightarrow CF_3\cdot + NO \qquad\qquad (9)$$

is known to be susceptible to homolytic cleavage and free radical attack
at the nitrogen atom (53). For example:

$$CF_3NO \rightarrow CF_3\cdot + NO$$

$$CF_3\cdot + CF_3NO \rightarrow (CF_3)_2NO\cdot \xrightarrow{NO} (CF_3)_2NONO$$

B. Poly(thiocarbonyl Fluoride)

Thiocarbonyl fluoride

$$(F-\overset{\overset{\displaystyle S}{\|}}{C}-F)$$

can be polymerized to form elastomeric materials. Poly(thiocarbonyl
fluorides) will be discussed in Part II, Chapter 9.C.

References

1. R. J. Plunkett, U.S. Pat. 2,230,654 (1941).
2. J. C. Montermoso, *Rubber Chem. Technol.*, **34**, 1521 (1961).
3. W. E. Mochel, L. F. Salisbury, A. L. Barney, D. D. Coffman, and C. J. Mityhton, *Ind. Eng. Chem.*, **40**, 2285 (1948).
4. J. C. Tatlow, *Rubber Plastic Age*, **39**, 33 (1958).
5. E. Tufts, *Rubber Age (NY)*, **84**, 963–67 (1959).
6. Engineering Guide to the Du Pont Elastomers.
7. E. I. du Pont de Nemours & Co., Rubber Chemicals Division, Report No. BL-262 (1954).
8. J. H. Simons, *Fluorine Chemistry*, Vol. L, Academic Press, New York, 1950, p. 414.
9. S. Dixon, D. R. Rexford, and J. S. Rugg, *Ind. Eng. Chem.*, **49**, 1687 (1957).
10. W. O. Teeters, H. J. Passino, and A. L. Dittman, U.S. Pat. 2,770,606 (1956).
11. J. R. Pailthorp and H. E. Schroeder, U.S. Pat. 2,968,649 (1961).
12. J. R. Pailthorp and J. F. Smith, U.S. Pat. 3,056,767 (1962).
13. D. R. Rexford, Brit. Pat. 789,786 (1958).
14. D. Sianesi, G. C. Bernardi, and G. Diotalleri, Belg. Pat. 626,289 (1963).
15. E. S. Lo, U.S. Pat. 3,178,399 (1965).
16. R. M. Adams and F. A. Bovey, *J. Polymer Sci.*, **9**, 481 (1952).
17. R. C. Ferguson, *J. Am. Chem. Soc.*, **82**, 2416 (1960).
18. G. Bier, R. Schaff, and K. H. Kahrs, *Angew. Chem.*, **66**, 291 (1954).
19. R. J. Orr and H. C. Williams, *Can. J. Chem.*, **33**, 1328 (1955).
20. E. S. Lo, U.S. Pat. 2,951,064 (1960).
21. A. N. Bolstad, U.S. Pat. 2,996,487 (1962).
22. F. J. Hohn, U.S. Pat. 2,962,484 (1960).
23. E. S. Lo, U.S. Pat. 2,970,988 (1961).
24. G. H. Crawford, U.S. Pat. 2,975,164 (1961).
25. F. J. Hohn, U.S. Pat. 3,116,269 (1963).
26. F. J. Hohn, U.S. Pat. 2,949,446 (1960).
27. A. N. Bolstad and E. S. Lo, U.S. Pat. 2,951,063 (1960).
28. E. S. Lo, U.S. Pat. 2,837,503 (1958).
29. E. S. Lo, U.S. Pat. 2,986,556 (1961).
30. A. N. Bolstad and J. M. Hoyt, U.S. Pat. 2,892,824 (1959).
31. G. H. Crawford, U.S. Pat. 2,836,583 (1958).
32. E. S. Lo and G. H. Crawford, U.S. Pat. 2,951,065 (1960).
33. H. Iserson, F. E. Lawler, and M. Hamptschein, U.S. Pat. 3,062,794 (1962).
34. E. S. Lo, U.S. Pat. 2,959,575 (1960).
35. E. E. Frisch, U.S. Pat. 3,202,643 (1965).
36. T. Gramstad and R. N. Haszeldine, *J. Am. Chem. Soc.*, **173** (1956).
37. J. F. Smith and G. T. Perkins, *Rubber Plastics Age*, **42**, 59 (1961).
38. J. F. Smith, *Rubber World*, **142**, No. 3, 102 (1960).
39. A. N. Lyubimov, A. S. Novikov, F. A. Galil, A. V. Gribacheva, and A. F. Varenik, *Vysokomolekul, Soedin.*, **5**, (5) 687–92 (1963).
40. R. D. Chambers, J. Hutchinson, and W. K. R. Musgrave, *Tetrahedron Letters*, 619 (1963).
41. T. G. Degteva, F. A. Galil, N. A. Slovokhtova, and T. N. Oyumaeva, *Vysoko-molekul. Soedin.*, **5** (10) 1485 (1963).

42. A. S. Noviokov, A. S. et al., *Vysokomolekul. Soedin.*, **4**, 423 (1962).
43. K. L. Paciorek, L. C. Mitchell, and C. T. Lenk, *J. Polymer Sci.*, **45**, 405 (1960).
44. M. I. Bro, *J. Applied Polymer Sci.*, **1**, 310 (1959).
45. A. S. Novikov, F. A. Galil, N. A. Slovokhotova, T. N. Oyumaeva, and V. A. Kargin, *Vysokomolekul. Soedin.*, **4**, 1799 (1962).
46. D. A. Barr and R. N. Haszeldine, *J. Chem. Soc.*, **1955**, 1881.
47. J. B. Rose, Brit. Pat. 787,254 (1958).
48. R. N. Haszeldine and C. J. Willis, Brit. Pat. 943,795 (1960).
49. G. H. Crawford, U.S. Pat. 3,072,592 (1963).
50. R. N. Haszeldine and C. J. Willis, U.S. Pat. 3,197,451 (1965).
51. R. N. Haszeldine and C. J. Willis, Brit. Pat. 901,206 (1962).
52. R. N. Haszeldine and C. J. Willis, Brit. Pat. 901,207 (1962).
53. D. A. Barr and R. N. Haszeldine, *J. Chem. Soc.*, **1956**, 3424.

CHAPTER 5

ELASTOMERS BY CATIONIC MECHANISMS

A. Polyisobutene and Butyl Rubber

J. P. KENNEDY

*Central Basic Research Laboratories, Esso Research
and Engineering Co., Linden, New Jersey*

Contents

I. POLYISOBUTENE

A. Introduction

Isobutene, one of the abundantly available, inexpensive monomers of petrochemical origin has a long and distinguished history in polymer

291

chemistry.   The "prehistoric" chemistry of isobutene may be traced back
at least until 1873 to Russia (1).   Continued scientific interest in cationic
isobutene polymerizations started at about 1940 when R. M. Thomas and
his co-workers from Standard Oil Development Company, published their
classic "The Preparation and Structure of High Molecular Weight Poly-
butenes" (2).   This paper was written several years after the discovery
of butyl rubber by the senior authors (3,4) and gave the first glimpse into
the tremendous complexity of this polymerization.   The early researchers
described the synthesis and characterization of high molecular weight
(up to 2–5 million) elastomeric polyisobutenes prepared by Friedel-
Crafts initiators, e.g., $AlCl_3$ and BF_3.   The significance of low temperatures for increased molecular weights in ionic polymerizations was investigated.   Also, the authors elucidated and proposed the overall structure
of polyisobutylenes, and discussed the effect of temperature, poisons, i.e.,
butenes and higher olefins, catalyst and monomer concentrations, etc.,
on product molecular weight and conversion.   This publication was a
fountainhead leading to further research.   For example, the early authors
found that the molecular weight of polymer increases with decreasing
polymerization temperature.   This observation has since been recognized
as a general phenomenon in polymer chemistry, and holds particularly
true in ionic polymerizations.   Flory, in 1953, replotted the early author's
molecular weight data in the log mol wt versus $1/°K$ fashion and obtained
a straight line (5).   This and other information in this publication was
the background for further investigations leading to the discovery of the
inversion temperature (6,7), to the discovery of discontinuity in the
molecular weight–temperature relation (8), to liquid phase polymerization
of isobutene at $-185°C$ (9), which is still the lowest temperature liquid
phase polymerization of any system reported to date, to the investigation
of poisoning activity of branched olefins by English workers (10), etc.
Quite recently the original observation that n-butenes and other higher
olefins are "polymerization poisons" has been further studied, and these
studies lead to an expanded general termination theory in cationic
polymerizations (11).

   The next important milestone in the polymer chemical history of
isobutene was the opening of the first commercial butyl rubber plant
in 1943 at Baton Rouge, Louisiana.   Shortly thereafter, important scientific publications concerning low temperature isobutene polymerizations
started to appear in England.   The English investigators studied in great
detail mechanistic and kinetic aspects of this polymerization reaction.
The influence of trace impurities was recognized and the concept of

cocatalysis was introduced (12,13), thermodynamic aspects of the polymerization were worked out (14), and in a continuous flow of publications many important facets of this very complex reaction have been elucidated. Excellent review articles and surveys are available on this subject, and the status of cationic polymerizations in general and isobutene polymerizations in particular until about 1962 have been discussed in great detail by many authors (15–19).

A comprehensive but, unfortunately, rather uncritical compilation of patents, technological developments, and industrially important aspects of isobutene polymers and copolymers covering the period up until 1959 also exists (20).

In view of the large amount of pertinent and critically surveyed scientific material, i.e., recent books (15,17), surveys, and review articles (16,18,19,21) available in this field, it would be unprofitable at this time to review basic knowledge of cationic isobutylene polymerizations. It would be more compatible with the purpose of this book to select and examine those aspects of the polymer chemistry of isobutylene which have a bearing on high molecular weight elastomeric polyisobutene and butyl rubber or on the synthesis and mechanism of high molecular weight rubbery polymers in general. In addition, recently published pertinent information will also be considered.

B. The Structure of Rubbery Polyisobutene

It is generally accepted that the major portion or the overall structure of high molecular weight polyisobutene can be represented by the following repeat unit:

$$-CH_2-\overset{\overset{\displaystyle CH_3}{|}}{\underset{\underset{\displaystyle CH_3}{|}}{C}}-$$

Those interested in details of structure determinations, structure versus property relationships, and related subjects should consult the authoritative reviews by Thomas and Sparks (22) and Buckley (23).

Russian workers claimed (24) that with certain Ziegler catalysts polyisobutenes having the following enchainment can be obtained:

$$-CH_2-\overset{\overset{\displaystyle }{|}}{\underset{\underset{\displaystyle CH_3}{|}}{CH}}-CH_2-$$

This contention has been completely refuted recently (25). For a more detailed discussion on this subject see Ref. 21.

Structural details of cationically obtained polyisobutenes have been investigated. Dainton and Sutherland (26) from a study of the infrared spectra of low molecular weights polyisobutenes prepared with $BF_3 \cdot H_2O$ catalyst concluded that polymer endgroups are predominantly of the

$$\overset{\displaystyle C}{\underset{\displaystyle |}{-C-}}C{=}C \text{ and also of the } -C{=}\overset{\displaystyle C}{\underset{\displaystyle |}{C}}-C \text{ type formed by proton expulsion,}$$

i.e., monomer transfer steps, and possibly of $-OH$ groups originating from the cocatalyst. These findings were corroborated by Flett and Plesch (27). Plesch also showed that with certain catalyst–cocatalyst systems, kinetic termination may result in the incorporation of catalyst residues into the polymer (27). Russian authors (28) studied the structure of high molecular weight polyisobutenes using thermal depolymerization technique. They concluded that the polymer chain consists mainly of head-to-tail monomer linkages; however, the presence of at least 2.1% tail-to-tail monomer units were indicated by the presence of diisocrotyl and 2,5-dimethyl-2-hexene among the degradation products. Recently, Slichter and Davis (29) showed by NMR spectroscopy that methyl group rotation in polyisobutene encounters relatively large hindrance.

Polyisobutene appears to have a completely linear structure and there is no evidence in the published literature to the contrary, i.e., branching. Very high molecular weight polyisobutylenes ($\sim$5–8 $\times$ 10^6) might require weeks of gentle solutioning to dissolve them. It is important that these extremely high molecular weight products be not shaken or agitated during dissolution because molecular weight breakdown might otherwise occur (29a).

C. The Synthesis of Rubbery Polyisobutenes

The factor which more than anything else determines the usefulness and application of polyisobutene is its molecular weight. Fortunately, polyisobutene is one of the very few high polymers whose molecular weights can be controlled relatively simply from the lowest oligomer level up to many millions. All these products are commercially available and find widespread use in various applications as determined by their molecular weight range. Thus, low molecular weight polyisobutenes ($\eta = < \sim 1.0$ dl/g) are tacky, very viscous semisolids, and can be used in adhesive compositions, binders, caulking and sealing agents, etc., whereas higher molecular weight products ($\eta = 1.0$–5.0 dl/g) find application as additives to waxes and polyethylene to improve their modulus, elasticity, low temperature flexibility, bend fracture, etc. A detailed discussion of the property-application relationship falls outside the scope of this chapter. General information on this subject is well presented in a commercial pamphlet (30) and a comprehensive but uncritical survey is in a recent book (20).

Technically, two methods exist to control the molecular weight of polyisobutene: by changing the polymerization temperature and by the use of transfer agents (commonly referred to as "poisons"). Some measure of molecular weight control can also be achieved by selection of the catalyst used. Thus $SnCl_4$, $AlBr_3$, $ZrCl_4$, VCl_4, and $FeCl_3$ give lower molecular weight products under comparable conditions than BF_3, $AlCl_3$, $AlEtCl_2$, or $TiCl_4$. Industrially, however, the inexpensive BF_3 and aluminum halides are used almost exclusively.

It is well known that polyisobutenes of extremely high molecular weights can be obtained at increasingly low temperatures. The first published report describing the synthesis of high molecular weight polyisobutylene is a German patent (31) in which Otto and Müller-Cunradi disclose the use of BF_3 catalyst at temperatures below $-10°C$. Highest molecular weight polymer was obtained at liquid ethylene temperature (32).

Today polyisobutylenes are manufactured by two companies: Badische Anilin und Soda-Fabrik A.G. and the Standard Oil Company (N.J). The trade name of Badische's polyisobutylenes is Oppanol, that of Jersey's is Vistanex. Both companies use a continuous polymerization process for the manufacture of their respective products. Badische manufactures low molecular weight polymers with BF_3 complexes as catalyst and isobutane as solvent at about $-10°C$, whereas Jersey uses aluminum chloride catalyst dispersed in hydrocarbon solvents and operates in butane solvent at -20 to $-40°C$.

Badische makes rubbery, high molecular weight Oppanols continuously on a moving conveyor belt: isobutylene and liquid ethylene at about 1:1 volumes are transported on a moving stainless steel trough and the catalyst, BF_3 in liquid ethylene, is introduced. Polymerization is almost instantaneous and the heat of polymerization is removed by the partial evaporation of the internal coolant ethylene. In contrast, the Standard Oil process for the manufacture of high molecular weight rubbery Vistanex employs a continuously operating vertical cylindrical overflow reactor. The monomer mixed with methyl chloride and the catalyst, aluminum chloride in methyl chloride, are injected simultaneously through small orifices at the bottom of the reactor. The reactor is cooled externally with liquid ethylene and, to improve cooling, liquid ethylene is circulated through cooling elements inside the reactor. The polymer is formed in the form of a fine slurry. The fine dispersion of the rubber in the methyl chloride–unreacted monomer diluent medium assures satisfactory heat exchange only when the temperature is below about −80°C. Above this temperature level the tiny rubber particles adhere to each other, they coalesce, subsequently large chunks of rubber are formed and the reactor becomes plugged ("fouling"). Under these conditions effective stirring, i.e., removal of reaction heat, becomes impossible and the reactor has to be stopped and cleaned, a very costly operation. The slurry overflows into a hot water tank, where the catalyst is quenched and the solvent and unreacted monomer are flashed off. The aqueous polymer slurry is filtered, dried, extruded to remove residual amounts of water, cut to suitable sizes, and packaged. Vistanex manufacture is very similar to butyl rubber manufacture (Section II-B).

D. The Problem of Cocatalysis in Isobutene Polymerization

1. Conventional Systems

Cocatalysis in cationic polymerizations in general and isobutene polymerizations in particular is a controversial and still largely unsettled problem. For reviews on this subject see Refs. 17 and 21. Since the early workers in this field recognized the significance of cocatalysis in ionic polymerizations, great progress has been made, and many systems have been investigated in detail from this point of view. It is interesting, however, that in spite of the large amount of fine work on cocatalysis, the problem of cocatalysis is still obscure in one of the most important industrial cationic polymerization systems, i.e., isobutene–AlCl₃. Perhaps the closest paper to this chemistry has recently been published by Czechoslovak authors who investigated the isobutene–AlBr₃ system (33). It was found that isobutene apparently can be polymerized at a high rate with AlBr₃ in heptane solvent in the absence of added cocatalyst, e.g., water. The absence of moisture in the system was indicated by showing that under substantially identical conditions BF₃ and TiCl₄ were inactive as polymerization catalysts and additions of water were necessary to initiate polymerization with these two Friedel-Crafts halides. However, it must be emphasized that these results do not necessarily apply for AlCl₃ and generalization from AlBr₃ to AlCl₃ should be made only with the utmost care. Only a carefully executed direct experiment in the complete absence of moisture or other impurities could provide the final answer whether or not AlCl₃ requires a cocatalyst for isobutene polymerization.

In a related series of investigations Beard, Plesch, and Rutherford (34) carried out polymerizations of isobutene with AlCl₃ catalyst in methylene dichloride solvent. It was found that polymerization occurred in this system in the apparent absence of water. Water did not affect the rate of polymerization but reduced the molecular weight. These workers believe that the solvent might be the cocatalyst in this system:

$$CH_2{=}C\overset{\displaystyle CH_3}{\underset{\displaystyle CH_3}{\big\langle}} + AlCl_3 + CH_2Cl_2 \rightarrow Cl{-}CH_2{-}CH_2{-}CH_2{-}C\overset{\displaystyle CH_3}{\underset{\displaystyle CH_3}{\big\langle}}{\oplus} \quad AlCl_4{\ominus}$$

Unfortunately, the undiluted isobutene–AlCl₃ experiment has not been carried out and we still do not know whether or not AlCl₃ in the absence of solvent requires the presence of water (or cocatalyst) to initiate isobutene polymerization.

It appears that the problem of cocatalysis in isobutene polymerizations becomes increasingly complex and involved (35). Some facets of this chemistry which either have not yet been discussed or have been considered only to a very limited extent will now be presented.

Very unusual observations have recently been published by Cheradame and Sigwalt (36) who studied the thoroughly purified isobutene–methylene dichloride–TiCl₄ system using a sealed H shaped reactor at −72°C. Experimentally, it was found that upon breaking the catalyst vial in the first arm of the reactor only partial conversion occurred and relatively low molecular weight (∼50,000) product formed. No further polymerization took place when the content of the first compartment was poured into the second compartment at −78° by tipping the reactor. However, strangely, further and almost total monomer conversion occurred and high molecular weight (∼270,000) product formed when the volatile components (i.e., unreacted monomer) in the first compartment were distilled over into the second one. Similar results have been obtained in the absence of solvent as well.

These workers suggest that small amounts of butene impurities in the system stopped the reaction at low conversions in the first compartment and after the "poison" had been consumed and the volatile monomer distilled over into the second compartment polymerization could resume and go to completion. Questions which still remain in this author's mind are: why is it necessary to physically separate the unreacted monomer from the finished macromolecules and "consumed" butene for polymerization to resume, what is responsible for reinitiating the polymerization in the second compartment in the absence of (non-volatile) catalyst, or, if this reaction is not reinitiation but polymerization resumption of "dormant" chains, how do these get over in the second compartment, etc.

This author has been interested for some time in the problem of cocatalysis in the isobutene–$AlCl_3$ system. This monomer–catalyst combination is somewhat unusual since the catalyst is insoluble in the monomer, i.e., the system is heterogeneous. Until recently, investigations involving cocatalysis in cationic polymerizations have been carried out largely in the homogeneous phase. Heterogeneity of a system might complicate the interpretation of results. For example, an inactive homogeneous system might become active in the presence of solid catalyst. This expectation is based on the following reasoning and experiment. The radioactive chloride exchange between labeled $AlCl_3$ and methyl chloride, i.e.,

$$\tfrac{1}{3}AlCl_3{}^* + CH_3Cl \rightleftharpoons \tfrac{1}{3}AlCl_3 + CH_3Cl^*$$

has been studied in these laboratories (37). Results showed that $AlCl_3$ rapidly exchanges its chlorine with methyl chloride in the presence of solid $AlCl_3$; however, this exchange becomes extremely slow in the absence of solid $AlCl_3$, i.e., after $AlCl_3$ has completely dissolved. It was proposed that in the presence of highly acidic $AlCl_3$ surfaces strong polarization of the methyl chloride might occur with the momentary formation of methyl carbonium ion-like transition states. Thus the ordinarily quite inactive methyl chloride molecule might provide an initiating species in the presence of solid $AlCl_3$. In general, then, in the presence of highly polar surfaces, the strong electrostatic field on the solid might induce some reation, which, in the proper environment, will give catalytic species, i.e., will be able to initiate polymerization, which, in the absence of such solids, could not occur. Such an event might be, for instance, the hydride abstraction by the solid from a suitable monomer so that the latter in fact provides its own initiating carbonium ion species. Such an initiation mechanism might, of course, also occur in suitable homogeneous media in the presence of very strong acids.

In this context it is of interest that the solubility of $AlCl_3$ in hydrocarbons increases significantly in the presence of growing polyisobutene chains. Thus when $AlCl_3$ in methyl cholride solution is injected into pure isobutene or into an isobutene solution in propane at $-50°C$ no visible precipitation occurs, and the system remains completely homogeneous. However, when the same catalyst solution is introduced into pure propane, immediate milkiness indicates $AlCl_3$ precipitation. Apparently the long growing polyisobutene carbonium ions are strongly associated with the gegenion ($AlCl_4^{\ominus}$ or $Al_2Cl_7^{\ominus}$?) so that the comparatively small inorganic moiety is being "pulled" into solution by the long hydrocarbon chain associated with it.

There is yet another aspect of cocatalysis which merits attention. It has been shown conclusively that in some systems polymerization stops because of cocatalyst exhaustion (16,35). In such systems reaction could be reinitiated by introducing additional amounts of cocatalyst. The fundamental equations describing this situation are:

$$\text{MeX}_n \; + \quad \text{HX} \quad \rightarrow \quad \text{H}^{\oplus}\text{MeX}^{\ominus}_{n+1} \tag{1}$$
$$\quad\text{Catalyst} \quad\; \text{Cocatalyst} \quad\;\; \text{Catalyst complex}$$

$$\text{H}^{\oplus}\text{MeX}^{\ominus}_{n+1} + \quad \text{M} \quad \rightarrow \quad \text{HM}^{\oplus} \quad\; + \text{MeX}^{\ominus}_{n+1} \tag{2}$$
$$\qquad\qquad\quad\; \text{Monomer} \quad\; \text{Growing chain} \quad\;\; \text{Gegenion}$$

$$\text{HM}^{\oplus} + \text{MeX}^{\ominus}_{n+1} \rightarrow \quad \text{HMX} \quad\; + \; \text{MeX}_n \tag{3}$$
$$\qquad\qquad\quad\;\; \text{"Dead" chain} \quad\;\; \text{Catalyst}$$

The catalyst reformed in the last reaction can then be "reactivated" by additional cocatalyst and the cycle starts again until more or all the monomer is consumed.

Isobutene polymerizations in the presence of small amounts of $AlCl_3$ catalyst do not go to completion, but stop at low conversions shortly after catalyst introduction (38). In this system polymerization cessation cannot be explained by assuming cocatalyst exhaustion. We have tested this possibility by introducing cocatalytic amounts of water, moist solvents, and hydrochloric acid to freshly synthesized polyisobutenes in a homogeneous system after it was ascertained that polymerization had stopped at low conversion. However, in no case did we notice polymerization resumption or an increase in the amount of polymer formed even after extended periods of time. Consequently, cocatalyst exhaustion does not seem to be the reason for polymerization termination in the isobutene–$AlCl_3$–cocatalyst(?) system.

Significantly, however, if the cocatalytic amount of HCl or moisture is introduced to the system before the $AlCl_3$ catalyst is added, the rate of polymerization is increased and more polymer is formed than in the absence of initially added HCl. However, the molecular weight of the

TABLE I

The Effect of HCl on the Amount of Polyisobutene Formed
and on its Molecular Weight[a]

Amount of HCl, mole/l	Polyisobutene, g	Mol wt, $\times 10^{-3}$
0	0.3650	951.4
0.001	0.4359	256.4
0.01	0.4211	157.3
0.05	0.5330	85.0
0.1	0.5731	91.0

[a] Polymerization conditions: $[M] = 3.14$ molar in n-pentane, system volume 28 ml; $[AlCl_3] = 9.4 \times 10^{-3}$ moles/l methyl chloride, 0.2 ml added at $-78°C$ Homogeneous polymerizations.

polymer formed in the presence of HCl is decreased considerably.   Table I show these data.

Qualitatively, the experimental observations could be explained by assuming that HCl has a dual function: it is a cocatalyst and a transfer agent.   It is a cocatalyst in the sense that the concentration of initiating species is increased in the presence of HCl provided HCl is present before catalyst introduction.   In other words, the catalyst must have reacted instantaneously with the HCl after its introduction into the system and the initiating species must have formed instantaneously before polymerization has taken place.   When HCl (or water) was added to the "dead" system after polymerization cessation at low conversions, no increase in yield occurred.   The fact that the polymerization does not reinitiate on cocatalyst introduction might indicate that the MeX_n catalyst reformed in the above reaction sequence (Eq. 3) is probably substantially different from that employed originally (Eq. 1).

According to the experiment (Table I) the molecular weight of polyisobutene is decreased if the polymerization is carried out in the presence of HCl, e.g., this halide is a transfer agent.   Molecular weight reduction may be explained by the following conventional mechanism:

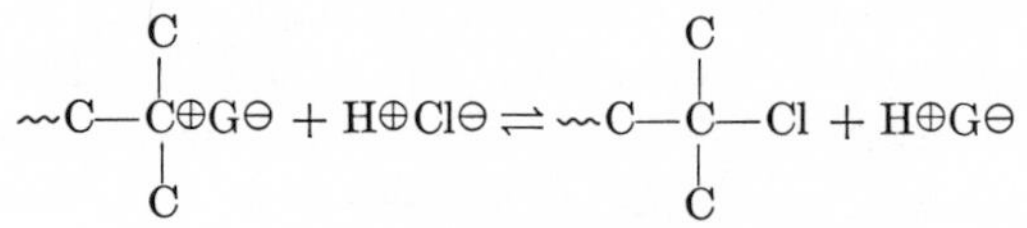

$$\sim C-\overset{\overset{\textstyle C}{|}}{\underset{\underset{\textstyle C}{|}}{C}}^{\oplus}G^{\ominus} + H^{\oplus}Cl^{\ominus} \rightleftharpoons \sim C-\overset{\overset{\textstyle C}{|}}{\underset{\underset{\textstyle C}{|}}{C}}-Cl + H^{\oplus}G^{\ominus}$$

$$H^{\oplus}G^{\ominus} + \text{Monomer} \rightarrow \text{H-Monomer}^{\oplus}G^{\ominus}$$

Thus, although we still do not know with certainty whether or not AlCl₃ requires the presence of a cocatalyst to initiate the polymerization of isobutene and only a direct experiment can furnish the final answer, these results seem to indicate that HCl is a rate promotor and transfer agent in this polymerization.

2. Aluminum Alkyl Systems

Whereas aluminum trichloride or bromide and aluminum alkyl dihalides are known to be efficient and very active cationic catalysts, it has been shown that aluminum dialkyl halides or aluminum trialkyls alone, i.e., in the absence of a cocatalyst, do not initiate the polymerization of certain olefins (39–41). Some time ago a new class of cationic catalyst–cocatalyst systems for the polymerization of isobutene (and other cationically initiable monomers) was discovered in our laboratories. We found that certain aluminum dialkyl halides or aluminum trialkyls in the presence of suitable cocatalysts are efficient cationic catalysts for isobutene polymerization (42,43).

A comprehensive discussion of the problem of cocatalysis in the presence of aluminum alkyls is an extremely complex subject and is outside the scope of this chapter. Reference is made among other papers to the work of Professors Furukawa and Saegusa in Kyoto (39) and Dr. Sinn in Munich (40,41), who investigated aspects of this subject. The advance in cationic isobutene polymerizations involving aluminum alkyl halides and cocatalysts as disclosed in the above-mentioned patents should now be summarized.

It is of interest to compare the Lewis acidity of the AlCl₃, AlRCl₂, AlR₂Cl, AlR₃ series. The two first compounds are the strongest Lewis acids and, for example, polymerize isobutene or vinyl ethers instantaneously without purposely added cocatalyst; AlR₂Cl is a much less strong Lewis acid: it efficiently polymerizes isobutene but only in the presence of a suitable cocatalyst; however, it is still able to polymerize isobutyl vinyl ether without extraneous cocatalyst; AlR₃ is the least active, it has limited catalytic activity for isobutene and does not polymerize isobutyl vinyl ether.

We have investigated in detail the catalytic activity of AlMe₃, AlEt₃, and AliBu₃ in the presence of tert-butyl chloride cocatalyst for the low temperature polymerization of isobutene in methyl chloride solvent (43). These aluminum alkyls are completely inactive in the absence of cocatalyst, e.g., tert-butyl chloride; however, they polymerize isobutene instantaneously when the cocatalyst is introduced. Figure 1 shows the

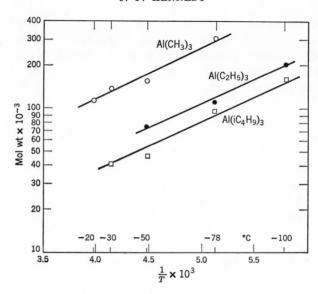

Fig. 1. Effect of temperature on the molecular weight of polyisobutene obtained with AlR$_3$ + *tert*-BuCl catalyst system in methyl chloride diluent.

effect of temperature on the molecular weight of polyisobutenes obtained with various aluminum trialkyls in the presence of *tert*-butyl chloride cocatalyst. Aluminum trimethyl seems to give the highest molecular weight product, but even they are much below those obtained with AlCl$_3$, AlEtCl$_2$, or AlEt$_2$Cl + cocatalyst systems under essentially identical experimental conditions. It is interesting that the log mol wt versus $1/T$ lines in Fig. 1 are parallel to each other, the highest being AlMe$_3$, and then in decreasing order AlEt$_3$ and AliBu$_3$, and that they are linear in the temperature range investigated from -20 to $-100°$C. These data indicate that the overall polymerization mechanism involved with these aluminum alkyls is probably the same. The molecular weight depressing effect of increasingly large alkyl groups is obscure.

Similarly, the low temperature polymerization of isobutene with AlEt$_2$Cl will not take place in the absence of a cocatalyst. Thus, isobutene can be stirred with AlEt$_2$Cl in various solvents over a wide temperature range (-30 to $-90°$C) without polymerization taking place. However, on the introduction of minute amounts of Brönsted acids or other suitable materials instantaneous and vigorous polymerization commences. Cocatalyst efficiencies have been studied by introducing various

TABLE II
Cocatalyst Efficiency of Various Brönsted Acids in
Conjunction with $AlEt_2Cl$

Cocatalyst	Approx. conc., moles	Time, min	Approx. yields, %	Remarks
None	—	—	—	No polymerization
HCl	10^{-4}	<1	100	Very fast reaction
HBr	10^{-4}	<1	100	Very fast reaction
HF	10^{-2}	~150	80	Medium reaction rate
H_2O	–	60	80	Medium reaction rate
CCl_3COOH	10^{-4}	90	20	Medium reaction rate
CH_3OH	10^{-3}	250	1	Slow reaction
CH_3OCH_3	10^{-2}	180	0.1	Extremely slow reaction

cocatalysts into identical isobutene–$AlEt_2Cl$ mixtures in methyl chloride solvent at $-50°C$ and determining the amount of polymer formed and polymerization time. Table II shows the findings.

This compilation of trends indicates a qualitative series of cocatalyst efficiencies in conjunction with $AlEt_2Cl$:

$$HCl, HBr > HF, H_2O > CCl_3COOH \gg CH_3OH > CH_3COCH_3$$

This sequence follows approximately the acidities of the materials involved. It is peculiar that water appears to be a more effective cocatalyst than CCl_3COOH. Acetone could possibly act as a proton donor through its enolic form $CH_3COH{=}CH_2$, or perhaps some unidentified trace of acidic impurity in it provided the cocatalyst.

The mechanism of initiation with Brönsted acid cocatalysts could be visualized by the following set of equations:

$$AlEt_2Cl + H^{\oplus}Cl^{\ominus} \rightleftharpoons H^{\oplus}AlEt_2Cl_2^{\ominus} \tag{4}$$

$$H^{\oplus}AlEt_2Cl_2^{\ominus} + CH_2{=}C\begin{smallmatrix}CH_3\\[2pt]\\CH_3\end{smallmatrix} \rightarrow CH_3{-}C^{\oplus}\begin{smallmatrix}CH_3\\[2pt]\\CH_3\end{smallmatrix}\ AlEt_2Cl_2^{\ominus} \tag{5}$$

$$CH_3{-}C^{\oplus}\begin{smallmatrix}CH_3\\[2pt]\\CH_3\end{smallmatrix}\ AlEt_2Cl_2^{\ominus} + M \rightarrow CH_3{-}C\begin{smallmatrix}CH_3\\[2pt]\\CH_3\end{smallmatrix}{-}{-}{-}{-}CH_2{-}C^{\oplus}\begin{smallmatrix}CH_3\\[2pt]\\CH_3\end{smallmatrix}\ AlEt_2Cl_2^{\ominus} \tag{6}$$

According to this concept, it is conceivable that HCl which is probably somewhat dissociated in methyl chloride solvent interacts with $AlEt_2Cl$ to form the hypothetical acid $H^{\oplus}AlEt_2Cl^{\ominus}$ (Eq. 4), which in the presence of isobutene is immediately consumed by polymerization initiation (Eq. 5). In this system, of course, chloride anions cannot exist, only stable complex anions in which the $Cl^{\ominus}$ is bound coordinatively.

Equation 5 suggests that the first propagating species or the first carbonium ion formed during initiation is the *tert*-butyl cation. Consequently, this equation suggests that polymerization of isobutene could be initiated with *tert*-butyl cations in conjunction with $AlEt_2Cl$, e.g.,

$$CH_3\!-\!\underset{\underset{\textstyle CH_3}{|}}{\overset{\overset{\textstyle CH_3}{|}}{C}}\!-\!Cl \;+\; AlEt_2Cl \;\rightleftharpoons\; CH_3\!-\!\underset{\underset{\textstyle CH_3}{|}}{\overset{\overset{\textstyle CH_3}{|}}{C}}\!\oplus \quad AlEt_2Cl_2\!\ominus \tag{7}$$

Experiments with *tert*-butyl chloride as cocatalyst confirmed this hypothesis. When $AlEt_2Cl$ and isobutene were stirred in methyl chloride diluent and small amounts of *tert*-butyl chloride were added to the quiescent system at low temperatures, instantaneous and sometimes even explosive polymerization occurred.

Basically, it can be postulated that the polymerization reaction of isobutene and other cationically responsive monomers could be initiated not only by the *tert*-butyl cation but by suitable carbonium ions in general in conjunction with the $AlEt_2Cl_2^{\ominus}$ (or similar) gegenion. These species could be created by:

$$RX + AlR_2Y \rightleftharpoons R^{\oplus} + AlR_2XY^{\ominus} \tag{8}$$

$$R^{\oplus} + C\!=\!C \rightleftharpoons R\!-\!C\!-\!C^{\oplus} \tag{9}$$

where RX is an organic halide and Y can be either halogen or hydrogen, and $C\!=\!C$ denotes a cationically initiable monomer.

The key problem is the proper selection of the cocatalyst, which under polymerization conditions will become the initiating carbonium ion. In the first approximation (minimizing changes in entropy, as well as the influence of solvation, dielectric constant, gegenion, temperature, etc.), it could be expected that polymerization initiation will be favored if the stability of the initiating cation, $R^{\oplus}$, is lower than the propagating carbonium ion, $R\!-\!C\!-\!C^{\oplus}$. This (dangerously) simple theory could be tested in a system in which only the initiating $R^{\oplus}$ could be varied keeping all other variables (monomer, gegenion, solvent, temperature, etc.) constant. Our $AlEt_2Cl + RCl$ system (Eq. 9) provided us with this possibility.

Isobutene was selected as "reference" monomer in the first part of our studies. Isobutene is the ideal monomer for research of this type because the molecule contains only primary or vinylidene type hydrogens which are relatively inactive under electrophilic conditions. Bothersome hydride migrations, rearrangements, etc., which occur with great ease with other simple monomers such as propylene or 3-methyl-1-butene, cannot interfere in the polymerization of isobutene.

Thus, we introduced certain amounts of various organic chlorides as potential carbonium ion sources into quiescent mixtures of isobutene-methyl chloride solvent and catalytic amount of $AlEt_2Cl$ stirred at $-50°C$. Table III shows our findings.

The extent of the reaction was judged by cocatalyst efficiency, i.e., gram polyisobutene formed per mole of cocatalyst employed. It was not possible to determine accurately the number of moles of polymer formed because only viscosity average molecular weights were available. The experiments summarized in the individual columns were executed under various conditions (changes in cocatalyst introduction rate, cocatalyst dilution, time of reaction, etc.), therefore, the data should not be treated quantitatively and can only be regarded as indicators of trends. Molecular weight reproducibility is very poor, the reason for which remains largely unexplained.

Although the data are far from being conclusive, they strongly suggest that a relation exists between carbonium ion availability and/or carbonium ion stability on the one hand, and cocatalytic activity on the other. In other words, cocatalytic activity seems to be determined by a balance between carbonium ion availability and carbonium ion stability.

Cocatalyst efficiency could be used as an indicator to estimate, or rather to compare, cocatalyst activities of the various halides employed. Figure 2 shows (the logarithm of) cocatalyst efficiencies as a function of carbonium ion stabilities. Since we cannot independently and quantitatively determine carbonium ion stabilities for the ions studied in these series of investigations, the horizontal axis of Fig. 2 merely indicates a qualitative sequence of carbonium ion stabilities dictated by organic chemical intuition. In other words, Fig. 2 is in fact a hystogram rather than a true plot. Carbonium ion availability, i.e., the ease of dissociation of the alkyl halide, seems to determine cocatalyst efficiencies among halides shown on the left-hand side of Fig. 2. Thus, methyl, ethyl,

TABLE
Polymerization of Isobutene with AlEt₂Cl

		1	2	3		4		5		6	
Cocatalyst chloride		Methyl	Ethyl	sec-Butyl		Isopropyl		Isobutyl		Allyl	
Probable initiating cation		C⊕	C—C⊕	C—C—C⊕ (with C above middle)		C—C⊕ (with C above)		C—C—C⊕ (with C above first)		C=C—C⊕	
iC₄⁻	ml	50	50	50	50	50	50	50	50	50	50
	g	35	35	35	35	35	35	35	35	35	35
	mole	0.625	0.625	0.625	0.625	0.625	0.625	0.6250	0.625	0.625	0.625
CH₃Cl	ml	50	50	50	50	50	50	50	50	50	50
AlEt₂Cl	ml		0.72								
	g		0.69								
	mole	0.00575	0.00575	0.00575	0.00575	0.00575	0.00575	0.00575	0.00575	0.00575	0.00575
Mole of cocat. added × 10⁻²		Any amount	5.57	3.3	1.22	7.17	11.4	0.237	0.0584	0.816	0.25
Mole cat. / Mole cocat.		>1	1	1.75	4.7	0.75	5.0	2.4	9.8	0.71	2.3
g PIB formed		0	0.802	29.0	4.5	27.5	30.3	17.0	2.4	19.0	3.7
% Conv.		0	2.3	82.8	12.8	78.5	86.5	48.5	6.9	54.3	10.5
Cocat. eff. g/mole Mol wt (visc.) × 10⁻³		0	1.4	880	369	384	2,680	7,180	4,130	2,330	1,480
		0	271	402	705	229	253	385	831	512	746

sec-butyl and isopropyl chlorides are rather poor cocatalysts probably because the C—Cl bond is too strong and no dissociation can take place or because the rate of dissociation is too slow. These halides would, of course, be highly efficient cocatalysts, as potential sources of aggressive (unstable) carbonium ions, if they could dissociate under the experimental conditions employed.

The complete inertness of methyl chloride as cocatalyst is significant on two counts, one practical, one theoretical. Practically, it is important because methyl chloride can thus be used as inert diluent for this polymerization. No other monohalogenated aliphatic hydrocarbon can be used as inert solvent, and our experiments indicate that probably most, if not all, halogen alkyls are more or less active cocatalysts with AlEt₂Cl.

The inertness of methyl chloride is theoretically significant because this fact has a certain bearing on the question whether there exist methyl carbonium ions or not. Since methyl chloride is the only halogen alkyl or alkene which does not seem to be a cocatalyst, it appears that at least in this system methyl carbonium ions do not form or form extremely slowly.

III

Catalyst in Conjunction with Various Cocatalysts

7	8	9		10			11	12	13	14
β-Methyl allyl	3-Chloro-1-butene	Crotyl		tert-Butyl			Benzyl	1-Chloro-ethyl benzene	Diphenyl chloro-methane	Triphenyl chloro-methane
C=Ċ—C⊕ (C above)	C=Ċ—C⊕ (C above)	Ċ=C—C⊕ (C above)		C—C⊕ (C above and below)			φ—CH₂⊕	φ—ĊH⊕ (C above)	φ—ĊH⊕ (φ above)	φ—Ċ—φ⊕ (φ above)
50	50	10	10	50	50	100	2.5	10	2.5	2.5
35	35	7	7	35	35	70	1.7	7	1.7	1.7
0.625	0.625	0.125	0.125	0.625	0.625	1.25	0.031	0.125	0.031	0.031
50	50	10	10	50	50	100	2.5	10	2.5	2.5
								0.13	0.030	0.030
								0.12	0.029	0.029
0.00575	0.00575	0.001	0.001	0.00575	0.00575	0.01	0.00024	0.001	0.00024	0.00024
0.25	0.05	0.00098	0.00073	0.0023	0.0007	0.0044	0.000037	0.00092	0.0069	0.018
2.3	11.5	102	137	250	820	228	650	109	3.5	0.014
28.0	3.4	0.783	0.993	23.0	2.0	6.4	0.195	0.295	0.934	0.83
80.0	9.7	11.1	14.2	65.6	5.7	9.2	11.4	4.2	55.0	48.7
11,300	6,800	80,000	137,000	1,030,000	295,000	145,000	527,000	31,900	13,500	4,600
220	685	294	604	447	667	865	784	325	294	314

Contrary to the situation with the simple aliphatic halides, carbonium ion stability seems to limit cocatalytic efficiency of aromatic halides as indicated on the right-hand side of Fig. 2. Thus, the comparatively low cocatalyst efficiency of triphenyl methyl chloride is probably due to the high stability of the triphenyl methyl carbonium ion. It is conceivable that large amounts of these ions exist in methyl chloride, but initiation is impeded because of energetic factors; i.e., the initiating carbonium ion is more stable than the propagating ion of the tert-butyl type.

It appears then that the best cocatalysts are halides shown in the middle of Fig. 2. This is in qualitative agreement with the postulate that cocatalyst efficiency is determined by a judicious balance of carbonium ion availability and carbonium ion stability. Empirically then, tert-butyl chloride and benzyl chloride give excellent cocatalysts. These halides have very high cocatalystic efficiencies (in one case 10^6 g PIB/mole cocatalyst was reached), and the molecular weights which they produce are excellent (particularly when we consider that these 700,000–800,000 molecular weights were obtained at −50°C).

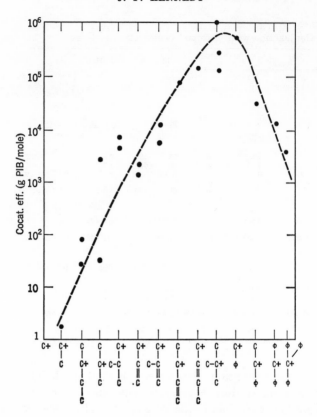

Fig. 2. Polymerization of isobutene with various initiating ions in conjunction with AlEt₂Cl.

E. Aspects of the Mechanism of Low Temperature Isobutene Polymerization with AlCl₃ Catalyst

This author has examined some of the fundamentals of the AlCl₃ initiated polymerization of isobutene and particularly the effect of temperature on the molecular weight under various experimental conditions (44–46). These investigations brought to light a new concept, i.e., the inversion temperature (47) which has already been reviewed (17,18), and therefore will not be examined further. However, the problem of termination in this system has not yet been explored in detail.

It has been shown that small quantities of AlCl₃ catalyst dissolved in methyl chloride solvent when introduced into an isobutene–n-pentane (a good solvent) system induces only a limited amount of polymerization (38). Also, incremental catalyst addition gives only an incremental amount of polymerization in this homogeneous system. This is also true in a bulk system without solvent (48) or in the presence of methyl chloride, a bad solvent for polyisobutene. The conversion versus catalyst amount curves are linear and, significantly, can be extrapolated back to the origin. This (and other evidence) indicates that impurities probably do not play a major role in termination. Since, as we have shown above, neither does cocatalyst exhaustion explain polymerization cessation the question arises, why do these polymerization reactions terminate?

The question remained unanswered even after the very interesting work by a Czechoslovak group (49) who studied the isobutene–AlCl₃ system in various solvents. This work was surveyed and criticized in detail by Plesch (17) and Erusalimskii (18).

A possibility for termination is hydride transfer from the monomer to the growing chain coupled with resonance stabilization of the allylic ion formed from the monomer:

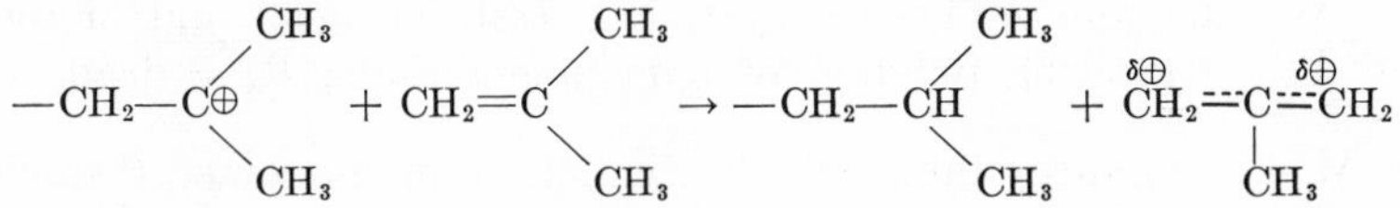

$$
\begin{array}{ccccccc}
& \overset{\displaystyle CH_3}{\diagup} & & \overset{\displaystyle CH_3}{\diagup} & & \overset{\displaystyle CH_3}{\diagup} & \\
-CH_2-C^{\oplus} & & + \ CH_2{=}C & & \rightarrow -CH_2-\overset{..}{C}H & & + \ \overset{\delta\oplus}{CH_2}{=\!\!=}C{=\!\!=}\overset{\delta\oplus}{CH_2} \\
& \diagdown & & \diagdown & & \diagdown & | \\
& CH_3 & & CH_3 & & CH_3 & CH_3 \\
\end{array}
$$

However, extended studies in these laboratories showed that this reaction is probably not very important (51).

It seems that at the present we do not know why polymerization stops in isobutene polymerizations catalyzed by AlCl₃. The least objectionable assumption is a spontaneous unimolecular termination probably involving the gegenion, i.e.,

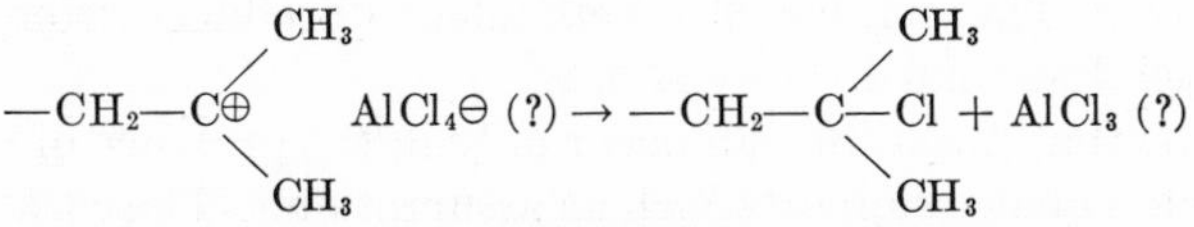

$$
\begin{array}{ccccc}
\overset{\displaystyle CH_3}{\diagup} & & & \overset{\displaystyle CH_3}{\diagup} & \\
-CH_2-C^{\oplus} & AlCl_4^{\ominus} \ (?) & \rightarrow & -CH_2-C{-}Cl + AlCl_3 \ (?) \\
\diagdown & & & \diagdown & \\
CH_3 & & & CH_3 & \\
\end{array}
$$

Research which was aimed to elucidate the termination mechanism operating in this system was, however, fruitful in uncovering many subtle effects governing this intriguing polymerization system (51). Advance in this area will now be outlined.

It is believed that the following is a useful kinetic model for the homogeneous polymerization of isobutene in hydrocarbon solvent using

AlCl$_3$ catalyst in methyl chloride, in the absence or presence of various terminators or transfer agents (ignoring solvation, the presence of gegenion and the influence of free ions or associated ion pairs):

Initiation
$$M + C \xrightarrow{k_1} M\oplus = P_1\oplus \qquad (10)$$

Propagation
$$P_n\oplus + M \xrightarrow{k_2} P\oplus_{n+1} \qquad (11)$$

Transfer

(a) With monomer
$$P_n\oplus + M \xrightarrow{k_3} P_n + M\oplus \qquad (12)$$

(b) With transfer agent
$$P_n\oplus + X \xrightarrow{k_4} P_n + X\oplus \qquad (13)$$

Termination

(a) Spontaneous
$$P_n\oplus \xrightarrow{k_5} P_n \qquad (14)$$

(b) Allylic, with monomer
$$P_n\oplus + M \xrightarrow{k_6} P_n + \overset{\oplus}{M} \qquad (15)$$

Allylic, with poison
$$P_n\oplus + X \xrightarrow{k_7} P_n + \overset{\oplus}{X} \qquad (16)$$

where M = monomer, C = catalyst, X = transfer agent and/or terminator, $P_n{}^{\oplus}$ = growing polymer of n monomer units, P_n = dead polymer, $\overset{\oplus}{M}$ = resonance stabilized allyl cation from monomer ("suicide" step), $\overset{\oplus}{X}$ = resonance stabilized allyl cation from terminating agent ("poisoning").

Initiation. The details of initiation in the isobutene–aluminum chloride system are still obscure, and after 30 years of intensive study we still do not know the exact nature of the actual catalytic species. Most disturbing is the fact that the "true" catalyst concentration is for some reason much less than what one would expect from the measured catalyst concentration. The number of kinetic chains produced from 1 mole of AlCl$_3$ is much less than 1.

It could be that initiation can occur only in the presence of cocatalyst, but no direct evidence supports such an assumption. The cocatalyst may be H$_2$O, HCl, or RX, and unless the experiment is carried out in high vacuum under extreme precautions, ubiquitous impurities may function as cocatalyst. Thus in our system C in Eq. 10 might be a complex, perhaps an ion pair formed from AlCl$_3$ and cocatalyst (i.e., H$_2$O, or HCl, or CH$_3$Cl) and could be visualized as any of the following species: AlCl$_3$·H$_2$O, AlCl$_3$·HCl, Al$_2$Cl$_6$·H$_2$O, H$^{\oplus}$AlCl$_3$OH$^{\ominus}$, H$^{\oplus}$AlCl$_4{}^{\ominus}$, H$^{\oplus}$Al$_2$Cl$_7$, CH$_3{}^{\oplus}$AlCl$_4{}^{\ominus}$, H$^{\oplus}$Al$_2$Cl$_6$OH$^{\ominus}$, etc.

For our treatment, however, it is safe and sufficient to assume that initiation (Eq. 10) is much faster than any other kinetic step. Thus the catalyst immediately reacts to form a "pool" of active initiating complexes whose concentration, which is initially equal to the "catalyst" concentration, decreases as the complexes are consumed by termination reactions. Thus the concentration of active complexes is changing with time and no steady state concentration is attained. If the original catalyst concentration is small, the termination process may use up the active complexes before the propagation reaction has time to consume all of the monomer, resulting in low conversions. A similar non-steady state polymerization scheme has been proposed by Burton and Pepper (50) to explain the polymerization of styrene by sulfuric acid.

Propagation. Propagation (Eq. 11) is visualized as a carbonium ion attack on the olefinic double bond with the simultaneous regeneration of propagating species $\sim C^{\oplus} + C{=}C \rightarrow \sim C\text{-}C\text{-}C^{\oplus}$. Whether free ions or associated carbonium ion–gegenion pairs or both are the true propagating species is not yet known.

Transfer. Transfer can occur either with the monomer or with transfer agents present in the system (Eq. 12 and 13). These processes do not interfere with the kinetic chain in the first approximation; however, they strongly affect the molecular weight of the product. The basic mechanism of chain transfer probably involves proton expulsion and reprotonation, i.e., $\sim C^{\oplus} \rightarrow \sim C^{=} + H^{\oplus}$ followed by $H^{\oplus} + C{=}C \rightarrow HC\text{-}C^{\oplus}$, where $C{=}C$ can be the monomer or transfer agent. The reaction is probably a concerted one (i.e., $\sim C^{\oplus} + C{=}C \rightarrow \sim C^{=} + C\text{-}C^{\oplus}$) and no protons, not even in the solvated state, have appreciable physical lifetimes in this system.

Figure 3 helps to visualize the situation; the control represents an uninhibited polymerization. As a rule, in cationic polymerizations, molecular weights are determined by transfer to the monomer and the kinetic chain by, probably, spontaneous termination. A pure chain transfer agent affects only the number of molecules leaving the polymer yield, i.e., the length of the kinetic chain, unchanged. A pure rate poison, however, reduces the yield of polymer and may also reduce the number of molecules by lowering the lifetime of the growing chain. By definition, the molecular weight should remain unaffected by a pure rate poison. Molecular weight depression by poisons can be visualized, however, if the last mole of polymer in the kinetic chain is terminated by the poison and not by chain transfer. This is shown in row 3 of Fig. 3. This "end effect" must be averaged in all molecules present in determining

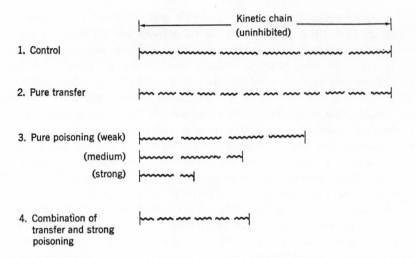

Fig. 3. Visualization of the effect of termination and chain transfer on number of moles of polymer.

its effect on molecular weight. Thus, the effect of a rate poison on molecular weight should increase as the yield decreases and consequently an extremely strong rate poison could drastically reduce the molecular weight without being a chain transfer agent.

Most poisons, as might be expected, cause both transfer and termination. In such a case, the number of moles may rise or fall depending on the relative magnitude of termination and transfer. In the particular case shown in row 4 of Fig. 3, these processes balance each other, causing the number of molecules to be the same as in the case of the control. In this case, the experimentally obtained molecular weights should be examined; if they are strongly depressed, this indication for simultaneous termination and transfer and the material is a strong poison as well as a transfer agent.

This kinetic model was used to develop poison coefficients (P.C.) and transfer coefficients (T.C.). The poison coefficient of a material shows quantitatively the extent of inhibition on the rate of polymerization (yield) caused by this material, whereas its transfer coefficient is a quantitative measure of its molecular weight depressing effect. The defining expressions for these coefficients are:

$$\frac{W_0}{W_p} = 1 + \frac{k_7}{k_5}X \quad \text{where P.C.} = \frac{k_7}{k_5}$$

and

$$\frac{MW_0}{MW_p} = 1 + \frac{k_4 + k_7}{k_3M + k_5} X \quad \text{where T.C.} = \frac{k_4 + k_7}{k_3M + k_5}$$

where W and MW are the weight of polymer formed and its molecular weight, respectively, and the subscripts 0 and p show the absence or presence of an agent which might affect the rate (terminator or "poison") and/or the molecular weights (transfer agent), respectively; M is the initial monomer concentration, X is the poison or transfer agent concentration, and the rest of the symbols are derived from the model. These equations are derived in Ref. 51.

Simple n-alkenes, e.g., propene, butene-1, pentene-1, hexene-1, were found to be "pure poisons," i.e., they only decreased the extent of polymerization but did not affect molecular weights. The effect of n-butenes on the polymerization has already been investigated (2). The mechanism responsible for termination could be hydride transfer coupled with allylic stabilization:

$$\sim CH_2 - C\overset{\oplus}{\underset{CH_3}{\overset{CH_3}{<}}} + CH_2 = CH - CH_2 - R \rightarrow$$

$$\sim CH_2 - C\overset{CH_3}{\underset{CH_3}{<}}H + \overset{\delta\oplus}{CH_2} \text{---} CH_2 \text{---} \overset{\delta\oplus}{CH} - R$$

Increasing poisoning activity with increasing molecular weights from propylene to 1-hexene might be due to the enhanced electron donating character of the larger alkyl groups which in turn will augment the stability of the allylic ions:

$$\overset{\delta\oplus}{CH_2} \text{===} CH \text{===} \overset{\delta\oplus}{CH_2} < \overset{\delta\oplus}{CH_2} \text{===} CH \text{===} \underset{CH_3}{\overset{\delta\oplus}{CH}} < \overset{\delta\oplus}{CH_2} \text{===} CH \text{===} \underset{C_2H_5}{\overset{\delta\oplus}{CH}} < \overset{\delta\oplus}{CH_2} \text{===} CH \text{===} \underset{C_3H_7}{\overset{\delta\oplus}{CH}}$$

2-Octene is the only n-olefin studied which exhibits a measurable molecular weight depressing effect. It is conceivable that in the presence of this olefin decreased molecular weights are due to chain transfer:

$$—CH_2—C^{\oplus}\underset{CH_3}{\overset{CH_3}{\big<}} + CH_3—CH=CH—CH_2—C_4H_9 \rightarrow$$

$$—CH_2—CH\underset{CH_3}{\overset{CH_3}{\big<}} + CH_3—CH=CH—\overset{\oplus}{CH}—C_4H_9$$

$$CH_3—CH=CH—\overset{\oplus}{CH}—C_4H_9 \rightarrow CH_3—CH=CH—CH=CH—C_3H_7 + H\oplus$$

$$H\oplus + M \rightarrow M\oplus \text{ etc.}$$

Interestingly, 3-methyl-1-butene did not affect yields or molecular weights. Perhaps the allylic H is shielded efficiently in this molecule by the neighboring vinyl and two methyl groups so that the attacking polyisobutene carbonium ion cannot reach it and no hydride transfer can occur. The structurally closely related vinyl cyclohexane has a measurable poisoning effect perhaps because the allylic H in this molecule is more available since the rotation of the two methylene groups shielding the allylic H is somewhat restricted. If the shielding around the allylic H is even partially removed, poisoning activity appears, for example, in 4-methyl-1-pentene.

3,3-Dimethyl-1-butene appears to be completely inert probably because no allylic hydrogen atoms for termination are available in this molecule.

α-Branched olefins are known to be strong molecular weight depressors, i.e., transfer agents (10). Transfer activity of α-branched olefins is probably due to the fact that these olefins may incorporate into the chain since they yield tertiary carbonium ions but these ions are sterically hindered, and are unable to propagate further. These crowded or "buried" ions probably stabilize themselves by deprotonation. The ejected proton then may reinitiate a chain which results in decreased product molecular weight:

$$—CH_2—C^{\oplus}\underset{CH_3}{\overset{CH_3}{\big<}} + CH_2=C\underset{R}{\overset{R}{\big<}} \rightarrow —CH_2—C\underset{CH_3}{\overset{CH_3}{\big|}}—CH_2—C^{\oplus}\underset{R}{\overset{R}{\big<}} \rightarrow$$

$$—CH_2—C\underset{CH_3}{\overset{CH_3}{\big|}}—CH=C\underset{R}{\overset{R}{\big<}} + H\oplus$$

The bulkier the substituents flanking the growing site, the more effective transfer activity becomes, i.e., 2,2,4-trimethyl-1-pentene is a more efficient transfer agent than 2-ethyl-1-hexene which in turn is more effective than 2-methyl-1-pentene.

Parenthetically it could be remarked that the poisoning activity of α-branched olefins can be readily explained by allylic hydride transfer and stabilization of the substituted allylic ion.

Conjugated dienes decrease both the yield and the molecular weight indicating that these materials have both appreciable poison and transfer coefficients. The poisoning effect of dienes has qualitatively been known since the discovery of butyl rubber. This rate retardation is probably connected to the incorporation of the diene in the chain and the formation of a comparatively stable terminally substituted allylic ion, for example, with isoprene:

This macrocation may further stabilize itself by proton ejection:

Such end conjugation has been found experimentally (49).

Incorporation of conjugated dienes into the polyisobutene chain of course is the important basic step in butyl rubber synthesis. This will be discussed later.

Among the halides allylic chlorides have pronounced transfer activities. Allyl chloride seems to be a mild but pure transfer agent (11). Tertiary butyl halides, however, are effective transfer agents but leave the polymerization rate practically unaffected. This effect of allyl and alkyl halides is well known (50).

II. BUTYL RUBBER

A. Introduction

Butyl rubber has established and assured its position among modern synthetic elastomers. It offers many excellent physical and chemical properties which make it of interest for many applications, e.g., inner

TABLE IV
Free World Rubber Consumption, 1955–1965 (000 of Long Tons)

Rubber	1955	1960	1965
SBR	860	1,450	2,060
PBR	0	10	230
PIR	0	0	50
EPT	0	0	15
Butyl	80	125	155
Neoprene	90	140	160
Nitrile	35	65	100
Other	0	10	30
Total Synthetic	1,065	1,800	2,800
Natural	1,730	1,625	1,800
Total New Rubber	2,795	3,425	4,600
% Synthetic	38%	52%	61%

tubes, mechanical molded goods, extruded articles, etc. Table IV shows recent consumption data for butyl and other synthetic rubbers (52). Continued healthy growth for butyl rubber during the years ahead can be expected.

Butyl rubbers offer the following outstanding properties:

Low price	Butyl is a low priced specialty elastomer available today in the rubber industry.
Low gas permeability	Butyl has the lowest permeability toward gases, i.e., highest retention of gas pressure among the commercially available elastomers. This renders it eminently suitable for inner tube or inner liner (for tubeless tires) applications.
Low resilience	Butyl gum has the lowest resilience among commercially available synthetic elastomers. This damping property can be put to use in vibration eliminators, motor and body mounts, truck or trailer springs, etc. Butyl tires ride "smooth" because of damping of mechanical shock.
Inertness	Butyl rubber, on account of its largely saturated nature, has excellent chemical resistance toward most acids, alkalies, and salt solutions. This property can be utilized in chemical tank linings, protective clothing, gloves, etc. Butyl offers excellent resistance to oxygenated organic compounds, e.g., alcohols, esters, ketones, and animal and vegetable oils.

Good weather (oxygen and ozone) resistance	Butyl, because of its low degree of unsaturation, possesses good resistance to the deteriorating effect of oxygen and ozone. Butyl is used in automotive rubber parts including window channels, gaskets, body shims, bellows, and hoses, and also in electrical insulation and proofed fabrics.
Good heat and age resistance	Butyl maintains its tear strength after extended heating, and does not harden (as natural rubber) on exposure to heat. Aging temperatures, of which vulcanizates of various polymers lose 50% of their initial tensile strength after 8 hr of heating are (in °F): 330 butyl, 323 silicone rubber, 277 natural rubber, 222 polyurethanes, 197 SBR, 170 nitrile rubber, 163 Hypalon, 142 neoprene.
Good tear resistance	Properly compounded butyl has excellent tear resistance. Moreover, butyl maintains high tear resistance after long exposure to air, atmospheric conditions, and high temperatures.
Good electrical properties	Butyl, as a largely saturated polymer free from electrolytes, has excellent electrical properties. Since it resists water, weathering, fungi, etc., it is an attractive material for electrical insulating.

B. The Manufacture of Butyl Rubber

Butyl rubber is produced by an extremely rapid cationic polymerization which is completed within a matter of less than a second in contrast with the many hours necessary for the copolymerization of styrene and butadiene by a radical mechanism employed in the production of SBR. Commercial grades of butyl are prepared by copolymerizing isobutene with 1.5–4.5% isoprene using $AlCl_3$ catalyst in methyl chloride solvent at about $-100°C$.

The simplified scheme of butyl synthesis is shown in Fig. 4.

The purity of raw materials used is of paramount importance in the manufacture of butyl rubber. For example, the n-butene concentration in isobutene must be below 0.5%, the methyl chloride solvent must be dried carefully, and the isoprene must also be of highest purity.

To control the violence of the reaction, to remove the heat of polymerization, and to obtain the rubber in the form of discrete small particles (slurry) rather than a solid mass, the reaction mixture is highly diluted with the methyl chloride solvent. Thus, the feed contains about 30 vol % mixed monomers and 70 vol % diluent. The feed and the catalyst solution, about 0.2 wt % $AlCl_3$ in methyl chloride, are introduced simultaneously at the bottom of a large vertical stainless steel reaction vessel. The reactor is cooled with liquid ethylene to about $-100°C$. Inside the reactor there are placed concentric cylindrical cooling units with circulating liquid ethylene providing additional cooling surfaces. The monomer feed and the catalyst solution are introduced continuously at the bottom in the middle of the inner cooling element close to the powerful stirrer. The monomer and catalyst solutions mix, a violent exothermic reaction occurs, and butyl rubber is formed almost instantaneously. The reaction mixture is continuously circulated with great speed upward in the center of the internal cooling element, then down between this cooling surface and the outer cooling jacket. The rate of production is controlled by diluting the catalyst solution or by diminishing the amount of injected catalyst solution. The flow sheet of a butyl plant is shown in Fig. 5.

The reaction mixture, consisting of rubber particles, unreacted monomer(s), and solvent, ascends in the reactor and passes through a short narrow tube ($\sim$3 inches) and then through a wider one into the top of a vertical cylindrical unit, the "flash tank." Here the reaction mixture is met by steam or hot water. Unreacted monomers and solvent immedi-

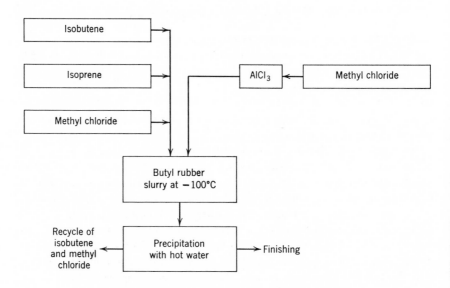

Fig. 4. Simplified scheme of butyl rubber synthesis.

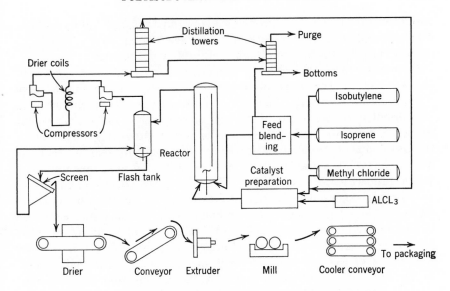

Fig. 5. Plant flowsheet for butyl production.

ately volatilize and the rubber precipitates into the water. The unreacted isobutene and methyl chloride solvent are recovered, purified, and
recycled. Small amounts of isoprene which remain after polymerization
are discarded. Zinc stearate and stabilizer are added to the slurry in
the flash tank. The rubber is then filtered, dried, compacted in an
extruder, sheeted on a mill, banded, and cut to suitable sizes for packaging.

C. The Structure and Important Characteristics of Butyl Rubber

Butyl rubbers are high molecular weight ($<$150,000 units by viscosity)
random copolymers of a major proportion of isobutene (up to 98.5 mole %)
and a minor amount of a diene (1–5%) obtained by cationic polymerization. The small amount of diene incorporated in the backbone of the
rubber is necessary to provide sites for vulcanization. As a matter of
fact, the basic invention of Thomas and Sparks (53,54) was the in-
coporation of conjugated dienes into the elastomeric polyisobutene chain.
This author has recorded a short history of the discovery of butyl rubber
(55).

From the point of view of rubber chemistry the discovery of butyl not
only meant the discovery of a new synthetic elastomer but, more importantly, the discovery of a new and, for practical purposes, extremely

important principle, i.e., the principle of low functionality.   Butyl rubber demonstrated that the presence of only a few double bonds (1–3%) in an elastomer is sufficient to obtain useful soft vulcanizates.

During the 1930's, before butyl rubber was discovered, it was generally believed that in order to render a material rubbery the presence of large amounts of C=C double bonds was necessary as in natural rubber or in its derivatives.   With the discovery of very high molecular weight rubbery polyisobutenes by R. M. Thomas and M. Otto in the old Standard Oil Development Company, scientists in this company realized that the number or even the presence of double bonds was not necessary to impart elastic properties to a structure.   Thus, Standard Oil investigators set out to render high molecular weight polyisobutene vulcanizable by introducing a limited number of double bonds into the macromolecule.   In the first crucial laboratory experiment in 1938, Thomas and Sparks used butadiene in conjunction with isobutene and polymerized this mixture with gaseous BF₃ in the presence of powdered dry ice as internal coolant. In an almost instantaneous strongly exothermic reaction they obtained a largely insoluble rubbery mass, the first butyl rubber.

Today the concept of low functionality for vulcanization is one of the most important concepts in rubber chemistry and technology.   The recently developed sulfur vulcanizable ethylene-propylene terpolymers are further important examples of the application of this principle.

Commercially available butyl rubbers are copolymers of isobutene and a conjugated diene, usually isoprene.   Some of the general properties of butyls are shown in Table V.   The viscosity average molecular weights of these products are in the range 350,000–450,000, the mole % isoprene in the chain is 0.6–2.5.   "Highly" unsaturated butyl rubber containing about 4 mole % unsaturation is discussed by Waddell et al. (56).

According to Rehner (57), the isoprene unit enters the chain in the *trans*-1,4 manner, and by chemical analysis there is little evidence for 1,2 and 3,4 isoprene entry.

$$
\begin{array}{ccc}
\overset{\displaystyle CH_3}{\underset{|}{}} & \overset{\displaystyle CH_3}{\underset{|}{}} & \\
-CH_2-C=CH-CH_2- & -CH_2-C- & -CH_2-CH- \\
 & \underset{|}{CH=CH_2} & \underset{|}{C=CH_2} \\
 & & \underset{|}{CH_3} \\
\textit{trans}\text{-}1,4 & \textit{trans}\text{-}1,2 & \textit{trans}\text{-}3,4
\end{array}
$$

Due to the highly saturated character of the butyl molecule, this synthetic elastomer is accordingly ozone resistant.   However, during vul-

TALBE V

General Properties of Enjay Butyl Polymers (23)

	Enjay Butyl 035		Enjay Butyl 150 and 165		Enjay Butyl 215		Enjay Butyl 217		Enjay Butyl 218[a] and 268		Enjay Butyl 325 and 365	
	Min.	Max.	Min.	Max.	Min.	Max.	Min.	Max.	Min.	Max.	Min.	Max.
Viscosity range												
ML 212°F at 8 min	41	49	41	49	41	49	61	70	71	—	41	49
Viscosity average mol wt[b]	350,000		350,000		350,000		400,000		450,000		350,000	
Mole % unsaturation[b]	0.6–1.0		1.0–1.4		1.5–2.0		1.5–2.0		1.5–2.0		2.1–2.5	
Tensile strength, psi												
40 min cure at 307°F	2600		2500		2400		2400		2400		2300	
Ultimate elongation, % minimum	700		650		550		550		550		500	
Modulus, psi at 400% elongation												
20 min cure at 307°F	400	590	590	770	770	950	770	950	770	950	950	1125
40 min cure at 307°F	650	900	900	1125	1125	1375	1125	1375	1125	1375	1375	1600

[a] Scorch recipe for Enjay Butyl 218

Polymer	100
Zinc oxide	2
Sulfur	2
Tetramethylthiuram disulfide	0.6

	min.	max.
Scorch time		
Minutes at 293°C	6	17

Base recipe for physical properties

Polymer	100
Channel black	50
Zinc oxide	5
Stearic acid	3
Sulfur	2
Benzothiazolyl disulfide	0.5
Tetramethylthiuram disulfide	1

[b] Not a specification; values shown are typical.

canization not all the double bonds are being consumed, and conventional butyls still slowly degrade under ozone attack. Recently, isobutene-cyclopentadiene copolymers have been described (58), which contain "protected" double bonds, and consequently are completely ozone resistant. The cyclopentadiene probably enters the copolymer in a 1,4 manner:

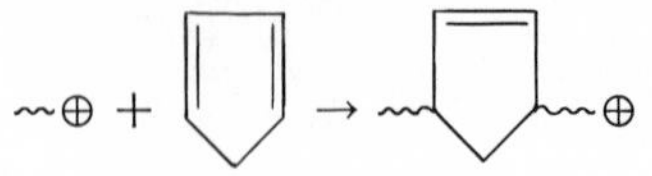

so that ozone attack, although cleaving the double bond, cannot cleave the chain as a whole.

This author showed that another completely ozone resistant butyl can be obtained by copolymerizing isobutene with small amounts of β-pinene (59). Isobutene-β-pinene copolymers and isobutene-2-methyl-1,5-hexadiene or 2-methyl-1,4-pentadiene have been patented (60,61). It is conceivable that the β-pinene unit in the chain has the following configuration:

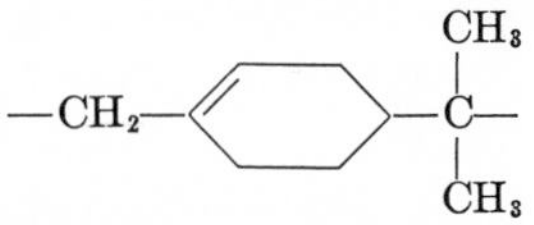

D. Aspects of the Chemistry of Isobutene-Diene Copolymerizations

While the homopolymerization of polyisobutene has been extensively investigated under a variety of conditions (see Section I), there is a dearth of reliable published information on isobutene copolymers. This is particularly true for isobutene-diene copolymers, the only isobutene copolymer systems which are of interest from the point of view of rubber chemistry at the present time.

The most important kinetic step in the synthesis of an isobutene-diene copolymer is the incorporation of the diene in the chain. This step is quantitatively expressed by copolymerization or reactivity ratios. Although the published reactivity ratios for isobutene–butadiene and isobutene–isoprene systems using $AlCl_3$ catalyst have been derived by making certain assumptions (62), a more exact reinvestigation of these data by this author showed that they were acceptable, i.e., r_1 (isobutene) $= 115 \pm 15$, r_2 (butadiene) $= 0.01 \pm 0.01$, and r_1 (isobutene) $= 2.5 \pm 0.5$, r_2 (isoprene) $= 0.4 \pm 0.1$. The reactivity ratios for copolymers prepared with $AlEtCl_2$ are almost identical to those obtained with $AlCl_3$ catalyst. In these experiments, mole % unsaturation (i.e., diene incor-

TABLE VI
The Effect of Temperature on Butadiene Incorporation
(AlEtCl$_2$ Catalyst)

Temperature			Mole % butadiene Incorporated[a]			
°C	°K	$1/T \times 10^3$	Total	*trans*-1,4	*trans*-1,2	$1/U - 1$
−30	243	4.12	15.6	12.5	3.1	5.4
−78	195	5.13	12.6	10.2	2.4	6.9
−100	173	5.78	9.2	7.0	2.2	9.8
−125	148	6.75	6.0	4.4	1.6	15.6

[a] Unsaturation by infrared method.

poration) was plotted as a function of conversion, and the linear lines obtained were extrapolated back to "zero conversion." This "instantaneous copolymer" is the copolymer which would form at infinitely low conversions, i.e., at initial monomer concentrations. The reactivity ratios obtained for AlEtCl$_2$ catalyst in methyl chloride solvent at −100°C were r_1 (isobutene) = 43.0, r_2 (butadiene) = 0, and r_1 (isobutene) = 2.17, r_2 (isoprene) = 0.5 (62a).

Interestingly, butadiene incorporation is strongly affected by polymerization temperature, whereas isoprene incorporation is not. Apparently butadiene incorporation decreases with decreasing temperatures. The data are shown in Table VI.

From these data one can calculate the activation energy for butadiene incorporation. According to the simple copolymerization theory, the rates of the four reactions in Scheme I govern copolymer composition (cf. Chapter 4.B).

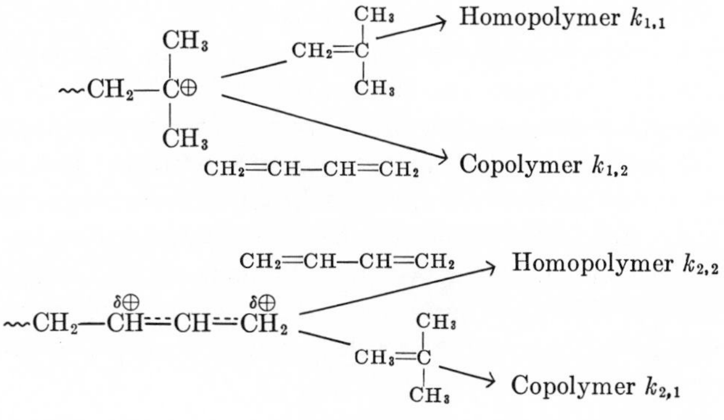

Scheme I

Since the reactivity ratio for butadiene in this system is close to zero ($r_2 \sim 0$), $k_{2,2}$ is also close to zero, so that we can ignore this rate constant which greatly simplifies our treatment. Thus, mole % butadiene incorporation U in the copolymer is given by:

$$U = \frac{k_{1,2}\,[\text{BD}]}{k_{1,2}\,[\text{BD}] + k_{1,1}\,[\text{iso-Bu}]} = \frac{k_{1,2}}{k_{1,2} + k_{1,1}\,[\text{iso-Bu}]/[\text{BD}]}$$

so that
$$U = 1/1 + r_1 F$$

where [BD] and [iso-Bu] are the concentrations of butadiene and isobutene in the charge, respectively, and F is the "feed ratio." Since r_1, the reactivity ratio for isobutene, is a ratio of rate constants $k_{1,1}/k_{1,2}$ one can write:

$$U = \frac{1}{1 + \dfrac{A_{1,1} \exp\{-E_{1,1}/RT\}}{A_{1,2} \exp\{-E_{1,2}/RT\}} \cdot F} \approx \frac{1}{1 + \exp\{(E_{1,2}-E_{1,1})/RT\}\,F}$$

and

$$\ln\left(\frac{1}{U} - 1\right) = \frac{E_{1,2} - E_{1,1}}{RT} + \ln F$$

The plot of $\ln(1/U - 1)$ versus $1/T$ gave a straight line with an intercept of 0.43 at $1/T = 0$. The number 0.43 is in excellent agreement with feed ratio used in these experiments, i.e., [BD]/[iso-Bu] = 30/70. The slope of this line gave 1.07 kcal/mole for $E_{1,2} - E_{1,1}$, indicating that the activation energy for butadiene incorporation is about 1.07 kcal higher than that for isobutene propagation.

It should be remarked that butadiene incorporation considered here is the total incorporation of this diene; since butadiene may enter the chain by 1,4 or 1,2 processes the rate constant for butadiene entry $k_{1,2}$ is a sum of two rate constants reflecting 1,4 and 1,2 incorporation.

In contrast to butadiene, isoprene incorporation seems to be unaffected by polymerization temperature in the temperature range from -30 to $-125°C$. This could mean that the overall activation energy for incorporating an isoprene molecule or an isobutene monomer in the chain is about equal. Since isobutene-isoprene copolymers have consistently lower molecular weights than isobutene homopolymers (44,63) obtained under essentially identical conditions, diminished copolymer molecular weights might be due to unfavorable entropy factors or perhaps some unknown kinetic influences. Research in this field is being continued and will be published (64).

E. New Developments

One of the most important molecular parameters determining the physical properties of butyl rubber is its molecular weight. The molecular weight in turn is largely determined by polymerization temperature. As has been shown recently (65), the most economical temperature to obtain highest molecular weight product with conventional Friedel-Crafts type catalysts is around $-100°C$. The linearly ascending log molecular weight versus $1/T$ curve shows a sharp break at this temperature level, and lowering the temperature below $\sim -100°C$ does not raise the molecular weight substantially (Fig. 6).

Since commercial butyl polymerizations are carried out at about $-100°C$, there is great incentive to raise the polymerization temperature and save the cost of refrigeration. However, by raising the temperature product molecular weights decrease, leading to unacceptable physical properties. The elementary step which is most likely responsible to lower the molecular weights at elevated temperatures is chain transfer to monomer. In general terms, chain transfer to monomer is the most important molecular weight determining event.

The rate of this step is a molecular constant, i.e., it is largely determined by the structure of the molecules involved. Thus, only by substantially changing the polymerization mechanism can this rate be changed in relation to the rate of molecular weight building event, the rate of propagation. Changing the nature of the catalyst would be a way to achieve this. Figure 6 shows the effect of three different Friedel-Crafts catalysts on the molecular weight of butyl rubber under substantially identical experimental conditions. At low temperatures (below $-100°C$) BF_3 seems to give highest molecular weight butyl rubber, whereas at higher temperatures (above $-100°C$) $AlEtCl_2$ and $AlCl_3$ seem to be better; however, the difference is not large. Consequently, these particular catalysts (gegenions) do not significantly alter the important molecular weight controlling events.

Recently a new catalyst system has been developed in our laboratories (42,43) which gives high molecular weight butyl rubber at comparatively high temperatures. The catalyst is a combination of a "modified" Friedel-Crafts halide (66), e.g., $AlEt_2Cl$, and a suitable cocatalyst. The

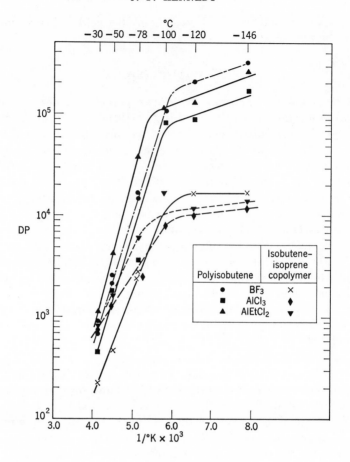

Fig. 6. The effect of temperature on the DP of polyisobutene and isobutene-isoprene copolymer.

catalyst alone does not polymerize isobutene, isobutene–diene mixtures, or other cationically initiable monomers. Only when a cocatalyst is added does polymerization commence. Suitable cocatalysts are proton sources, e.g., HCl, or carbonium ion sources, e.g., *tert*-butyl chloride. The molecular weights of a typical butyl rubber obtained with this catalyst at $-30°$ and $-60°C$ are $\sim90,000$ and $\sim260,000$, respectively, whereas those obtained with a conventional $AlCl_3$ under identical conditions are $\sim25,000$ and $\sim65,000$.

In a typical experiment using *tert*-butyl chloride as cocatalyst at $-50°C$, we obtained 580,000 mol wt and 1.81 mole % unsaturation (isoprene) in the rubber. This material gave the following physical properties (cured at 307°F with 50 p carbon black, 10 p oil, 5 p ZnO, 2 p sulfur, 1 p stearic acid, 1 p PBN, 1 p Methyl Tuad, 1 p Altax):

Cure time, min	Tensile, psi	300% Modulus, psi	Elongation, %
10	2030	240	890
30	2350	590	690
60	2410	830	610

References

1. A. Butlerov and V. Goryaniov, *Ber.*, **6**, 561 (1873).
2. R. M. Thomas, W. J. Sparks, P. K. Frolich, M. Otto, and M. Mueller-Cunradi, *J. Am. Chem. Soc.*, **62**, 276 (1940).
3. U.S. Pat. 2,356,127, JASCO (Dec. 29, 1937).
4. Standard Oil Development Co., Austral. Pat. 109,187 (Nov. 23, 1938).
5. P. J. Flory, in *Principles of Polymer Chemistry*, Cornell Univ. Press, Ithaca, N.Y., 1953.
6. J. P. Kennedy and R. M. Thomas, *J. Polymer Sci.*, **55**, 311 (1961).
7. J. P. Kennedy, I. Kirshenbaum, and R. M. Thomas, *J. Polymer Chem.*, A1, 331 (1963).
8. J. P. Kennedy and R. M. Thomas, *Advan. Chem. Ser.*, No. **34**, 111 (1962).
9. J. P. Kennedy and R. M. Thomas, *J. Polymer Sci.*, **45**, 229 (1960).
10. C. Horrex and F. T. Perkins, *Nature*, **163**, 486 (1949).
11. J. P. Kennedy and R. G. Squires, *J. Macromol. Sci. (Chem.)*, A1, 995 (1967).
12. A. G. Evans, M. Polanyi, D. Holden, H. A. Skinner, P. Plesch, and M. A. Weinberger, *Nature*, **157**, 102 (1946).
13. P. H. Plesch, M. Polanyi, and H. A. Skinner, *J. Chem. Soc.*, **1947**, 252.
14. P. H. Plesch, *Ric. Sci.*, **25a**, 140 (1955).
15. P. H. Plesch, Ed., *Cationic Polymerization and Related Complexes*, Hefner, Cambridge, 1953.
16. D. C. Pepper, *Quart. Rev.*, **8**, 88 (1954).
17. P. H. Plesch, Ed., *The Chemistry of Cationic Polymerization*, Macmillan, New York, 1963.
18. B. L. Erusalimskii, *Russian Chem. Rev. English Transl.*, **32**, 651 (1963).
19. D. C. Pepper, in *Friedel-Crafts and Related Reactions*, Chapter 30, G. A. Olah, Ed., Interscience, New York, 1964, p. 1293.
20. H. Güterbock, *Polyisobutylen und Isobutylen Mischpolymerisate*, Springer, Berlin, 1959.
21. J. P. Kennedy and A. W. Langer, *Fortschr. Hochpolymer-Forsch.*, **3**, 508 (1964).
22. R. M. Thomas and W. J. Sparks, in *Synthetic Rubber*, Chapter 24, G. S. Whitby, Ed., Wiley, New York, 1954.

23. D. J. Buckley, *Rubber Chem. Technol.*, **32**, 1475 (1959).

24. A. V. Topchiev, B. A. Krentsel, N. F. Bogolomova, and Y. Y. Goldfarb, *Dokl.*, *Akad. Nauk. SSSR*, **111**, 121 (1956).

25. R. Bacskai and S. J. Lapporte, *J. Polymer Sci.*, **A1**, 225 (1963).

26. F. S. Dainton and G. B. B. M. Sutherland, *J. Polymer Sci.*, **4**, 37 (1949).

27. M. St. C. Flett and P. H. Plesch, *J. Chem. Soc.*, **1952**, 3355.

28. Y. M. Slobodin and N. I. Matusevich, *Vysokomolekul. Soedin.*, **5**, 774 (1963); *Chem. Abstr.*, **59**, 4041b (1963).

29. W. P. Slichter and D. D. Davis, *Bull. Am. Phys. Soc.*, **9**, 221 (1964).

29a. R. M. Thomas, J. C. Zimmer, L. B. Turner, R. Rosen, and P. K. Frölich, *Ind. Eng. Chem.*, **32**, 299 (1940).

30. *Vistanex, Polyisobutylene, Properties and Applications*, Bulletin V-01, Enjay Chemical Company, New York (1960).

31. M. Otto and M. Müller-Cunradi, Ger. Pat. 641,284 (July 26, 1931) (to I. G. Farbenindustrie).

32. M. Otto and M. Müller-Cunradi, U.S. Pat. 2,203,873 (June 1, 1937).

33. M. Chmelir, M. Marek, and O. Wichterle, *Intern. Symp. Macromol. Chem.*, *Prague, 1965, Preprint* **P110**.

34. J. H. Beard, P. H. Plesch, and P. P. Rutherford, *J. Chem. Soc.*, **1964**, 2566.

35. R. H. Biddulph, P. H. Plesch, and P. P. Rutherford, *J. Chem. Soc.*, **1965**, 275.

36. H. Cheradame and P. Sigwalt, *Compt. Rend.*, **259**, 4273 (1964).

37. J. P. Kennedy and F. P. Baldwin, *Polymer*, **6**, 237 (1965).

38. J. P. Kennedy and R. M. Thomas, *J. Polymer Sci.*, **46**, 233 (1960).

39. T. Saegusa, H. Imai, and J. Furukawa, *Makromol. Chem.*, **79**, 207 (1964), and references herein.

40. H. Sinn, H. Winter and W. Tirpitz, *Makromol. Chem.*, **48**, 59 (1961).

41. H. Sinn and Tirpitz, *Makromol. Chem.*, **85**, 280 (1965).

42. J. P. Kennedy, Belg. Pat. 663,319 (April 30, 1965).

43. J. P. Kennedy and F. P. Baldwin, Belg. Pat. 663,320 (April 30, 1965).

44. J. P. Kennedy and R. G. Squires, *Polymer*, **6**, 579 (1965), and previous papers in this series.

45. J. P. Kennedy and R. M. Thomas, *J. Polymer Sci.*, **45**, 229 (1960).

46. J. P. Kennedy and R. M. Thomas, *J. Polymer Sci.*, **46**, 481 (1960).

47. J. P. Kennedy, I. Kirshenbaum, D. C. Murray, and R. M. Thomas, *J. Polymer Sci.*, **A1**, 331 (1963).

48. J. P. Kennedy and R. M. Thomas, *J. Polymer Sci.*, **55**, 311 (1961).

49. Z. Zlamal and A. Kazda, *J. Polymer Sci.*, **A1**, 3199 (1963), and references herein.

50. R. E. Burton and D. C. Pepper, *Proc. Roy. Soc. (London)*, **A263**, 58 (1961).

51. J. P. Kennedy and R. G. Squires, *J. Makromol. Sci. (Chem.)*, **A1**, 805 (1967). This is the first of seven papers and other publications in this series.

52. R. E. Workman, *Rubber Age*, **97**, 49, Aug. 1965.

53. R. M. Thomas and W. J. Sparks, U.S. Pat. 2,356,128 (Aug. 22, 1944).

54. R. M. Thomas and W. J. Sparks, U.S. Pat. 2,356,130 (Aug. 22, 1944).

55. J. P. Kennedy, in *Copolymerization*, Chapter 5, G. Ham, Ed., Interscience, New York, 1964.

56. H. H. Wadell, J. V. Fusco, and L. T. Eby, *Rubber World*, **146**, 57 Aug. 1962.

57. J. Rehner, *Ind. Eng. Chem.*, **36**, 46 (1944).

58. S. Minckler, Jr. and A. B. Small, *Ind. Eng. Chem. Prod. Res. Develop.* **1**, 216 (1962).

59. J. P. Kennedy, unpublished results, 1963.
60. E. Ott, U.S. Pat. 2,373,706 (April 17, 1945) (to Hercules Powder Co.).
61. W. J. Sparks and R. M. Thomas, U.S. Pat. 2,384,975 (Sep. 6, 1941) (to Jasco, Inc.).
62. F. M. Lewis, C. Walling, W. Cummings, E. R. Briggs, and F. R. Mayo, *J. Am. Chem. Soc.*, **70**, 1519 (1948).
62a. J. P. Kennedy and N. H. Carter, *J. Polymer Sci.*, *A1*, **5**, 2455 (1967).
63. V. I. Anosov and A. A. Korotkov, *Vysokomolekul. Soedin.*, **2**, 354 (1960).
64. J. P. Kennedy and N. H. Canter, *J. Polymer Sci.*, *A1*, **5**, 2712 (1967) and *Trans. N.Y. Acad. Sci.*, in press.
65. J. P. Kennedy and R. M. Thomas, *Advan. Chem. Ser.*, **34**, 111 (1962).
66. G. Natta, G. Dall'Asta, G. Mazzanti, V. Giannini, and S. Cesca, *Angew. Chem.*, **71**, 205 (1959).

CHAPTER 5

ELASTOMERS BY CATIONIC MECHANISMS

B. Poly(vinyl Ethers)*

JOGINDER LAL

The Goodyear Tire & Rubber Company, Akron, Ohio

Contents

* Contribution No. 350 from the Research Division, The Goodyear Tire & Rubber Company, Akron, Ohio.

I. INTRODUCTION

Vinyl alkyl ethers, in view of their ease of synthesis from acetylene and an alcohol via Reppe (1) chemistry, offer an interesting and useful class of monomers. By a judicious choice of the alkyl groups in the monomers, they are candidates for preparing a wide variety of polymers. Quite recently, Imperial Chemical Industries (2) have developed a less expensive process for making these monomers using ethylene instead of the traditional acetylene. Details of the process have not been revealed as yet. The more attractive prices of the monomers will, undoubtedly, stimulate further interest in their polymers and copolymers.

Vinyl alkyl ethers, particularly the lower alkyl members of the series, are extremely reactive. They are readily susceptible to attack by electrophilic reagents. This is due to resonance between the vinyl group and the ether oxygen and may be formally represented as follows:

$$\overset{\ominus}{C}H_2-CH=\overset{\oplus}{O}R \leftrightarrow CH_2=CH\ddot{O}R$$

$$(1) \qquad\qquad (2)$$

The alkoxy group can also withdraw electrons from the double bond by the inductive effect but the resonance effect predominates. The double bond is, therefore, electron-rich and favors reaction with electron acceptors and acidic reagents (Friedel-Crafts halides, Lewis acids (3)) but resists attack by anionic catalysts and free radicals. In vinyl aryl ethers, electrons are fed into the ring by the resonance effect resulting in increased susceptibility to attack by free radicals.

II. FRIEDEL-CRAFTS HALIDES, BORON TRIFLUORIDE AND ITS COMPLEXES, AND OTHER WELL-KNOWN POLYMERIZATION CATALYSTS

A. Historical and Early Developments

The first polyvinyl ether was reported by Wislicenus (4). The addition of iodine to vinyl ethyl ether resulted in a violent reaction and a "balsamous" material. Reppe and Schlichting (5) disclosed that a variety of vinyl ethers undergo polymerization by contacting them with small amounts of anhydrous inorganic acidic agents such as $AlCl_3$, $SnCl_4$, $SnCl_2$, $ZnCl_2$, BF_3, complexes of boron halides with ethers, ketones, or alcohols, $SiCl_4$, $HgCl_2$, $KHSO_4$, H_2SO_4, H_3PO_4, and acidic bleaching earths. The products obtained were viscous liquids or solids depending on the chemical structure of the monomer and on the degree of polymerization. Shostakovskii and coworkers (6,7) employed metal halides, iodine, and BF_3 etherate as catalysts for polymerizing several vinyl ethers at room temperature and below $-15°C$ and also obtained low molecular weight materials. Chalmers (8) investigated the polymerization of a number of vinyl ethers and reported differences in their polymerizability. Mueller-Cunardi and Pieroh (9) were successful in obtaining solid, high molecular weight, rubbery poly(vinyl isobutyl ether) by polymerizing purified vinyl isobutyl ether at -40 to $-80°C$ with BF_3 catalyst. These extremely rapid polymerizations were carried out in solvents to which solid carbon dioxide had been added or in bulk with the addition of solid carbon dioxide alone. Otto et al. (10) have described a process for continuously polymerizing vinyl isobutyl ether on an endless steel belt. Boron trifluoride dissolved in ethylene or propane was used as a catalyst to produce semisolid to solid products. The product was marketed in Germany as Oppanol C. Bunn and Howells (11) reported that Oppanol C had an identity period of 6.60 Å along the fiber axis. This demonstrates that this product had sufficient structural regularity for crystallization on stretching. Oppanol E, a solid high polymer of vinyl isopropyl ether, was also reported (12). Polymers from long chain alkyl vinyl ethers, for instance, poly(vinyl octadecyl ether), are waxlike solids even when the polymer viscosity is low (13).

B. Stereoisomerism

Schildknecht et al. (14–17) demonstrated conclusively that by controlling the conditions of polymerization two types of polymers could be obtained from vinyl isobutyl ether. When liquid propane ($\sim -70°C$) containing boron trifluoride catalyst (0.01% or more) was rapidly added to cooled vinyl isobutyl ether diluted with liquid propane, nearly instantaneous or "flash polymerization" took place. The polymer obtained was tacky, rubberlike and amorphous. On the other hand, when boron trifluoride etherate was used at -80 to $-60°C$, polymerization proceeded slowly and the polymer grew around the catalyst as a separate phase. This type of polymerization was termed (17) "polyphase" or "proliferous polymerization." The polymer was a nontacky solid which gave a crystalline x-ray diffraction pattern in the unstretched condition and a sharp fiber diagram on stretching. It showed distinct differences from Oppanol C in solubility behavior and stability to milling (14). According to Muthana and Mark (18) polymer–solvent interactions were different for the crystalline and rubbery polymers of vinyl isobutyl ether. In toluene, the polymer molecules of the former type favor greater segment–segment interaction as compared with segment–solvent interaction, presumably due to a more regular structure than in the latter polymer. Schildknecht (19) has reviewed the polymerization of vinyl ethers and properties of the polymers.

The weight average molecular weights of vinyl isobutyl ether polymers obtained by the flash and polyphase polymerizations are of the order of 200,000. The difference in the properties of the two types of polymers were attributed by Schildknecht et al. (15) to differences in the stereoisomeric arrangement of the alkoxy groups along the polymer carbon chain. The crystalline polymer was assumed to have a *dldl* alternation of side groups, and the rubberlike polymer to contain a random succession of *d* and *l* groups. Natta and coworkers (20) have assigned it an isotactic (21,22) structure, i.e., the side groups have the same *ddddd* or *lllll* steric arrangement. The identity period along the fiber axis was found to be 6.50 ± 0.05 Å. The polymer chains have a 3/1 helical conformation. This value of identity period is in close agreement with the value of 6.60 Å reported by Bunn (11) but is higher than the value of 6.20 Å reported by Schildknecht and coworkers (15). In terms of Natta's (23) definition, the structure originally proposed by Schildknecht and coworkers would be called syndiotactic. Schildknecht and Dunn (24) have favored a block copolymer structure of *d* and *l* isotactic sequences as well as amorphous segments, rather than whole polymer molecules having asymmetric carbon atoms of identical configuration.

The addition of boron trifluoride etherate failed to catalyze the polymerization of vinyl methyl ether containing liquified propane and solid carbon dioxide. The presence of an "activator" such as chloroform in the above system induced a slow, smooth polymerization to give a nontacky, crystalline polymer (17,25).

The polymerization of vinyl isopropyl ether differs from that of vinyl isobutyl ether (26). Under similar conditions of proliferous polymerization with boron trifluoride etherate (liquid propane, −78°C), vinyl isopropyl ether polymerizes about three times as fast as vinyl isobutyl ether and gives a polymer of significantly higher viscosity. The polymers of vinyl isopropyl ether prepared with boron trifluoride etherate and other complexes of boron trifluoride are noncrystalline. On stretching, x-ray fiber patterns are obtained from several polymers indicating the presence of stereoregular segments in the polymer chains.

High molecular weight poly(vinyl n-butyl ether), prepared by proliferous polymerization in propane at −78°C, is tacky, rubberlike, and normally noncrystalline (15,26). At high elongations, it crystallizes to give x-ray fiber patterns. Mild conditions obtained by a proper choice of reaction temperature and catalyst concentration were found necessary by Zoss (27) for the formation of form-stable, rubberlike polymers of vinyl isopropyl ether. Temperature control is the determining parameter in the preparation of form-stable, rubberlike poly(vinyl n-butyl ether) (28). Form stable, microcrystalline polymers of vinyl alkyl ethers containing lower alkyl groups, which yield x-ray fiber diagrams in stretched condition have been claimed by Zoss (29). Schildknecht (30) has reported the preparation of "uniform" high molecular weight poly(vinyl alkyl ethers) by carrying out the polymerization in the presence of a suitable diluent such as liquified propane. The high molecular weight polymer is insoluble in the diluent which is a selective solvent for the concurrently formed low molecular weight fraction. In all the patents mentioned above, boron trifluoride etherate was the preferred catalyst.

C. The Side Chain Effect

Approximate threshhold temperatures above which the vinyl alkyl ethers, in the absence of special activators, will polymerize readily with boron trifluoride etherate in ether solution as catalyst are given in Table I (17,29). Nakano (31) reported that in the polymerization of vinyl methyl ether the threshold concentration of boron trifluoride etherate increases on (1) decreasing polymerization temperature, (2) increasing the monomer concentration, and (3) by the use of ethers as diluents.

Qualitatively, polymerization tendencies of these five monomers containing different alkyl groups increases in the following order: iso-

TABLE I
Threshhold Temperatures for Polymerization of Vinyl Alkyl Ethers (17,29)

Alkyl group in vinyl alkyl ether	Approx. threshhold temp., °C
Methyl	−30 to −25
Ethyl	−53 to −50
n-Butyl	−80 to −70
Isobutyl	< −80
Isopropyl	< −100

propyl > isobutyl > n-butyl > ethyl > methyl. Similar results have been reported recently by Fishbein and Crowe (32). The effect of the alkyl group in vinyl alkyl ether on polymerization rates with boron trifluoride etherate and iodine catalysts has been discussed in terms of (1) electron donating capacity of the alkyl group to the olefinic double bond, thereby favoring the formation of π complex with the Lewis acid catalyst, and (2) steric blocking of the ether oxygen or olefinic double bond (17,32–34).

D. Homogeneous Polymerization Systems

Okamura and co-workers (35–43) have extensively studied the polymerization of vinyl alkyl ethers by boron trifluoride etherate at −74 to −78°C. They have examined the characteristics of polymers prepared in "homogeneous" and "heterogeneous" systems, i.e., in reaction media in which the polymer is soluble and insoluble, respectively (36). Polymerization of vinyl isobutyl ether in n-hexane-chloroform (both solvents for the polymer) mixture at −78°C was heterogeneous if boron trifluoride etherate was added to the monomer–solvent mixture, irrespective of the composition of the solvent. However, polymerization proceeded homogeneously if the monomer was added dropwise to the catalyst–solvent combination, even when the solvent was pure n-hexane. Apparently, heterogeneous polymerization takes place because polymer formation around the catalyst is faster than the dissolution rate (36). Polymers obtained in homogeneous systems had lower molecular weights than those obtained in heterogeneous systems. Isotactic crystalline polymers were obtained in both types of polymerizations of vinyl methyl ether and vinyl isobutyl ether.

Polymers of vinyl ethyl, isopropyl, and n-butyl ethers showed lower crystallinity (37). To what extent these differences are due to packing of side groups and/or stereoregularity of the polymer molecules is not clear.

Okamura and co-workers (38–43) have also investigated the influence of polymerization conditions on the properties of poly(vinyl alkyl ethers). In general, the molecular weights of polymers decreased with increasing temperature, concentration of polar solvents, and concentration of chain transfer agents such as methanol and acetic acid. Higashimura and coworkers (42,43) have emphasized that the presence of a counterion in the vicinity of the growing polymer chain end is essential for obtaining isotactic polymer. In a solvent of high dielectric constant, wherein the counterion may become sufficiently separated from the growing chain end, the steric repulsion between substituents on the growing chain end and that of the monomer may dominate stereospecific control and lead to the formation of a syndiotactic polymer as in low temperature free radical (44) polymerization. Higashimura et al. (45) have further proposed that due to the presence of a counterion near the charged carbon atom of the last monomer unit in the growing polymer chain, the electron configuration of this carbon may not be adequately represented by sp^2 (trigonal) hybrids but by sp^3 (tetrahedral) hybrids, particularly in solvents where solvation of the ion pair is not pronounced, i.e., where dissociation of the ion pair is not appreciable. Cram and Kopecky (46) and Bawn and Ledwith (34) have discussed the mechanism of stereospecific polymerization of vinyl isobutyl ether with boron trifluoride etherate to isotactic polymer.

Goodman and Fan (47) have synthesized syndiotactic poly(α-methyl vinyl methyl ether) by carrying out polymerizations at $-78°C$ under homogeneous conditions with various cationic catalysts. Their mechanism for the polymerization is similar to that of Cram and Kopecky.

E. Kinetics of Polymerization

Eley and co-workers investigated in detail the kinetics of polymerization of several vinyl alkyl ethers with different catalysts under a variety of conditions (48). The ionic nature of reaction in vinyl ether polymerization was clearly demonstrated (49) by the accelerating effect (50) of solvents of high dielectric constant. Eley and Richards (49) proposed a basic scheme of elementary reactions in the polymerization of vinyl n–butyl ether with stannic chloride. This scheme has been found applicable in later studies on vinyl ether polymerizations (See Scheme I). C, M, and P_n are catalyst, monomer, and dead polymer, while M^+ and M_n^+ represent initial and propagating cations and Y^- is the gegenion (counterion) which is assumed to accompany the cation in solvents of low dielectric constant (51). The catalyst may already be an ion pair or capable of producing an ion pair by reacting with monomer; k_i, k_p, k_t and k_m are rate constants for the various reaction steps.

Initiation:

$$C + M \xrightarrow{k_i} M^+Y^-$$

Chain propagation:

$$M^+Y^- + M \xrightarrow{k_p} M_2^+Y^-$$

$$M^+_{n-1}Y^- + M \xrightarrow{k_p} M_n^+Y^-$$

Chain termination:

$$M_n^+Y^- \xrightarrow{k_t} P_n$$

Transfer to monomer:

$$M_n^+Y^- + M \xrightarrow{k_m} P_n + M^+$$

Scheme I

The transfer to monomer may involve a proton transfer from the growing chain:

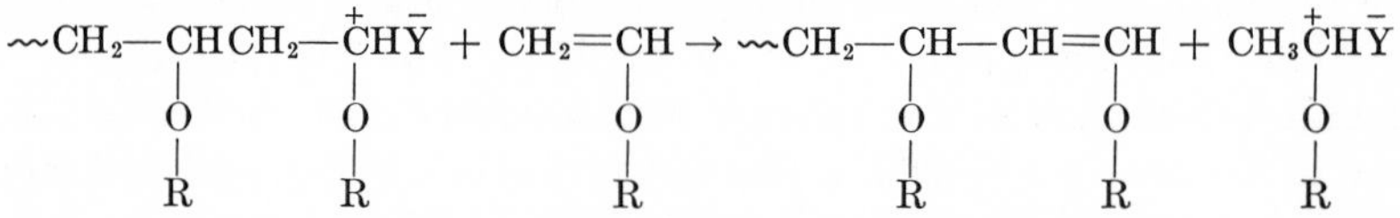

$$\sim CH_2-CHCH_2-\overset{+}{C}H\overset{-}{Y} + CH_2{=}CH \rightarrow \sim CH_2-CH-CH{=}CH + CH_3\overset{+}{C}H\overset{-}{Y}$$

with O—R substituents below each chain unit.

This reaction, since its early proposal by Eley and Richards (52), has been accepted as one of the most important reactions that control the molecular weight of polymers in cationic polymerizations. Solvents and other added agents may also participate in transfer reactions. Except for the fact that the poly(vinyl alkyl ethers) possessed unsaturation, no definite information on the structure of endgroups was available. Recently, Imanishi and co-workers (53,54) have concluded that monomer transfer involved ether cleavage of monomer by the growing ion pair to give an acetal endgroup (**3**). During spontaneous unimolecular termination the formation of a tetrahydropyran endgroup (**4**) free from unsaturation was similarly postulated.

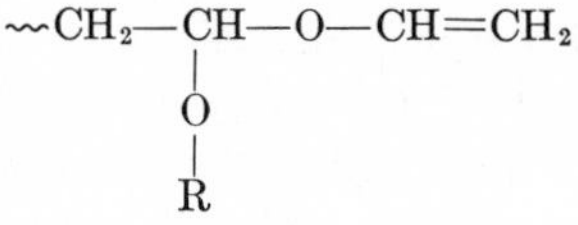

$$\sim CH_2-CH-O-CH{=}CH_2$$

with O—R substituent below

(**3**)

tetrahydropyran structure:

$$\sim CH_2CH \begin{smallmatrix} & CH_2 & \\ \diagup & & \diagdown \\ & & CHOR \\ O & & CH_2 \\ \diagdown & & \diagup \\ & CH & \\ & OR & \end{smallmatrix}$$

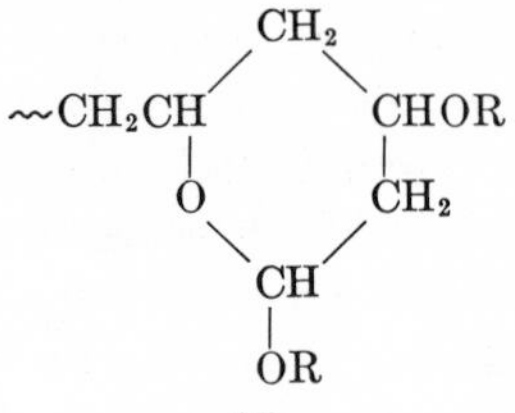

(**4**)

Coombes and Eley (55) have investigated the polymerization of vinyl n-butyl ether and vinyl 2-chloroethyl ether in sealed dilatometers at 25°C with boron trifluoride etherate in diethyl ether solvent. The initial rates of polymerization for both monomers followed the kinetics

$$-d[\mathrm{M}]/dt = k[\mathrm{M}]^2[\mathrm{C}]$$

The reaction of the catalyst with the monomer was presumed to give an ion pair:

$$\mathrm{BF_3 \cdot OEt_2} + \mathrm{CH_2{=}CH{-}OR} \xrightarrow{k_i} [\mathrm{Et\,CH_2\overset{+}{C}H{-}OR}][\mathrm{BF_3OEt}]^-$$

The chain termination process most likely involves transfer of a proton to gegenion. Kennedy (56) has shown that in the polymerization of vinyl isobutyl ether with boron trifluoride–diethyl-1-C^{14}-ether complex at $-75°C$ in liquid propane only 0.71% of polymer chains contain labeled ethyl groups. Although the mechanism of initiation by $[\mathrm{BF_3OC_2H_5}]^-$ $[\mathrm{C_2H_5}]^+$ suggested for isobutylene polymerization could be operative, initiation by chain transfer is dominant in the case of vinyl isobutyl ether. Eley (48) has stated that in this case traces of water might be assisting in initiation and transfer process.

Eley and Seabrooke (57) have extended the polymerization of vinyl n-butyl ether to four additional boron trifluoride etherates using the same ether as in the etherate as polymerization solvent. The ethers were: di-n-butyl ether, diisopropyl ether, anisole, and tetrahydrofuran (THF). The overall rate of polymerization decreased with the etherates in the order:

anisole > diisopropyl ether > diethyl ether (55) > di-n-butyl ether > THF

Eley and Johnson (58) have demonstrated that under extremely dry conditions the system vinyl n-butyl ether–boron trifluoride–hexane shows no signs of polymerization at -78 and 25°C and conclude that this monomer cannot be its own cocatalyst.

The kinetics of low temperature (-78 and $-40°C$) polymerization of vinyl isobutyl ether with boron trifluoride etherate in 50:50 n-hexane-toluene solvent have been further investigated by Blake and Eley (59,60). The catalyst was completely dissolved in the solvent mixture before the monomer was introduced. From the reaction kinetics obtained at $-78°C$, it appears that the normal steady state approximation applies. The DP increased on increasing the initial monomer concentration but decreased very slightly with increase in the catalyst concentration. The degree of isotacticity of the polymer decreased with increasing polymerization temperature and initial monomer and catalyst concentration.

Polymerization at −40°C gave atactic polymer. The kinetics of polymerization, the limited yields of polymer, and the decrease in the molecular weight on increasing the catalyst concentration were shown to fit a non-stationary mechanism, involving fast initiation and slow propagation. The situation is analogous to that reported by Burton and Pepper (61) for the polymerization of styrene with sulfuric acid.

F. Other Metal Halides as Catalysts

Grosser (62) has reported the preparation of rubbery polymers from vinyl isopropyl ether by employing a mild catalyst such as gallium trichloride (0–50°C, homogeneous systems in chloroform). Iwasaki and coworkers (63) have determined the reactivity of $MgCl_2$, $MgBr_2\cdot3THF$, $MgBr_2\cdot2$ dioxane, $MgI_2\cdot2Et_2O$, MgF_2, and $MgF_2\cdot HF$ for polymerizing vinyl methyl, ethyl, and isopropyl ethers. The polymerizations were carried out in bulk at 0–80°C. It was necessary to disperse the catalysts uniformly due to their poor solubility in reaction media. The molecular weight of poly(vinyl isobutyl ether) prepared with different catalysts increased in the order:

$$MgBr_2\cdot2\text{ dioxane} > MgF_2\cdot HF > MgCl_2 > MgBr_2\cdot3THF > MgI_2\cdot2Et_2O$$

III. METAL OXIDES AS CATALYSTS

Iwasaki (64) reported the stereospecific polymerization of vinyl isobutyl ether with preheated CrO_3 at 80°C in toluene solvent (heterogeneous system). The polymer possessed $[\eta] = 2.1$ dl/g and gave 33% MEK-insoluble fraction (isotactic). In CrO_3, the chain structure of tetrahedra linked together at corners were believed to be related to the stereoregular polymerization (65). Furukawa (66) has proposed a multicentered coordinated polymerization with water as a cocatalyst.

Amorphous polymers of vinyl isobutyl ether were also obtained with MgO, Al_2O_3, NiO, MoO_3, and V_2O_5 (64,67). Iwasaki and co-workers (67) have attributed the activity of magnesium oxide, normally considered basic, to active protons on magnesium oxide.

IV. ZIEGLER-NATTA TYPE CATALYSTS

A. Cationic Aspects of Coordination Polymerization; Modified Friedel-Crafts Catalysts

Lal (68) reported that the slow polymerization of vinyl isobutyl ether at −78°C with a titanium tetrachloride–trialkylaluminum catalyst system gave high molecular weight, isotactic, crystalline polymer which resembled the boron trifluoride etherate catalyzed polymer in its physical properties. It was also shown that these Ziegler catalysts polymerized vinyl allyl ether in the same manner as reported by Butler (69) for other similar compounds using the etherate catalyst. These data indicate that the mechanism of polymerization of vinyl ethers by the Ziegler catalyst is cationic in nature. This appears to have been the first report in the literature which showed that Ziegler catalysts could also polymerize a suitable monomer by cationic mechanism. Further evidence on the cationic mechanism of polymerization by the Ziegler catalyst was obtained by Lal and coworkers (70) by the polymerization of β-vinyloxy-ethyl methacrylate. By infrared analysis and glass transformation temperature measurements it was shown conclusively that the polymer obtained with the Ziegler catalyst (Al/Ti = 1.2:1) at −20°C possessed the same structure as that obtained with stannic chloride (71), a typical cationic catalyst. Amorphous polymer formed when the temperature during polymerization of vinyl isobutyl ether with the titanium tetrachloride–triisobutylaluminum catalyst was allowed to rise above −20°C (72). Highest inherent viscosity was obtained at the Al/Ti ratio of 1.5 in this series of experiments (73).

Bogdanova (74) has also reported the polymerization of vinyl isopropyl, n-butyl, isobutyl, cyclohexyl, and β-decalyl ethers with titanium tetrachloride–triethylaluminum catalyst at about −78°C. Roch and Saunders (75) have obtained crystalline poly(vinyl isobutyl ether) using titanium tetrachloride–dibenzene chromium catalyst.

Natta and coworkers (76) confirmed the cationic mechanism of some Ziegler catalysts during vinyl alkyl ether polymerization. Stereospecific polymerization of vinyl isobutyl ether at −78°C was obtained with the insoluble catalyst derived from triethylaluminum and titanium tetrachloride at Al/Ti molar ratio of about 1. Whereas triethylaluminum was ineffective in polymerizing vinyl isobutyl ether, crystalline polymers were obtained with diethylaluminum choride and ethylaluminum dichloride. These data suggest that in the trialkylaluminum–titanium tetrachloride system for stereospecific polymerization of vinyl alkyl ethers the alkyl-aluminum halides *in situ* are the true catalytically active species. Natta and co-workers have coined the term "modified Friedel-Crafts catalysts" to designate halogen compounds of polyvalent metals on the highest valence level, in which some of the halogen atoms are replaced by organic groups such as acetyl, alkoxy, or alkyl. Dall'Asta and co-workers (77, 78) have used "modified Friedel-Crafts" catalysts containing aluminum or titanium for the stereospecific polymerization of vinyl isopropyl, neopentyl and n-butyl ethers at $-78°C$. Using soluble bis(cyclopentadienyl) titanium dichloride/triamylaluminum catalyst in the polymerization of vinyl isobutyl ether in toluene solvent the polymer was crystalline if the temperature of polymerization was below $-30°C$, whereas the polymer prepared above this temperature was amorphous (76). During polymerization with soluble catalysts, the stereospecific control was attributed to the formation of a doubly anchored complex between the catalyst and the vinyl ether monomer in which both the vinyl double bond and the ether oxygen participate. Natta et al. (76) have also employed well-defined, soluble bimetallic catalyst complexes having the structural formula $(C_5H_5)_2TiCl_2AlX_2$, where C_5H_5 is cyclopentadienyl and X is a chloride or ethyl group:

$$(C_5H_5)_2Ti \underset{Cl}{\overset{Cl}{\diagdown}} Al \underset{X}{\overset{X}{\diagup}}$$

It was elegantly demonstrated that the cationic activity of the complexes for stereospecific polymerization of vinyl isobutyl ether at $-78°C$ diminishes by substituting the chloride groups bound to aluminum by ethyl groups. The catalytic activity of these complexes for polymerizing ethylene followed the reverse pattern, i.e., increased. Thus, variation in the composition of complexes in this series altered the polymerizing ability from cationic-coordinated to anionic-coordinated. Neither alkyl groups nor cyclopentadienyl groups initially present in the catalyst complex were detected in the poly(vinyl isobutyl ether), indicating a cationic mechanism (79) of polymerization. However, terminal cyclopentadienyl groups were found in polyethylene when both X groups in the catalyst were chlorides and ethyl groups were detected in this polymer when at least one of the X groups was a chloride.

Natta and coworkers (80–83) have extended the stereospecific polymerization to β-chlorovinyl alkyl ethers and 1-alkenyl alkyl ethers which contain internal double bonds. Crystalline *threo* and *erythro* diisotactic polymers were obtained from several monomers. Independently, Heck and Breslow (84) have reported the polymerization of alkenyl alkyl ethers to give crystalline materials. In those cases which were investigated, the polymers from the *cis* and *trans* isomers had essentially the same melting points.

B. Vandenberg's Modified Ziegler-Type Catalysts

Vandenberg et al. (85) briefly reported on the highly crystalline polymers of vinyl ethers, obtained at room temperature with Vandenberg's

(86) modified Ziegler-type catalysts or with other highly stereospecific, specially prepared metal sulfates (87–89). The vinyl ethers containing ethyl and higher n-alkyl groups gave essentially amorphous, reasonably high molecular weight rubbery materials (89). Vinyl isobutyl ether was polymerized predominantly to a tacky rubber of low crystallinity. This is in contrast to the studies with boron trifluoride etherate by Schildknecht et al. who found vinyl isobutyl ether to give the most stereospecific polymerization (15–17). Allyl ethyl ether and allyl tert-butyl ether practically failed to polymerize, suggesting that the vinyl ethers polymerize by a special cationic mechanism rather than the coordinated anionic mechanism of 1-olefin polymerization. Vandenberg (89) has proposed a cationic insertion mechanism for the polymerization of vinyl ethers with his vanadium based stereospecific catalyst.

C. Russian Investigations

The stereospecific polymerization of vinyl alkyl ethers at room temperature (20–30°C) with modified Ziegler catalysts has been reported by Russian investigators (90–92). The catalysts used were prepared from: (1) $TiCl_4$, LiC_3H_7, and Al (iso-C_4H_9)$_3$ in the molar ratio 1:1:4 (heterogeneous system) (90), (2) Al(iso-C_4H_9)$_2$Cl and Al(iso-C_4H_9)$_3$ in the preferred molar ratio 1:3 (homogeneous catalyst) (91), and (3) $VOCl_3$, LiC_3H_7, and Al(iso-C_4H_9)$_3$ in the molar ratio 1:1.5:3 (heterogeneous system) (92). The polymers of vinyl isobutyl, n-butyl, cyclohexyl, and β-decalyl ethers prepared by the above catalysts had reasonably high molecular weights. The heterogeneous catalyst based on vanadium oxychloride was particularly suitable for polymerizing vinyl n-butyl ether. The homogeneous catalyst was found (91) to be most effective in copolymerizing vinyl alkyl ethers with each other.

V. OTHER CATALYSTS BASED ON ORGANOMETALLIC COMPOUNDS

Kray (93) has reported that vinyl isobutyl ether polymerizes at 80°C with in situ n-butylmagnesium bromide in cyclohexane at a slow, controlled rate to yield a high molecular weight polymer melting at 104°C. Vinyl methyl, ethyl, and n-butyl ethers were also polymerized with Grignard reagent. Kray speculated that the mechanism of polymerization may be cationic and may involve the formation of a complex between Grignard reagent and vinyl alkyl ether. Martin (94) has polymerized vinyl ethyl, n-butyl, 2-ethylhexyl, and 2-butyloctyl ethers with Grignard reagents at 0–64°C and obtained products which varied from tacky, viscous liquids to polymers of extremely high molecular weights. Grignard reagent freed from ether was an extremely active catalyst.

Russian investigators (95) have also polymerized vinyl ethyl, n-propyl, isopropyl, n-butyl, and isobutyl ethers at 20°C in the presence of butyl, ethyl, or phenyl magnesium bromides. Except for poly(vinyl ethyl ether), these polymers contained high proportions of crystalline fractions.

Bruce and Farren (96) have repeated Kray's work and found that vinyl ethers fail to polymerize with Grignard reagent if atmospheric contaminants are excluded. Iwasaki and co-workers (67) independently found that Grignard reagents and diethylmagnesium were catalytically inactive for vinyl ethyl ether polymerization. Oxygen, acetaldehyde, benzaldehyde, acetone, and metallic oxides (V_2O_5, NiO, HgO) were effective cocatalysts for polymerizing vinyl isobutyl ether at 0–50°C with Grignard reagents. Poly(vinyl isobutyl ether) prepared at 80°C in carbon tetrachloride with the aid of ethylmagnesium bromide had a rather high molecular weight. It was amorphous, like the high molecular weight polymers of vinyl isopropyl and ethyl ethers prepared in this fashion. The mechanism of polymerization probably involved cationic initiation.

Triethylaluminum (Et_3Al) does not polymerize vinyl isobutyl ether in hydrocarbon solvents (97). Saegusa et al. (98) have reported that α-chlorodimethyl ether activates the polymerization of vinyl isobutyl ether with Et_3Al or Et_2Zn in methylene chloride at −78°C. They proposed the following cationic initiation mechanism:

$$Et_3Al + CH_3OCH_2Cl \rightleftharpoons [CH_3OCH_2]^+[Et_3AlCl]^-$$

Ternary systems based on an alkylmetal, H_2O, and a cocatalyst have also been reported by the same authors (99). The alkylmetals were triethylaluminum and diethylzinc and active chlorine compounds such as α-chlorodimethyl ether, acetyl chloride, and $tert$-butyl chloride as well as acetic anhydride were cocatalysts. The principal function of water was considered to be that of a modifier, i.e., to convert alkylmetal into another Lewis acid.

$$Et_3Al + H_2O \rightleftharpoons Et-\underset{\underset{Et}{|}}{(Al-O-)_n}AlEt_2 \quad (n = 1, 2, \ldots)$$

Probably, the real active species is an unstable complex (99) derived from the modified Lewis acid and the cocatalyst

$$Et_2AlOAlEt_2 + CH_3OCH_2Cl \rightleftharpoons [CH_3OCH_2]^+[Et_2AlOAlEt_2Cl]^-$$

(In the polymerization of styrene with $Et_3Al-H_2O-CH_3COCl$, the mechanism of polymerization initiated by acetyl cation was established by end-group analysis.)

Kern et al. (100) have reported that aluminum fluoride based catalysts are superior to boron trifluoride in forming stereoregular poly(vinyl alkyl ethers). These polymers contained highly stereoregular, crystalline fractions. Recently, Kern and Calfee (101) published details of the catalyst systems previously used for polymerizing vinyl methyl ether. A cationic mechanism of propagation was visualized, the anionic counterions being anchored in the catalyst gel. Precoordination of vinyl ether on sites in the aluminum fluoride gel led to stereospecific control during polymerization. Alkylaluminum fluorides and alkylaluminum chlorofluorides have been used by Kern (102) for polymerizing vinyl methyl and isobutyl ethers at about -40 to $-14°C$ to crystalline materials. Calfee (103) has reported the stereospecific polymerization of vinyl methyl, isobutyl, and 2-ethylhexyl ethers with aluminum fluoride and aluminum chlorofluorides at -10 to $25°C$. These catalysts had average particle radii of the order of 200 Å and were prepared by the action of fluorinating agents upon aluminum chloride under anhydrous conditions.

Takeda and coworkers (104) have polymerized vinyl isobutyl ether at Dry Ice temperature with the binary catalyst $(C_2H_5)_3Al-SnCl_4$ which was reported earlier by Nakano et al. (65). The polymer obtained at Al/Sn ratio of 1 had the highest stereoregularity. From various physicochemical measurements in nonpolar solvents it was concluded that the reaction between $(C_2H_5)_3Al$ and $SnCl_4$ in equimolar amounts produces $AlCl_3$ and $(C_2H_5)_3SnCl$ which form a complex. The complex was further believed to exist as multiple ions such as

$$[\overset{-}{Al}Cl_4 \cdots \overset{+}{Sn}(C_2H_5)_3]_m \overset{-}{Al}Cl_4 \quad \text{and} \quad [(C_2H_5)_3\overset{+}{Sn} \cdots \overset{-}{Al}Cl_4]_n(C_2H_5)_3\overset{+}{Sn}$$

in favor of the hypothesis (66) of the multicenter coordinated ionic catalyst for stereospecific polymerization.

VI. METAL SULFATE–SULFURIC ACID COMPLEX AS CATALYST

The polymerization of vinyl alkyl ethers with aluminum hexahydrosulfate heptahydrate catalyst at $10°C$ or above to give high molecular weight polymers was reported by Mosley (105). These polymers were not characterized for crystallinity, directly or indirectly. There is no indication in the patent that the author visualized any stereospecific control during polymerization with this catalyst. The stereospecific polymerization of vinyl ethyl ether and of vinyl n-butyl ether with a metal sulfate–sulfuric acid complex catalyst was first reported by Lal (106). It was shown that the reaction product of ferric sulfate hydrate with sulfuric acid not only polymerizes these vinyl alkyl ethers at about room temperature to high molecular weight materials, but that these materials also crystallize on stretching and give fiber diagrams on x-ray diffraction. It was also reported (107) that aluminum sulfate–sulfuric acid complex catalyst (105) similarly gave crystallizable polymers from vinyl alkyl ethers. It was further shown that poly(vinyl n-butyl ether) was more crystalline if polymerization was carried out in carbon disulfide instead of pentane. The catalytically active aluminum complex gave an essentially amorphous x-ray diffraction pattern (108). On standing in air, several crystalline peaks appeared. Finally, a crystalline pattern typical of catalytically inactive aluminum sulfate octadecahydrate was obtained.

The metal sulfate–sulfuric acid complex catalysts offer a combination of advantages over many other catalysts for preparing poly(vinyl alkyl ethers):

1. It is not necessary to use monomers and solvents of high purity.
2. These catalysts have low sensitivity toward oxygen and moisture.
3. They do not require low temperatures in order to obtain high molecular weight polymers.
4. They offer the possibility of obtaining highly crystalline polymers.
5. They can be used in very small amounts to obtain reasonable polymerization rates.
6. A good process control and uniformity of product are achieved (109).
7. It is possible to obtain vulcanizable copolymers of vinyl alkyl ethers by the use of such catalysts. (See Section IX on Vulcanizable Copolymers.)
8. They are economical.

A. Mechanism of Polymerization

These catalysts are insoluble in the polymerization medium. Steric considerations associated with the heterogeneous nature of the catalyst presumably play a key role in the various stages involved in the stereospecific polymerization. A coordinated cationic mechanism (Fig. 1) has been proposed by Lal and McGrath (110). Polymerization is initiated by a proton bound to a large heterogeneous counterion derived from the metal sulfate complex. The metal center coordinates with the ether

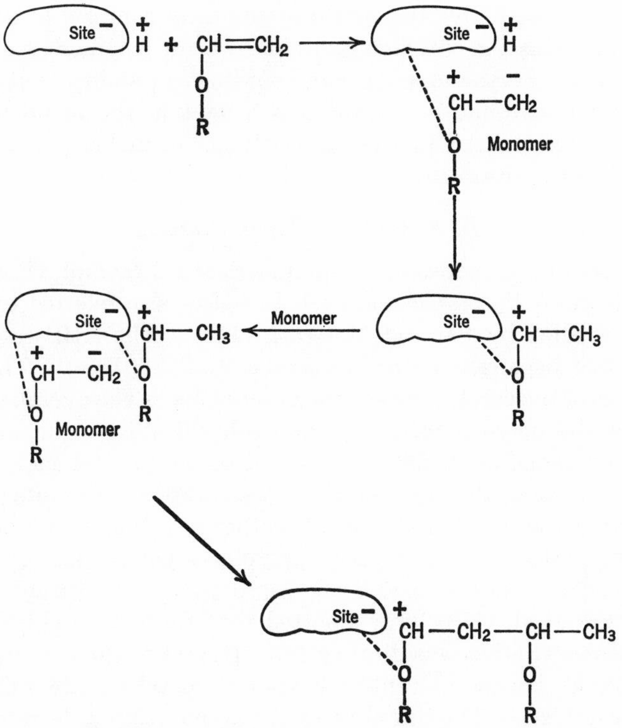

Fig. 1. Lal and McGrath's (110) mechanism of the cationic coordinated polymerization of vinyl alkyl ether with a metal sulfate–sulfuric acid complex catalyst.

oxygen of the vinyl ether carbonium ion and thereby exercises further steric control for the incoming monomer. The ether oxygen of the incoming monomer is also presumed to coordinate with the metal on the catalyst surface. The ability of these heterogeneous catalysts to yield high molecular weight polymers at somewhat elevated temperatures suggests that transfer to monomer and termination reactions are suppressed by the heterogeneous complex counteranion. Furukawa (66) has suggested that such a complex catalyst should possess at least two coordination centers. Okamura and coworkers (42,111–113) have also investigated the polymerization of vinyl alkyl ethers with a variety of metal sulfate-sulfuric acid complex catalysts. As a general trend, an increase in bonding between metal sulfate and sulfuric acid increases the stereoregularity of the polymer and its molecular weight. A strong coordinating ability of the catalyst stabilizes the bisulfate anion by complex formation and leads to the formation of a high molecular weight polymer even at higher temperatures. Furthermore, a high coordinating ability of the catalyst facilitates the coordination of the substituent in the monomer on the catalyst surface thereby producing a polymer of higher stereoregularity even at higher temperatures.

B. Kinetics of Polymerization

The effect of temperature on polymerization rate of vinyl n-butyl ether with the Al^{3+} complex catalyst, hereafter also referred to as AHS, has been studied by Lal and McGrath (110). The AHS catalyst was used as a fine dispersion in dried mineral oil (0.2–2% w/v). The average particle size of the catalyst was of the order of 5 μ. The overall activation energy for the polymerization of vinyl n-butyl ether in heptane in the temperature range of 30–58°C was found to be 9.7 kcal/mole. During the polymerization, the polymer did not separate as a discrete phase.

The polymerization of vinyl isobutyl ether in a dilatometer or a stirred flask using AHS suspension gave polyphase polymerization, i.e., the polymer could be seen as a discrete phase around the catalyst particle. Higashimura et al. (113) have reported that the rate of polymerization of vinyl isobutyl ether was directly proportional to the catalyst concentration (30°C, hexane solvent) for various metal sulfate–sulfuric acid complex catalysts. The initial rates of polymerization decreased in the following order:

$$Al^{3+} > Cr^{3+} > Fe^{3+} > Mg^{2+} > Fe^{2+} > Co^{2+}$$

Data on the particle size of the catalyst in these studies are not available. Qualitatively, the catalyst derived from aluminum sulfate was more active (72) than the catalyst prepared from ferric sulfate for polymerizing vinyl n-butyl ether as well as vinyl isobutyl ether.

C. The Side Chain Effect

Vinyl ethers having a linear alkyl group polymerize considerably faster than monomers having a branched alkyl group with the same number of carbon atoms (110) (Table II). Under identical conditions the apparent first order rate constant for the polymerization of vinyl n-butyl ether at 36°C is about 25 times the corresponding value for vinyl isobutyl ether. This lower rate is partly due to the formation of threadlike agglomerates of the catalyst during the polymerization of vinyl isobutyl ether. The bulkiness of the isobutyl group of this monomer might also

TABLE II

Apparent First Order Rate Constants for the Polymerization of Vinyl
Alkyl Ethers with AHS Catalyst (110)[a]

Alkyl group in monomer	Polym. temp., °C	Apparent first order rate constant, $min^{-1} \times 10^{3b}$
Ethyl	0.5	4.0
n-Butyl	0.5	0.55
n-Butyl	36	42
Isobutyl	36	1.7
n-Hexyl	36	0.54
n-Octyl	36	0.54

[a] Catalyst suspension in mineral oil. The average particle size of the catalyst was about 5 μ. During polymerization of vinyl n-butyl ether, the catalyst particles remained unchanged in size (microscopic examination).

[b] Heptane solvent; [M] = 0.71 mole/liter; [catalyst] = 12.1 mg/liter.

hinder the approach of the monomer to the reactive centers of the heterogeneous catalyst, thereby reducing the rate of polymerization. The influence of the length of the linear alkyl group on the retardation rate of polymerization of vinyl ethers is indicated by the following sequence:

$$ethyl > n\text{-butyl} > n\text{-hexyl} = n\text{-octyl}$$

D. Molecular Weight of Polymers

Using complexes of various metal sulfates and sulfuric acid as catalysts, Okamura and co-workers (111–113) have reported that the effect of increasing vinyl isobutyl ether and catalyst concentration on the intrinsic viscosities was generally not pronounced at 30°C in n-hexane. The molecular weight of the polymer was influenced by the nature of the metal ion in the complex and decreased in the following order (113):

$$Al^{3+} > Cr^{3+} \approx Fe^{3+} \approx Mg^{2+} > Fe^{2+} > Co^{2+}$$

An increase in the concentration of vinyl n-butyl ether from 5 to 20% slightly reduced the yield and intrinsic viscosity with the aluminum complex catalyst (114).

A noteworthy feature of polymerization of vinyl isobutyl ether with these heterogeneous catalysts in hydrocarbon solvents is that on increasing the polymerization temperature from −10°C to +50°C the intrinsic viscosities remain essentially unchanged, except in the case of the aluminum complex, which produces the highest molecular weight polymer at

20–25°C (111–113). In the polymerization of vinyl n-butyl ether with the aluminum complex in the 0–40°C range, the intrinsic viscosities of the polymers also showed a slight increase; the polymer having the highest intrinsic viscosity was obtained at 30°C (114). However, with the aluminum complex in carbon disulfide solvent, the inherent viscosities of the vinyl n-butyl and isobutyl ether polymers generally decreased when the temperature was increased from -20 to 50°C (110,115). Vinyl n-butyl ether yielded polymers of higher inherent viscosity (3–9.3 dl/g) than those obtained from vinyl isobutyl ether (0.5–4.7 dl/g) under similar conditions (115).

E. Stereoregularity of Polymers

The influence of various factors on the stereospecific polymerization of vinyl alkyl ethers has been investigated by Okamura and co-workers (42,112,113). In the polymerization of vinyl isobutyl ether at 30°C in hexane solvent, the degree of stereoregularity decreased on increasing monomer concentration. This was attributed to the increase in dielectric constant of the polymerization medium due to increasing concentration of the polar monomer and is similar to the behavior observed (38,39) in the case of "homogeneous" cationic polymerization.

The nature of the metal ion in the complex affected the stereospecific polymerization of vinyl isobutyl ether; the stereospecific control decreased in the following order (112):

$$Al^{3+} \approx Cr^{3+} > Fe^{3+} > Fe^{2+} > Mg^{2+}$$

Okamura and coworkers concluded that the stereoregularity of poly-(vinyl isobutyl ether) obtained with various metal sulfate complexes in n-hexane generally increases slightly when the temperature of polymerization is increased from -40 to 50°C. However, Lal and coworkers (115) have shown that increasing the polymerization temperature of vinyl isobutyl ether from -20 to 0°C decreased the crystallinity of the MEK-insoluble fraction and that, in the 0–50°C range, temperature has virtually no effect on crystallinity. These polymers were prepared with the Al^{3+} complex in n-heptane and carbon disulfide solvents. Data on poly(vinyl n-butyl ether) prepared in carbon disulfide also demonstrated that as the temperature was increased from -30 to 25°C, the stereoregularity and crystallinity of the polymers showed no significant difference. A lower degree of crystallizability was suggested for the polymer prepared at 50°C. The coordination of the ether oxygen atoms of the vinyl ether monomer and of the growing end of the polymer molecule to the metal centers in the heterogeneous counteranion is expected to decrease with increasing temperature of polymerization and thereby lead to the formation of less stereoregular polymer.

F. Metal Alkoxide (or Alkyl)–Sulfuric Acid Catalyst and Related Systems

Christman and Vandenberg (87) have polymerized several vinyl ethers at 0–30°C to high molecular weight polymers by employing as catalyst the reaction product of sulfuric acid with a metal alkoxide or alkylmetal. The metal in these compounds could belong to groups IIIA, IVA, IVB–VIB and VIII of the periodic table. The catalyst was activated by the addition of a metal alkoxide or alkylmetal to the monomer–diluent mixture prior to the addition of the catalyst. The molar ratio of metal alkoxide or alkylmetal to sulfuric acid was greater than one in all the examples cited in the patent. Polymerizations were carried out with these heterogeneous catalysts in diethyl ether, methylene chloride, ethyl acetate or chlorobenzene diluent. Thus, addition of the reaction product of aluminum isopropoxide and sulfuric acid (molar ratio 4.6:1) to a vinyl isopropyl ether–ethyl acetate solution activated with triisobutylaluminum–tetrahydrofuran complex resulted in the polymerization of vinyl isopropyl ether. The crude polymer was extracted with hexane to isolate the hexane insoluble, high molecular weight crystalline fraction. The mechanism of polymerization is probably similar to the one proposed for the stereospecific polymerization of vinyl ethers with Ziegler-type catalysts (89). Crystalline polymers could not be obtained from vinyl-2-ethylhexyl and isopropenyl ethyl ethers. The catalyst system of Christman and Vandenberg has been used by Lorenzi and co-workers for polymerizing optically active vinyl alkyl ethers to optically active polymers. (See Section VII.)

Vinyl methyl ether was also polymerized at 0–25°C by Vandenberg (116) to high molecular weight crystalline materials. The catalyst system of Christman and Vandenberg was modified by substituting sulfuric acid with hydrogen fluoride or boron trifluoride etherate. Nakano et al. (65) have reported that the reaction products of triisobutylaluminum with hydrogen chloride or phosphoric acid produced amorphous polymers from vinyl isobutyl ether.

Methyl, isopropyl, isobutyl, *tert*-butyl, and trifluoroethyl vinyl ethers have also been polymerized to high molecular weight polymers by Heck and Vandenberg (88). They used as catalyst various metal sulfates in combination with a metal alkoxide or a trialkylmetal or dialklmetal hydride or an alkylmetal alkoxide. Polymerizations were carried out at 0–30°C in methylene chloride, diethyl ether, or heptane diluent. The preferred ratio of the metal sulfate to the metal alkyl or alkoxide or alkyl alkoxide was about 1:0.1 to 1:5. In most preparations some amorphous material is produced along with the crystalline polymer. An analogous catalyst system based on anhydrous aluminum selenate and aluminum alkyl, alkoxide, or alkyl alkoxide was used by Chiang (117) to polymerize vinyl methyl ether to a mixture of amorphous and crystalline polymers

Nakano et al. (65) have qualitatively compared various types of catalysts for the polymerization of vinyl isobutyl ether at room temperature and at low temperature to produce stereoregular polymers. Vinyl methyl ether was also polymerized in a few cases. Complexes of metal halides, chromyl chloride, chromic oxide, reaction products of organometallic compounds and Lewis acids, trialkylaluminum–sulfuric acid systems, and metal sulfate–sulfuric acid complexes were found to catalyze the polymerization of vinyl alkyl ethers at room temperature to stereoregular polymers. Nakano et al. postulated that tetrahedral compounds which have one active edge, preferably of shorter length, are especially useful for stereospecific polymerization at room temperature. The proposed mechanism of stereospecific polymerization involves interaction of the active edge of the counteranion and sp^2-type structure of the terminal carbon atom of the growing polymer chain, coupled with restricted entry of the incoming monomer in a particular mode.

VII. OPTICALLY ACTIVE POLYMERS

Pino et al. (118) have reported the polymerization of vinyl(S)-2-methylbutyl ether in toluene with aluminum bromide at −78°C and with a mixture of diisobutylaluminum chloride and isobutylaluminum dichloride at 0 and −78°C. All the polymers exhibited optical activity, but only the polymers prepared at −78°C had some crystallinity. The optical activity of these polymers was low, comparable to that of the monomer.

Lorenzi et al. (119) have polymerized vinyl(S)-1-methylpropyl ether, vinyl(S)-2-methylbutyl ether, and the corresponding racemic monomers. Triisobutyl aluminum–sulfuric acid* and aluminum isopropoxide–sulfuric acid† catalyst systems (87) were employed, using diethyl ether and ethyl

---

* Molar ratio of triisobutylaluminum to sulfuric acid was 2.1:1. Polymerizations were activated by additional triisobutylaluminum.

† Molar ratio of aluminum isopropoxide to sulfuric acid was 4.5:1. Polymerizations were activated by addition of triisobutylaluminum–tetrahydrofuran complex.

acetate solvents, respectively.   With the exception of racemic vinyl(R) (S)-2-methylbutyl ether, these monomers were also polymerized with isobutylaluminum dichloride in toluene or a mixture of toluene and propylene.   Among the three catalyst systems investigated, the aluminum isoperoxide–sulfuric acid system was the most stereospecific.   Polymers of a relatively high degree of stereoregularity, were not obtained with the soluble catalyst isobutylaluminum dichloride even at −78°C.   Since the solvents were not the same in these experiments it is difficult to assess their influence.

VIII. GLASS TRANSFORMATION TEMPERATURES

The influence of pendant groups on the glass transformation temperature (T_g) of polymers has been reported in the literature for several series of polymers.   In these studies, the chemical nature, length, and size of the side groups were varied without altering the polymer backbone. T_g values for a few poly(vinyl alkyl ethers) have been determined by Schmieder and Wolf (120) using a torsion pendulum method.   Some low temperature properties on this series have also been reported by Fishbein and Crowe (32).   Lal and Trick (121) have measured dilatometrically T_g values of several high molecular weight poly(vinyl alkyl ethers). As shown in Table III, the glass transformation temperatures of poly-(vinyl n-alkyl ethers) decrease as the length of the n-alkyl side chain increases from methyl to n-octyl.   The lowest T_g value (-80°C) in this series was obtained from poly(vinyl n-octyl ether).   On lengthening the side chain, 14% of the specific volume increase represents an increase in free volume, i.e., within experimental error the same value as was found for poly(n-alkyl methacrylates) (122).

The data also show the influence of substitution and branching in the alkyl group on the T_g values of poly(vinyl alkyl ethers).   Poly(vinyl isopropyl ether), wherein substitution occurs on the carbon atom adjacent to the ether linkage, has a T_g about 40°C higher than the linear poly(vinyl ethyl ether).   Similarly, poly(vinyl sec-butyl ether) has a T_g 29°C higher than the estimated T_g value for poly(vinyl n-propyl ether).   Poly(vinyl 2-ethylhexyl ether), in which the substitution is two carbon atoms away from the ether group, has a T_g 11°C higher than poly(vinyl n-hexyl ether).

The influence of branching in the alkyl group on the T_g values can also be seen by comparing the T_g value of -55°C for poly(vinyl n-butyl ether) with -20°C for poly(vinyl sec-butyl ether) and -19°C for poly(vinyl isobutyl ether).   Similarly, the T_g value of -66°C for poly(vinyl 2-ethylhexyl ether) is 14°C higher than the T_g value of -80°C for

TABLE III

Effect of Substitution and Branching in the Alkyl Group on the T_g
Values of Poly(vinyl Alkyl Ethers) (121)

Alkyl group in poly(vinyl alkyl ether)		
• Name	Structure	T_g, °C
Methyl	$-CH_3$	-31
Ethyl	$-CH_2CH_3$	-42
Isopropyl	$-CH-CH_3$ $\quad\vert$ $\quad CH_3$	-3
n-Propyl	$-CH_2CH_2CH_3$	-49
sec-Butyl	$-CHCH_2CH_3$ $\quad\vert$ $\quad CH_3$	-20
n-Hexyl	$-CH_2CH_2(CH_2)_3CH_3$	-77
2-Ethylhexyl	$-CH_2CH(CH_2)_3CH_3$ $\qquad\vert$ $\qquad CH_2CH_3$	-66
n-Pentyl	$-CH_2CH_2(CH_2)_2CH_3$	-66
n-Butyl	$-CH_2CH_2CH_2CH_3$	-55
Isobutyl	$-CH_2CHCH_3$ $\qquad\vert$ $\qquad CH_3$	-19
n-Octyl	$-CH_2CH_2(CH_2)_5CH_3$	-80

polyl(vinyl n-octyl ether). The T_g value of poly(vinyl 2-ethylhexyl ether) is identical to the T_g value of poly(vinyl n-pentyl ether).

The influence of the T_g on dynamic mechanical properties of vulcanizates of poly(vinyl alkyl ethers) is discussed in Section XI. It will be seen that the T_g dominates the dynamic resilience behavior of a homologous series of elastomers having the same network structure.

IX. VULCANIZATION

A. Vulcanizable Copolymers

The homopolymers of vinyl alkyl ethers prepared by cationic polymerization contain insufficient unsaturation, believed to be mostly terminal, to yield suitable vulcanizates on curing. Therefore, vinyl alkyl ethers have been copolymerized with suitable monomers which make available chemically reactive groups for forming a three-dimensional network. Several comonomers have been used for this purpose.

1. Copolymerization with Dienes

By analogy with butyl rubber, a copolymer of vinyl alkyl ether with a diene should be vulcanizable with an accelerated sulfur curing recipe. Young and Sparks (123) copolymerized vinyl alkyl ethers with isoprene, 2,3-dimethyl-1,3-butadiene, and 2-methyl-1,3-pentadiene at low temperature using Friedel-Crafts catalysts. For instance, a mixture of equal parts by weight of vinyl isobutyl ether and diene was polymerized with a 3.9% solution of boron trifluoride in ethyl chloride in the presence of solid carbon dioxide. The resulting copolymer had a reasonably high molecular weight and was compounded as follows, all ingredients expressed as weight parts per 100 parts of the copolymer: zinc oxide, 5; stearic acid, 1; carbon black, 9; tetramethylthiuram monosulfide, 0.4; and sulfur, 1.5. A sample cured at 287°F (~142°C) for 30 min had a tensile strength of 2600 psi at 585% elongation at break and 300% modulus (stress at 300% elongation) of 1400 psi. Schildknecht (30) polymerized a mixture of vinyl isopropyl ether with butadiene at −80°C using boron trifluoride etherate catalyst. Although a "copolymer" was claimed, no data on unsaturation or vulcanizability of the product were given.

Nagasawa and co-workers (124) copolymerized vinyl isopropyl ether and cyclopentadiene (80:20 mixture) at −65°C with boron trifluoride etherate catalyst in chloroform. The copolymer was compounded with carbon black and an accelerated sulfur curing recipe and vulcanized. A cured sample exhibited tensile strength of 1180 psi and 460% elongation at break.

Butadiene or its alkyl derivatives are less reactive than vinyl alkyl ethers in cationic copolymerization systems. In contrast, 1-alkoxy-1,3-butadienes are very reactive toward cationic catalysts. Lal (125) has copolymerized vinyl n-butyl ether with 2–3% of 1-methoxy-1,3-butadiene at 5°C using Mosley's (105) aluminum hydrosulfate hydrate catalyst. The copolymers had inherent viscosities of 4.2–5.9 dl/g. They were vulcanized using an accelerated sulfur recipe. The solubility of a vulcanizate in benzene was only 3% and indicated that vulcanizable, true copolymers were obtained. Heck (126) has copolymerized vinyl methyl ether with 1-methoxyl-1,3-butadiene to obtain vulcanizable copolymers. The catalyst system of Christman and Vandenberg (87) was employed in preparing these copolymers, which were partially crystalline if the amount of 1-methoxy-1,3-butadiene in them did not exceed 10%. In this polymerization system, 1-methoxy-1,3-butadiene was much less reactive than vinyl methyl ether. Earlier, Lendle (127) had disclosed that the copolymerization of vinyl ethers with 1-methoxy-1,3-butadiene at −5 to 15°C using Friedel-Crafts catalysts in the presence of diluents yields highly viscous resins.

Some peculiarities of copolymerization of vinyl alkyl ether with dienes have been reported by Russian investigators. Ushakov et al. (128) copolymerized vinyl isopropyl ether with butadiene in carbon tetrachloride at 65°C using free radical initiators. The vinyl ether predominated in the copolymer which was attributed to a cationic polymerization mechanism. Similarly, in the copolymerization of vinyl isopropyl ether with chloroprene using dibenzoyl peroxide or azobisisobutyronitrile as initiator the vinyl ether component predominated in the product (129). Contact with air increased the vinyl ether content in the copolymer. It was postulated that traces of water in chloroprene led to the formation of HCl. Copolymerization by both free radical and cationic mechanisms took place but copolymers from the cationic mechanism predominated. It was found possible to alter polymerization conditions so as to produce copolymers of mol wt 60,000–90,000 containing up to 95–98% vinyl ether or coplymers of mol wt 60,000 containing 80–90% chloroprene.

2. Copolymerization with Other Vinyl Ethers, $CH_2\!\!=\!\!CHOR$, Containing Reactive R Groups

Herrle et al. (130) copolymerized vinyl alkyl ethers with small amounts of vinyl allyl ether or 1-vinyloxy-2-allyloxyethane to produce high molecular weight materials using principally Mosley's (105) catalyst. Boron trifluoride dihydrate and gaseous boron trifluoride were also used as catalysts in some cases. The copolymers were subsequently vulcanized with an accelerated sulfur recipe or with dibenzoyl peroxide. A vulcanizate containing carbon black filler had a tensile strength of about 1910 psi at 692% elongation at break. Crosslinked products have also been obtained from these unsaturated copolymers by reacting them with free radical initiators in the presence of a polymerizable monomer such as styrene or an acrylic monomer (131).

Vulcanizable copolymers of vinyl alkyl ethers with vinyl-2-chloroethyl ether and with β-vinyloxyethyl methacrylate (VEM) have been prepared by Lal (72) using Mosley's catalyst. The copolymers containing pendant chloroethyl groups were vulcanized with ditertiary amines or polytertiary amines. A copolymer prepared from vinyl n-butyl ether and VEM (molar ratio 97:3, 5°C), having an inherent viscosity of 6.2 dl/g, was compounded with 50 phr (weight parts per 100 parts of rubber) HAF carbon black and an accelerated sulfur recipe. Curing for 60 min at

290°F (143°C) gave an elastomer having a tensile strength of 1100 psi at 320% elongation at break. Solubility of the vulcanizate in benzene was 6.7%. This copolymer of VEM was also vulcanized with triethylene tetramine via addition of amine groups across the double bonds in the pendant methacryloxyethoxy groups. The reactivity ratios of vinyl n-butyl ether (M_1) and VEM (M_2) during copolymerization (Mosley's catalyst, pentane, 26°C) were reported by Lal and McGrath (110). Values of $r_1 = 0.82$ and $r_2 = 0.004$–0.3 (approximate) were obtained.

Low molecular weight copolymer of vinyl isobutyl ether with VEM (BF$_3$-etherate, CHCl$_3$, 0°C) has been prepared by Schreiber (132). Conjugated soybean, conjugated linseed, and nonconjugated linseed vinyl ethers have been copolymerized (stannic chloride catalyst, room temperature, benzene) with vinyl allyl ether, 1,2-divinyloxyethane, and VEM to yield viscous oils for subsequent film formation (133).

B. Vulcanization of Homopolymers

1. Radiation Curing

The crosslinking of poly(vinyl alkyl ethers) with high energy electrons was reported by Miller and coworkers (134). Duffy (135) investigated the effect of various fillers on the strength properties of poly(vinyl methyl ether) cured with ^{60}Co γ radiation or with an electron beam. Carbon black and magnesia were found to be effective fillers. Large doses of radiation, of the order of 40 megarads, are needed to give cross-linked materials with tensile strengths of about 1500–1700 psi. The effect of sulfur during radiation curing of high molecular weight poly(vinyl ethyl ether) has been reported by Lal and McGrath (136). The polymer was mixed with 50–55 phr of HAF carbon black in a Banbury before irradiation with ^{60}Co γ-rays. Chemically bound sulfur was found in samples irradiated in the presence of elemental sulfur or dicyclopenta-methylenethiuram tetrasulfide. The fact that the network released portion of the combined sulfur by reaction with triphenylphosphine indicates that some of the crosslinks were polysulfidic, i.e., contained three or more sulfur atoms in the crosslink. In the presence of small amounts of sulfur higher tensile strengths were obtained. This is probably due to the formation of disulfidic and/or polysulfidic crosslinks and generally higher crosslink densities compared to samples prepared in the absence of sulfur. Furthermore, the maximum tensile strength of the sulfur-containing samples was attained at a lower radiation dose than in the corresponding control experiments. For a given swelling ratio, which may be taken as a measure of crosslink density, a higher tensile strength was generally obtained for samples irradiated in the presence of sulfur.

2. Vulcanization with Peroxides

Schildknecht (137) reported that unfilled poly(vinyl n-butyl ether) can be rendered "substantially insoluble" in benzene by vulcanizing with dibenzoyl peroxide in the presence of sulfur. No quantitative data on swelling and per cent solubility of the vulcanizates were given. The vulcanization of poly(vinyl alkyl ethers) has been investigated in considerable detail in the Goodyear Research Laboratories (138–139). Lal and Scott (138) have demonstrated that attempts to vulcanize high molecular weight poly(vinyl n-butyl ether) containing 50 phr HAF carbon black using dibenzoyl peroxide or dilauroyl peroxide, either alone or in the presence of sulfur, produced poorly cured samples. However, vulcanizates exhibiting tensile strength of 1560 psi at 470% elongation were obtained with 10 phr of dicumyl peroxide (40% active) in the presence of 0.25 phr sulfur. These vulcanizates had high gel contents, low swelling ratios and low per cent solubility in benzene. These

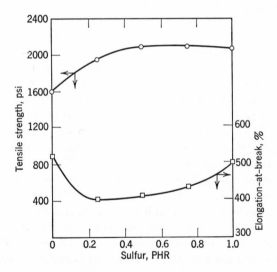

Fig. 2. Effect of sulfur concentration on tensile strength and per cent elongation-at-break of ISAF-loaded poly(vinyl ethyl ether) vulcanizates. ○, tensile strength; □, per cent elongation-at-break. ISAF, 55 phr; Di-Cup 40 C, 10 phr; sulfur, 0–1 phr, cure: 30 min/310°F. Data of Lal and McGrath (139).

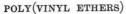

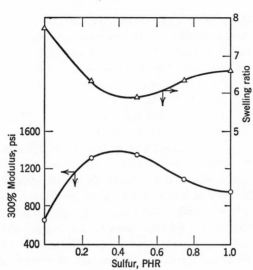

Fig. 3. Effect of sulfur concentration on 300% modulus and swelling ratio of ISAF-loaded poly(vinyl ethyl ether) vulcanizates. △, swelling ratio; ○, 300% modulus, HAF, 55 phr; Di-Cup 40 C, 10 phr; sulfur 0–1 phr, cure: 30 min/310°F. Data of Lal and McGrath (139).

data show that vulcanizates of practical usefulness can be obtained from homopolymers of vinyl alkyl ethers.

In the vulcanization of high molecular weight poly(vinyl alkyl ethers) with dicumyl peroxide, the addition of small amounts of sulfur in the compounding recipes gives higher degrees of crosslinking than in the absence of sulfur (139). Larger amounts of sulfur are less effective, probably because acidic impurities in sulfur waste increasing amounts of the peroxide by ionic (140) decomposition. However, these gum vulcanizates are weak rubbers. Carbon black reinforced vulcanizates, cured with dicumyl peroxide and small amounts of sulfur, exhibit higher tensile strength, lower elongation at break, higher modulus, and lower swelling ratio than control samples containing no sulfur. (Figs. 2, 3) The highest tensile strengths for vulcanizates of three HAF-loaded poly(vinyl alkyl ethers) were in the order:

$$ethyl \cong isobutyl > n\text{-butyl}$$

Sulfur becomes chemically bound to the crosslinked network during vulcanization. In the poly(vinyl ethyl ether) vulcanizates prepared with 4 phr of DiCup (95% active) and 0.75–2.0 phr of sulfur, a portion of the chemically bound sulfur can be removed by reacting these vulcanizates with triphenylphosphine. This decrease in the amount of bound sulfur is dependent on the value of the combined sulfur originally present and is not accompanied by crosslink scission to any appreciable degree. In the vulcanizate prepared with 2 phr of sulfur, the crosslinks are predominately of the disulfide and polysulfide type. However, there are many crosslinks of the C—C or C—S—C variety when the amount of sulfur in the above recipe is 0.5 phr or lower. Rehner et al. (141) have reported that the principal high boiling reaction products of 2,5-dimethylpentane, sulfur, and di-*tert*-butyl peroxide at 160–180°C are disulfides and polysulfides, most of which are formed from the isoparaffin starting material. They have emphasized that sulfur plays a major role and does not merely prevent minor side reactions as was suggested by Natta et al. (142). Based on model compound work, Wei and Rehner (143) have proposed a mechanism for the vulcanization of saturated hydrocarbon elastomers by the peroxide–sulfur system. It is reasonable to assume that most of the chemical reactions outlined by these authors (141,143) take place also in the vulcanization of poly(vinyl alkyl ethers) with dicumyl peroxide plus sulfur. In the presence of carbon black these reactions are likely to become even more complex.

Small amounts of sulfur in the curing recipe *decrease* the ratio of the fraction of structural repeat units in poly(vinyl n-butyl ether) undergoing scission p to the fraction of repeat units undergoing crosslinking q and increases the efficiency of crosslinking Z, i.e., the number of chemical crosslinks produced per molecule of dicumyl peroxide (139). For isomeric poly(vinyl butyl ethers), p, q, and Z decreased as follows (144):

$$n\text{-butyl} > \text{isobutyl} > sec\text{-butyl}$$

However, the ratio, p/q, was essentially unchanged. Lengthening the n-alkyl group from ethyl to n-butyl did not significantly change p, q, or Z values. Increasing the length from n-butyl to n-octyl almost doubled the Z value and halved the p value.

The chemistry of vulcanization of poly(vinyl alkyl ethers) has been studied by Lal and McGrath (145). The principal volatile products formed during the vulcanization of poly(vinyl ethyl ether) with dicumyl peroxide were ethane and acetaldehyde and to a lesser extent methane. In the presence of sulfur there was (*1*) a sharp increase in the proportion of ethyl alcohol, (*2*) a large increase in methane and, (*3*) a large decrease in ethane. These changes due to the presence of sulfur in the curing recipe can be explained by assuming side chain scission of the polymer. The length and/or shape of the alkyl group in the poly(vinyl alkyl ether) determine the composition of the volatile products during vulcanization. These data agree with the postulated chemistry of vulcanization of these polymers.

Kirk (146) reported the vulcanization of amorphous and crystalline poly(vinyl methyl ether) with dicumyl peroxide or its homologs. The degree of cure improved by the addition of quinone dioxime dibenzoate or triallyl cyanurate for gum and HAF black reinforced vulcanizates. The highest tensile strength for vulcanizates containing 50 weight parts of HAF black per 100 parts of amorphous poly(vinyl methyl ether) was about 1800 psi.

Rodriguez and Lynch (147) have also vulcanized poly(vinyl ethyl ether) with dicumyl peroxide. The influence of lubricants, fillers, and antioxidants on the properties of the vulcanizates was explored. Stearic acid, silicones, or polyethylene glycol (Carbowax 4000) were effective lubricants. Channel black and silica were outstanding with respect to tensile strength. Incroporation of HAF black in the compounding recipe gave vulcanizates with a good balance of properties. Retention of tensile strength and elongation at break in aged vulcanizates was greatly improved by incorporating trimethyl dihydroquinoline (Agerite Resin D) in the compounding recipe.

X. DYNAMIC PROPERTIES

Schmieder and Wolf (120) have measured at different temperatures the dynamic shear modulus and logarithmic decrement of *unvulcanized* poly(vinyl alkyl ethers) in which the pendant alkyl groups were methyl, ethyl, *n*-propyl, *n*-butyl, isobutyl, and *tert*-butyl and determined the glass transformation temperatures of these polymers. Dynamic properties of vulcanized poly(vinyl alkyl ethers), which are of considerable practical interest, have not been reported until recently. Lal, McGrath, and Scott (148) have systematically varied the side chain alkyl group in poly(vinyl alkyl ethers) and evaluated quantitatively the influence of these variations on the dynamic properties of gum vulcanizates of these polymers. The homopolymers were vulcanized with a dicumyl peroxide–sulfur curing recipe. Vulcanized cylindrical samples ($\frac{1}{2}$ in. long, $\frac{1}{2}$ in. diameter) were used for measuring dynamic mechanical properties with the Goodyear Forced Resonance Vibrotester (149). The frequency of the tester was 60 cps. The samples were subjected to 8% static compression and 2% dynamic longitudinal strain. Measurements were made

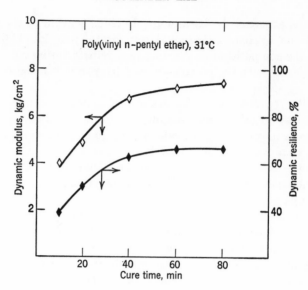

Fig. 4. Dependence of (◆) dynamic resilience and (◇) dynamic modulus of poly(vinyl n-pentyl ether) vulcanizates at 31°C on cure time (at 143°C). Data of Lal et al. (148).

at three temperatures: 31, 60, and 93°C.   In Fig. 4, the dynamic resilience and the dynamic modulus have been plotted as a function of cure time for vulcanizates of poly(vinyl n-pentyl ether).   The dynamic resilience has been defined as the ratio of the energy in two successive cycles of free vibration, but is commonly expressed as percentage.   The percent dynamic resilience is equal to 100 exp $(-2 \tan \delta)$, where $\tan \delta$ is the mechanical loss factor.   In the range of cure times used, the upper limit of which represents about 2.5 half-lives for the decomposition of dicumyl peroxide (150), the dynamic modulus and the dynamic resilience increase with cure time.   The data of Fig. 4 are consistent with the expectation that the rate controlling step in the crosslinking of poly(vinyl alkyl ethers) is the rate of decomposition of dicumyl peroxide.

Figure 4 indicates that the dynamic modulus and resilience of this rubber depend on its degree of crosslinking.   It also suggests that at this frequency and temperature of measurement, only a limited range of dynamic modulus and resilience values are permissible, i.e., the dynamic modulus and resilience values may not vary independently of each other as there may be a functional relation between the two.   This functional relation may be seen by eliminating the independent variable of cure time from Fig. 4 by plotting the dynamic resilience vs. the dynamic modulus of samples each at a different degree of cure as in Fig. 5. The least crosslinked sample is represented by the point with the lowest dynamic modulus and resilience values. Samples with increasing degrees of cross-linking are located further along the curve at higher dynamic modulus and resilience values. While Fig. 5 again indicates that no single dynamic modulus value and no single dynamic resilience value can be considered as characteristic of a rubber crosslinked to an unspecified degree, it does indicate that for vulcanizates of a given rubber the dynamic modulus and resilience values are limited to a rather narrow range of permissible values as indicated by the curve of Fig. 5. The dynamic modulus vs. dynamic resilience plot is a convenient way of comparing and categorizing dynamic mechanical properties of crosslinked rubbers. It also suggests the contribution of molecular weight and crosslinking to dynamic mechanical properties.

In Fig. 6 dynamic resilience values at 93°C are plotted against dynamic moduli for samples of seven poly(vinyl alkyl ethers), each at a different

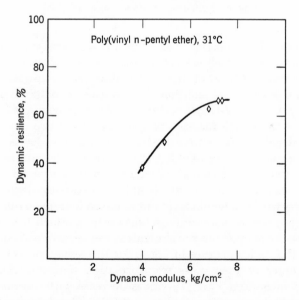

Fig. 5. Relationship between dynamic resilience and dynamic modulus of poly(vinyl n-pentyl ether) vulcanizates at 31°C. Data of Lal et al. (148).

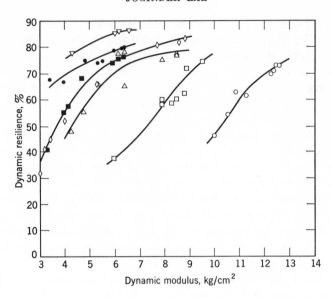

Fig. 6. Relationship between dynamic resilience and dynamic modulus of poly(vinyl alkyl ether) vulcanizates at 93°C: ○, ethyl; △, *n*-butyl; □, isobutyl; ◇, *n*-pentyl; ●, *n*-hexyl; ▽ *n*-octyl; ■, 2-ethylhexyl. Data of Lal et al. (148).

degree of cure.   Each polymer has its own characteristic locus of points in such a plot and in each case an increase in cure time, i.e., degree of crosslinking, has the effect of moving the data points upward along this characteristic curve.   In addition, in the case of poly(vinyl *n*-pentyl ether) and poly(vinyl 2-ethylhexyl ether) the data points fall on a common curve even though for each of these polymers two materials with different molecular weights were used and two different levels of dicumyl peroxide were used for crosslinking the samples.   The upward progression of the data points along the characteristic curve for a polymer with either increasing molecular weight or increasing degree of crosslinking agrees with similar observations made on four diene rubbers (151). With the exception of poly(vinyl isobutyl ether), which has the highest T_g among the seven polymers investigated, the relative position from left to right of the various curves in Fig. 6 for the remaining polymers is in increasing order of their glass transformation temperatures.   As was pointed out in Section VIII, T_g values decrease with increasing length of the *n*-alkyl group from ethyl to *n*-octyl.   The data for poly(vinyl *n*-pentyl and 2-ethylhexyl ethers), which have the same T_g, fall practically on a common curve. Similar data at 31 and 60°C have been reported by Lal and co-workers (148). The following features are noteworthy. The curves in Fig. 6 at the test temperature of 93°C maintain their positions relative to each other at these lower test temperatures. However, poly(vinyl isobutyl ether)shifts to the right of the curve for poly(vinyl ethyl ether) as the test temperature decreases to 31°C. These data demonstrate that the T_g apparently is a major factor in correlating the dynamic mechanical behavior of this homologous series of elastomers. The influence of the size and shape of the alkyl group is to be seen primarily in the T_g value. There is very little direct dependence of the dynamic mechanical properties on the structure of the alkyl side groups.

From Fig. 6 and the corresponding figures at the lower temperatures it is observed that at a given temperature of measurement the dynamic modulus for each rubber is limited to a relatively small range of values. It is preferable to consider only the 93°C data in Fig. 6 since this temperature is sufficiently above the T_g of each of these rubbers to minimize any of several difficulties that one might expect at temperatures closer to the T_g of a rubber. On the basis of rubber elasticity theory, one expects the dynamic modulus to reflect the contributions of both chemical and physical crosslinks in these rubbery networks. Since the chemistry of crosslinking for these poly(vinyl alkyl ethers) should be similar, about the same order of chemical crosslinking should be present in all these rubbers for equal cure times for a given amount of dicumyl peroxide used. Hence, the different dynamic modulus values exhibited by these rubbers primarily represent differences in degrees of physical crosslinking for the rubbers. These differences in physical crosslinking most likely are due to differences in the amount of chain entanglements occurring in these polymers.

Figure 7 shows data for poly(vinyl ethyl, n-butyl, and isobutyl ethers) at three temperatures. Poly(vinyl isobutyl and ethyl ethers) show a larger temperature coefficient of resilience than the other polymers of vinyl alkyl ethers. The relative position of the curves for poly(vinyl isobutyl ether) and poly(vinyl ethyl ether) is reversed then the temperature of measurement is increased from 31 to 93°C. Poly(vinyl isobutyl ether) has a high internal friction or dynamic viscosity at 31°C. Its modulus is about 15 kg/cm² and its resilience is less than 1% for various degrees of cure. At 60°C, the curve shifts considerably to the left in relation to the curves for the other poly(vinyl alkyl ethers). This large shift should be expected since poly(vinyl isobutyl ether) has the highest T_g. At 93°C, the curve for poly(vinyl isobutyl ether) lies to the left of the

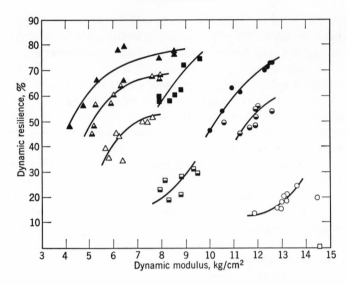

Fig. 7. Relationship between dynamic resilience and dynamic modulus of poly(vinyl alkyl ether) vulcanizates at three test temperatures: ○, ethyl, 31°C; △, n-butyl, 31°C; □, isobutyl, 31°C; ◐, ethyl, 60°C; ▲, n-butyl, 60°C; ◪, isobutyl, 60°C; ●, ethyl, 93°C; ▲, n-butyl, 93°C; ■, isobutyl, 93°C. Data of Lal et al. (148).

curve for the poly(vinyl ethyl ether). This position is anomalous with respect to the order of the T_g values of the polymers. It is evident that the curve for poly(vinyl ethyl ether) also shifts considerably to the left with increase in temperature, much more than the curves for the polymers with higher n-alkyl side chains. In general, the amount of displacement with temperature of the various curves for the polymers increases as their glass transformation temperatures increase and approach the test temperature. This agrees with the general temperature dependence observed for other viscoelastic materials (152).

The vulcanizates of vinyl alkyl ether polymers have low dynamic moduli as compared to accelerated sulfur gum vulcanizates of either SBR or natural rubber (Fig. 8). By suitably choosing the length of the alkyl group in these vinyl ether polymers a series of elastomers can be obtained having desirable combinations of dynamic modulus and dynamic resilience values. It is apparent from Fig. 8 that the glass transformation temperature is not the only factor controlling dynamic mechanical properties when chemical structural differences larger than those between polymers of a homologous series are involved. The curves of natural rubber ($T_g = -72°C$) and SBR ($T_g = -56°C$) are displaced markedly to the right of the curves for poly(vinyl alkyl ethers) having similar T_g values. Apparently, structural changes in the main chain of a polymer have a marked influence on dynamic behavior.

The effect of temperature on the dynamic modulus of three poly(vinyl alkyl ethers) is shown in greater detail in Fig. 9. These data were obtained on samples cured for 80 min at 143°C and it was assumed that the various samples have about the same degree of chemical crosslinking. Poly(vinyl isobutyl ether) begins to stiffen below 30°C and poly(vinyl ethyl ether) begins to stiffen starting at about 0°C. In the limited temperature range of 0–93°C, the dynamic modulus of poly(vinyl n-hexyl ether) increased with temperature at about two-thirds the rate predicted by the kinetic theory of rubberlike elasticity. The effect of temperature on the dynamic resilience of these same vulcanizates is presented in Fig. 10. Over the entire temperature range the curves are lined up in the order of the T_g of the polymers. The curves are rather similar in shape with the dynamic resilience being only 5% at temperatures about 60°C higher than the dilatometric T_g of the polymer and

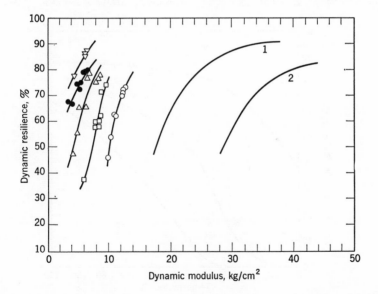

Fig. 8. Dynamic properties of rubber vulcanizates at 93°C for various poly(vinyl alkyl ethers): ○, ethyl; △, n-butyl; □, isobutyl; ●, n-hexyl; ▽, n-octyl; (1) natural rubber; (2) SBR. Data of Lal et al. (148).

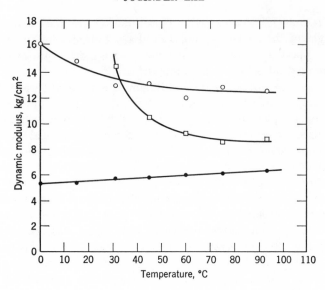

Fig. 9. Effect of temperature on the dynamic modulus of poly(vinyl alkyl ether) vulcanizates: ○, ethyl; □, isobutyl; ●, n-hexyl.   Data of Lal et al. (148).

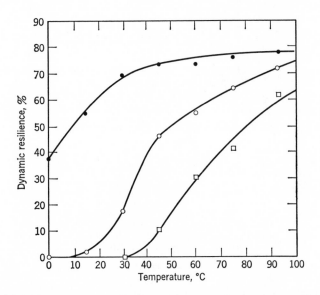

Fig. 10. Effect of temperature on the dynamic resilience of poly(vinyl alkyl ether) vulcanizates: ○, ethyl; □, isobutyl; ●, n-hexyl.   Data of Lal et al. (148).

then increasing with temperature to a limiting value (for this degree of vulcanization) of 70–80% at temperatures which are 100–140°C above the T_g. The curves appear to be superimposable by translation parallel to the temperature axis.

In Fig. 11 the dynamic resilience values for the polymers shown in Fig. 10 and also for all of the other vinyl alkyl ether polymers are plotted against $T - T_g$ values (where T is the test temperature in °C). Only vulcanizates cured for 80 min at 143°C using 4 phr dicumyl peroxide (95%) and 0.5 phr sulfur in the compounding recipe were used. It was assumed that the dynamic resilience values of these vulcanizates have not been greatly affected by the differences in chemical crosslink density and degree of polymerization that exists between these samples. Apparently, the T_g dominates the dynamic resilience behavior of these elastomers from a homologous series. This is in accord with what one would expect from the WLF equation (152) (cf. Chapter 1, pp. 9 and 17) and other observations on viscoelastic dependent properties (153). Lal and coworkers (148) have presented evidence in support of their assumption that the

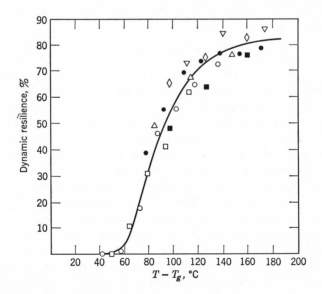

Fig. 11. Relationship between dynamic resilience and $(T - T_g)$ for poly(vinyl alkyl ether) vulcanizates: ○, ethyl; △, n-butyl; □, isobutyl; ◇, n-pentyl; ●, n-hexyl; ▽, n-octyl; ■, 2-ethylhexyl; T = test temp., °C. Data of Lal et al. (148).

data of Fig. 11 are not seriously affected by the differences in chemical crosslink density or degree of polymerization that exist between these particular samples. The dynamic resilience behavior of the vulcanizates of four poly(vinyl alkyl ethers) (alkyl group = ethyl, n-butyl, isobutyl, and n-octyl) having the same concentration of chemical crosslinks and essentially the same degree of polymerization indicates that the T_g dominates the dynamic resilience behavior of a homologous series of elastomers having the same network structure.

In Fig. 11, poly(vinyl isobutyl ether) exhibits a normal dependence of resilience on T_g; hence, the anomalous position of the poly(vinyl isobutyl ether) curve in Fig. 6 is to be attributed to the anomalous dependence of its dynamic modulus (and hence low degree of chain entanglements) on T_g. This is the only polymer of the homologous series discussed here that has shown any unusual dependence of dynamic mechanical properties on structure, as the other polymers all had dynamic mechanical properties that could be correlated with their T_g. (A plot of dynamic modulus at 93°C of the 80 min cure samples vs. T_g suggests that the anomaly in Fig. 6 is due somewhat more to the fact that the dynamic modulus of poly(vinyl isobutyl ether) is low with respect to its T_g rather than to the fact that the dynamic modulus of poly(vinyl ethyl ether) is high with respect to its T_g.)

XI. SUMMARY

A variety of cationic catalysts are now available for polymerizing vinyl alkyl ethers to high molecular weight materials in which the degree of stereoregularity extends over a wide range. Polymers having a low degree of stereoregularity are essentially amorphous, whereas others possessing a high degree of stereoregularity are highly crystalline. Semicrystalline polymers are characterized by block copolymer structure of d and l sequences as well as amorphous segments. In most polymerizations, some amorphous polymer is produced together with semicrystalline and/or crystalline fractions. The degree of stereoregularity in poly(vinyl alkyl ethers) prepared with common Friedel-Crafts type catalysts is generally quite low, even when low temperatures are employed during polymerization. Highly crystalline polymer fractions are produced when vinyl alkyl ethers are polymerized at about 30°C with metal sulfate–sulfuric acid and related catalyst systems, with Vandenberg's modified Ziegler-type catalyst, and with complex aluminum fluorides. These polymerizations are examples of coordinated cationic processes in which polymerization takes place via an insertion mechanism. The relationship of these catalysts to other stereospecific cationic catalysts is similar to that of the Ziegler-Natta type coordination anionic catalysts to other stereospecific anionic catalysts. Apparently, a heterogeneous catalyst is not essential for formation of an isotactic polymer during the polymerization of vinyl alkyl ethers. In homogeneous polymerizations, low temperatures are necessary to obtain a reasonable degree of stereoregularity and crystallinity in the polymer. Polymerization at the surface of a heterogeneous catalyst may, under favorable conditions, be beneficial in increasing the stereospecific control. However, this may not always be the case. Bulkiness of the side groups in the monomer and the growing chain end may interfere in the orientation of the incoming monomer and lead to the formation of a less isotactic polymer. Higashimura and associates have reported that the heterogeneous, stereospecific Mosley catalyst failed to give a crystalline polymer from vinyl *tert*-butyl ether. Likewise, Vandenberg's heterogeneous, modified Ziegler-type catalyst performed best with vinyl methyl ether.

The interaction between a growing cation and a counteranion is dependent on their nature and the polarity of the medium. An increase in the temperature of polymerization at the site of growth weakens interaction forces which govern the orientation of the incoming monomer in a preferred conformation at the site of growth and which minimize inversion of the growing carbonium ion before another monomer unit has been added. The stronger these interaction forces are, the less is the effect of temperature on the stereoregularity of the polymer.

The glass transformation temperatures of poly(vinyl *n*-alkyl ethers) decrease as the length of the *n*-alkyl side chain increases from methyl to *n*-octyl. On lengthening the side chain, 14% of the specific volume increase represents an increase in free volume.

Homopolymers of vinyl alkyl ethers can be effectively crosslinked with dicumyl peroxide, Co^{60} γ-radiation, or with an electron beam. Gum vulcanizates of these polymers are weak rubbers. It is necessary to incorporate a suitable reinforcing agent in compounding recipes to obtain vulcanizates with appreciable tensile strengths. The addition of small amounts of sulfur to the recipes results in the attainment of higher tensile strength and generally higher crosslink density than when sulfur is omitted. Data on the dynamic mechanical properties of gum vulcanizates of poly(vinyl alkyl ethers) demonstrate that the glass transformation temperature dominates the dynamic resilience behavior of a homologous series of elastomers having the same network structure. By suitably choosing the length of the alkyl group in these vinyl ether polymers a series of elastomers can be obtained having desirable combinations of dynamic modulus and dynamic resilience values.

References

1. W. Reppe, U.S. Pat. 1,959,927 (May 22, 1934) and many subsequent references; W. Reppe and co-workers, *Ann. Chem.*, **601**, 81 (1956).
2. *Chemical and Engineering News*, **43**, 42 (March 22, 1965).
3. G. N. Lewis, *J. Franklin Inst.*, **226**, 293 (1938).
4. J. Wislicenus, *Ann. Chem.*, **92**, 106 (1876).
5. W. Reppe and O. Schlichting, U.S. Pats. 2,104,000, 2,104,101, and 2,104,102 (Dec. 28, 1937).
6. A. E. Favorski and M. F. Shostakovskii, *J. Gen. Chem. USSR (English Transl.)*, **13**, 1 (1943); *Chem. Abstr.*, **38**, 330 (1944).
7. M. F. Shostakovskii and F. P. Sidel'kovskaya, *J. Gen. Chem., USSR (English Transl.)*, **13**, 428 (1943); *Chem. Abstr.*, **38**, 3387 (1944).
8. W. Chalmers, *Can. J. Res.*, **7**, 464 (1932).
9. M. Mueller-Cunardi and K. Pieroh, U.S. Pat. 2,061,934 (Nov. 24, 1936); Ger. Pat. 634,295 (Aug. 22, 1936); *Chem. Abstr.*, **30**, 8438 (1936).
10. M. Otto, H. Gueterbock, and A. Hellemans, U.S. Pat. 2,311,567 (Feb. 16, 1943).
11. C. W. Bunn and E. R. Howells, *J. Polymer Sci.*, **18**, 307 (1955).
12. K. Pieroh, U.S. Dept. Commerce, OTS PB 11,405.
13. L. Kollek and M. Jahrstorfer, U.S. Pat. 2,045,393 (June 23, 1936).
14. C. E. Schildknecht, A. O. Zoss, and C. McKinley, *Ind. Eng. Chem.*, **39**, 180 (1947).
15. C. E. Schildknecht, S. T. Gross, H. R. Davidson, J. M. Lambert, and A. O. Zoss, *Ind. Eng. Chem.*, **40**, 2104 (1948).
16. C. E. Schildknecht, S. T. Gross, and A. O. Zoss, *Ind. Eng. Chem.*, **41**, 1998 (1949).
17. C. E. Schildkneckt, A. O. Zoss, and F. Grosser, *Ind. Eng. Chem.*, **41**, 2891 (1949)
18. M. S. Muthana and H. Mark, *J. Polymer Sci.*, **4**, 531 (1949).
19. C. E. Schildknecht, *Vinyl and Related Polymers*, Wiley, New York, 1952.
20. G. Natta, I. W. Bassi, and P. Corradini, *Makromol. Chem.*, **18/19**, 455 (1956).
21. G. Natta, P. Pino, P. Corradini, F. Danusso, E. Mantica, G. Mazzanti, and G. Moraglio, *J. Am. Chem. Soc.*, **77**, 1708 (1955).
22. G. Natta, *J. Polymer Sci.*, **16**, 143 (1955).
23. G. Natta, *Chim. Ind. (Milan)*, **37**, 888 (1955); G. Natta and P. Corradini, *J. Polymer Sci.*, **20**, 251 (1956).
24. C. E. Schildknecht and P. H. Dunn, *J. Polymer Sci.*, **20**, 597 (1956).
25. A. O. Zoss, U.S. Pat., 2,616,879 (Nov. 4, 1952).
26. C. E. Schildknecht, *Ind. Eng. Chem.*, **50**, 107 (1958).
27. A. O. Zoss, U.S. Pat. 2,555,179 (May 29, 1951).
28. A. O. Zoss, U.S. Pat. 2,609,364 (Sept. 2, 1952); Brit. Pat. 609,999 (Oct. 8, 1948).
29. A. O. Zoss, U.S. Pat. 2,799,669 (July 16, 1957).
30. C. E. Schildknecht, U.S. Pat. 2,513,820 (July 4, 1950).
31. S. Nakano, *Kobunshi Kagaku*, **19**, 276 (1962).
32. L. Fishbein and B. F. Crowe, *Makromol. Chem.*, **48**, 221 (1961).
33. D. D. Eley and J. Saunders, *J. Chem. Soc.*, **1952**, 4167.
34. C. E. H. Bawn and A. Ledwith, *Quart. Rev. (London)*, **16**, 361 (1962).

35. S. Okamura, T. Higashimura, and H. Yamamoto, *Kobunshi Kagaku*, **16**, 45 (1959).
36. S. Okamura, T. Higashimura, and H. Yamamoto, *J. Polymer Sci.*, **33**, 510 (1958); *Kogyo Kagaku Zasshi*, **61**, 1636 (1958).
37. T. Higashimura, Y. Sunaga, and S. Okamura, *Kobunshi Kagaku*, **17**, 257 (1960).
38. S. Okamura, T. Higashimura, and I. Sakurada, *J. Polymer Sci.*, **39**, 507 (1959).
39. T. Higashimura, T. Kodama, and S. Okamura, *Kobunshi Kagaku*, **17**, 163 (1960).
40. T. Kodama, T. Higashimura, and S. Okamura, *Kobunshi Kagaku*, **18**, 267 (1961).
41. S. Okamura, T. Kodama, and T. Higashimura, *Makromol. Chem.*, **53**, 180 (1962).
42. T. Higashimura, T. Watanabe, K. Suzuoki, and S. Okamura, *J. Polymer Sci.*, **4C**, 361 (1963).
43. T. Higashimura, K. Suzuoki, and S. Okamura, *Makromol Chem.*, **86**, 259 (1965).
44. J. W. L. Fordham, *J. Polymer Sci.*, **39**, 321 (1959).
45. T. Higashimura, T. Yonezawa, S. Okamura, and K. Fukui, *J. Polymer Sci.*, **39**, 487 (1959).
46. D. J. Cram and K. P. Kopecky, *J. Am. Chem. Soc.*, **81**, 2748 (1959).
47. M. Goodman and Y. Fan, *J. Am. Chem. Soc.*, **86**, 4922, 5712 (1964).
48. D. D. Eley, in *The Chemistry of Cationic Polymerization*, P. H. Plesch, Ed., Pergamon Press, New York, 1963.
49. D. D. Eley and A. W. Richards, *Trans. Faraday Soc.*, **45**, 425 (1949).
50. D. C. Pepper, *Nature*, **158**, 789 (1946).
51. P. H. Plesch, *Research (London)*, **2**, 267 (1949).
52. D. D. Eley and A. W. Richards, *Research (London)*, **2**, 147 (1949).
53. Y. Imanishi, T. Higashimura, and S. Okamura, *Kobunshi Kagaku*, **19**, 154 (1962).
54. Y. Imanishi, H. Nakayama, T. Higashimura, and S. Okamura, *Kobunshi Kagaku*, **19**, 565 (1962).
55. J. D. Coombes and D. D. Eley, *J. Chem. Soc.*, **1952**, 4167.
56. J. P. Kennedy, *J. Polymer Sci.*, **38**, 263 (1959).
57. D. D. Eley and A. Seabrooke, *J. Chem. Soc.*, **1964**, 2226.
58. D. D. Eley and A. F. Johnson, *J. Chem. Soc.*, **1964**, 2238.
59. G. J. Blake and D. D. Eley, *J. Chem. Soc.*, **1965**, 7405.
60. G. J. Blake and D. D. Eley, *J. Chem. Soc.*, **1965**, 7412.
61. R. E. Burton and D. C. Pepper, *Proc. Roy. Soc. (London)*, **A263**, 58 (1961).
62. F. Grosser, U.S. Pat. 2,457,661 (Dec. 28, 1948).
63. K. Iwasaki, H. Fukutani, Y. Tsuchida, and S. Nakano, *J. Polymer Sci. A*, **1**, 1937 (1961).
64. K. Iwasaki, *J. Polymer Sci.*, **56**, 27 (1962).
65. S. Nakano, K. Iwasaki, and H. Fukutani, *J. Polymer Sci. A*, **1**, 3277 (1963).
66. J. Furukawa, *Polymer*, **3**, 487 (1962).
67. K. Iwasaki, H. Fukutani, Y. Tsuchida, and S. Nakano, *J. Polymer Sci. A*, **1**, 2371 (1963).
68. J. Lal, *J. Polymer Sci.*, **31**, 179 (1958).
69. G. B. Butler, *J. Am. Chem. Soc.*, **77**, 482 (1955).
70. J. Lal, E. F. Devlin, and G. S. Trick, *J. Polymer Sci.*, **44**, 523 (1960).
71. H. C. Haas and M. S. Simon, *J. Polymer Sci.*, **17**, 421 (1955).
72. J. Lal, unpublished results.
73. J. Lal, Ital. Pat. 606, 672 (July 18, 1960).
74. A. V. Bogdanova, *Vysokomolekul. Soedin.*, **2**, 576 (1960).

75. K. M. Roch and J. Saunders, *J. Polymer Sci.*, **38**, 554 (1959).
76. G. Natta, G. Dall' Asta, G. Mazzanti, U. Giannini, and S. Cesca, *Angew. Chem.*, **71**, 205 (1959).
77. G. Dall'Asta and I. W. Bassi, *Chim. Ind. (Milan)*, **43**, 999 (1961).
78. G. Dall'Asta and N. Oddo, *Chim. Ind. (Milan)*, **42**, 1234 (1960).
79. G. Natta and G. Mazzanti, *Tetrahedron*, **8**, 86 (1960).
80. G. Natta, M. Farina, and M. Peraldo, *Chim. Ind. (Milan)*, **42**, 255 (1960); *Makromol. Chem.*, **38**, 13 (1960).
81. G. Natta, M. Farina, M. Peraldo, P. Corradini, G. Bressan, and P. Ganis, *Atti. Accad. Nazl. Lincei, Rend. Classe Sci. Fis., Mat. Nat.*, **28**, 442 (1960).
82. G. Natta, M. Peraldo, M. Farina, and G. Bressan, *Macromol. Chem.*, **55**, 139 (1962).
83. G. Natta, *Soc. Plastics Engrs. Trans.*, **3**, 99 (1963).
84. R. F. Heck and D. S. Breslow, *J. Polymer Sci.*, **41**, 520 (1959).
85. E. J. Vandenberg, R. F. Heck, and D. S. Breslow, *J. Polymer Sci.*, **41**, 519 (1959).
86. E. J. Vandenberg, Ital. Pat. 571,741 (Jan. 14, 1958); Brit. Pat. 820,469 (Sept. 23, 1959).
87. D. L. Christman and E. J. Vandenberg, U.S. Pat. 3,025,282 (March 13, 1962).
88. R. F. Heck and E. J. Vandenberg, U.S. Pat. 3,205,283 (March 13, 1962).
89. E. J. Vandenberg, *J. Polymer Sci. C*, **1**, 207 (1963).
90. M. F. Shostakovskii, A. V. Bogdanova, A. V. Golovin, and S. Shamakhmudova, *Izv. Akad. Nauk SSSR, Otd. Khim. Nauk*, **1962**, 1813.
91. A. V. Bogdanova, M. F. Shostakovskii, and S. Shamakhmudova, *Izv. Akad. Nauk SSSR, Otd. Khim. Nauk*, **1964**, 543.
92. M. F. Shostakovskii, A. V. Bogdanova, and S. Shamakhmudova, *Izv. Akad. Nauk SSSR, Otd. Khim. Nauk*, **1964**, 363.
93. R. J. Kray, *J. Polymer Sci.*, **44**, 264 (1960).
94. F. E. Martin, Brit. Pat. 860,413 (Feb. 1, 1961).
95. M. F. Shostakovskii, A. M. Khomutov, and A. P. Alimov, *Izv. Akad. Nauk SSSR, Otd. Khim. Nauk*, **1963**, 1843.
96. J. M. Bruce and D. W. Farren, *Polymer*, **4**, 407 (1963).
97. J. Furukawa, T. Tsuruta, and S. Inoe, *J. Polymer Sci.*, **26**, 234 (1957).
98. T. Saegusa, H. Imai, and J. Furukawa, *Makromol. Chem.*, **64**, 224 (1963).
99. T. Saegusa, H. Imai, and J. Furukawa, *Makromol. Chem.*, **79**, 207 (1964).
100. R. J. Kern, J. J. Hawkins, and J. D. Calfee, *Makromol. Chem.*, **66**, 126 (1963).
101. R. J. Kern and J. C. Calfee, paper presented to the Division of Polymer Chemistry, Winter Meeting, American Chemical Society, Phoenix, Arizona, January 1966; *Polymer Preprints*, **7**, 191 (1966).
102. R. J. Kern, U.S. Pat. 3,231,554 (Jan. 25, 1966).
103. J. D. Calfee, Can. Pat. 722,244 (Nov. 23, 1965).
104. Y. Takeda, T. Okuyama, T. Fueno, and J. Furukawa, *Makromol. Chem.*, **76**, 209 (1964).
105. S. A. Mosley, U.S. Pat. 2,549,921 (April 24, 1951).
106. J. Lal, U.S. Pat. 2,984,656 (May 16, 1961); Brit. Pat. 846,690 (August 31, 1960).
107. J. Lal, U.S. Pat. 3,062,789 (Nov. 6, 1962).
108. J. Lal and R. N. Thudium, unpublished results.
109. N. B. Duffet and H. F. Wakefield, *Adhesives Age*, **3**, 28 (1960).
110. J. Lal and J. E. McGrath, *J. Polymer Sci. A*, **2**, 3369 (1964).

111. H. Yamaoka, T. Higashimura, and S. Okamura; *Kobunshi Kagaku*, **18**, 561 (1961).
112. S. Okamura, T. Higashimura, and T. Watanabe, *Makromol. Chem.*, **50**, 137 (1961).
113. T. Higashimura, T. Watanabe, and S. Okamura, *Kobunshi Kagaku*, **20**, 680 (1963).
114. M. F. Shostakovskii, A. M. Khomutov, and A. P. Alimov, *Izv. Akad. Nauk SSSR, Otd. Khim. Nauk*, **1964**, 1848.
115. J. Lal, J. E. McGrath, and G. S. Trick, *J. Polymer Science* (in press); *Polymer Previews*, **2**, 375 (1966).
116. E. J. Vandenberg, U.S. Pat. 3,159,613 (Dec. 1, 1964).
117. R. Chiang, U.S. Pat. 3,014,014 (Dec. 9, 1961).
118. P. Pino, G. P. Lorenzi, and S. Previtera, *Atti Accad. Nazl. Lincei Rend. Classe Sci. Fis., Mat. Nat.*, **29**, 562 (1960).
119. G. P. Lorenzi, E. Benedetti, and E. Chiellini, *Chim. Ind. (Milan)*, **46**, 1474 (1964).
120. K. Schmieder and K. Wolf, *Kolloid-Z.*, **134**, 149 (1953).
121. J. Lal and G. S. Trick, *J. Polymer Sci.*, *A*, **2**, 4559 (1964).
122. S. S. Rogers and L. Mandelkern, *J. Phys. Chem.*, **61**, 985 (1957).
123. D. W. Young and W. J. Sparks, U.S. Pat. 2,462,703 (Feb. 22, 1949).
124. F. Nagasawa, S. Nakano, A. Kageyama, K. Takiyuchi, T. Ichiriki, and I. Sukegawa, Japan. Pat. 1440 (Feb. 27, 1957); *Chem. Abstr.*, **52**, 7757 (1958).
125. J. Lal, U.S. Pat., 3,038,889 (June 12, 1962).
126. R. F. Heck, U.S. Pat., 3,025,276 (March 13, 1962).
127. A. Lendle, Ger. Pat. 847,349 (Aug. 25, 1962); *Chem. Abstr.*, **52**, 7775 (1958).
128. S. N. Ushakov, S. P. Mitsengendler, and V. N. Krasulina, *Izv. Akad. Nauk SSSR, Otd. Khim. Nauk*, **1957**, 366; *Chem. Abstr.*, **51**, 13459 (1957).
129. S. P. Mitsengendler, V. N. Krasulina, and L. B. Trukhmanova, *Izv. Akad. Nauk SSSR, Otd. Khim. Nauk*, **1956**, 1120; *Chem. Abstr.*, 51, 3178 (1957).
130. K. Herrle, H. Fikentsher, and H. P. Siebel, U.S. Pat. 2,825,719 (March 4, 1958).
131. H. P. Siebel, U.S. Pat. 2,830,032 (April 8, 1958).
132. H. Schreiber, U.S. Pat. 2,692,256 (April 11, 1957).
133. L. E. Gast, R. J. Stenberg, W. J. Schneider, and H. M. Teeter, *Offic. Dig., Federation Soc. Paint Technol.*, **32**, 1091 (1960).
134. A. A. Miller, E. J. Lawton, and J. S. Balwit, *J. Polymer Sci.*, **14**, 503 (1954).
135. D. Duffy, *Ind. Eng. Chem.*, **50**, 1267 (1958).
136. J. Lal and J. E. McGrath, *Rubber Chem. Technol.*, **36**, 248 (1963); J. Lal, Fr. Pat. 1,322,556 (Feb. 18, 1963).
137. C. E. Schildknecht, U.S. Pat. 2,429,587 (Oct. 21, 1947).
138. J. Lal and K. W. Scott, Ital. Pat. 641,220 (June 15, 1962); Brit. Pat. Spec. 903,943 (Aug. 22, 1962).
139. J. Lal and J. E. McGrath, *Rubber Chem. Technol.*, **36**, 1159 (1963).
140. A. R. Robinson, J. V. Marra, and L. O. Amberg, *Rubber Chem. Technol.*, **35**, 1083 (1962).
141. J. Rehner, Jr., G. G. Wanless, and P. E. Wei, *Rubber Chem. Technol.*, **35**, 118 (1962).
142. G. Natta, G. Crespi, E. DiGiulio, G. Ballini, and M. Bruzzone, *Rubber Plastics Age*, **42**, 53 (1961).
143. P. E. Wei and J. Rehner, Jr., *Rubber Chem. Technol.*, **35**, 133 (1962).

144. J. Lal and J. E. McGrath, paper presented at the International Symposium on Macromolecular Chemistry, Prague, Aug. 30–Sep. 4, 1965. *Preprint* No. P616.

145. J. Lal and J. E. McGrath, *Polymer Previews*, **2**, 382 (1966).

146. D. Kirk, U.S. Pat. 2,984,655 (May 16, 1961).

147. F. Rodriguez and S. R. Lynch, *Ind. Eng. Chem., Prod. Res. and Develop.*, **1**, 206 (1962).

148. J. Lal, J. E. McGrath, and K. W. Scott, *J. Appl. Polymer Sci.*, **9**, 3471 (1965).

149. S. D. Gehman, D. E. Woodford, and R. B. Stambaugh, *Ind. Eng. Chem.*, **33**, 1032 (1941).

150. D. K. Thomas, *J. Appl. Polymer Sci.* **6**, 613 (1962).

151. K. E. Gui, Figure 26 in S. D. Gehman, *Rubber Chem. Technol.*, **30**, 1202 (1957).

152. J. D. Ferry, *Viscoelastic Properties of Polymers*, Wiley, New York, 1961.

153. T. L. Smith and A. B. Magnusson, *J. Appl. Polymer Sci.*, **5**, 218 (1961).

CHAPTER 5

ELASTOMERS BY CATIONIC MECHANISMS

C. Elastomers from Cyclic Ethers

A. Ledwith and C. Fitzsimmonds

Donnan Laboratories, University of Liverpool

Contents

The barriers to single bond rotation in molecules containing $C{\overset{\displaystyle O}{\diagup\diagdown}}C$ linkages are much lower than those in comparable groupings having $C{\overset{\displaystyle C}{\diagup\diagdown}}C$ linkages (1). Consequently, introduction of C—O—C linkages into a macromolecule increases the rate with which it can adopt a large number of different conformations. Elastomeric properties are consequent on the ability of macromolecules to undergo stretching to statistically less favored conformations and relaxation to random conformations. The presence of $C{\overset{\displaystyle O}{\diagup\diagdown}}C$ linkages in a macromolecule might, therefore, be expected to confer it with elastomeric characteristics (cf. Chapter 1, p. 4).

Increased flexibility in the macromolecule will be most significant when ether linkages are introduced into the backbone skeleton. However, ether linkages are necessarily polar with respect to C—C or C—H bonds and this will give rise to an increase in intermacromolecular cohesion forces. Interaction between these two factors, molecular flexibility and intermolecular cohesion forces, is demonstrated by the variation in crystalline melting points of linear polymethylene oxides (1a). Figure 1

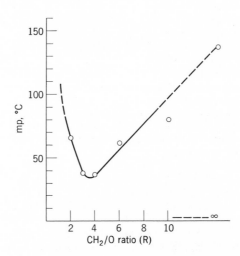

Fig. 1. Crystalline melting points of linear polyethers —[$(CH_2)_R$—O—]$_n$.

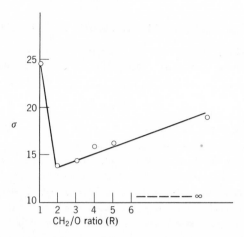

Fig. 2. Conformation parameters for linear polyethers —[(CH$_2$)$_R$—O—]$_n$.

shows the variation in crystalline melting points with increasing values of the ratio [R = CH$_2$/O], for a series of unsubstituted polymers having ether linkages in the backbone. The very polar polymer, polyoxymethylene with R = 1, has a high but indeterminate melting point (it decomposes before melting). At the other end of the scale we have linear polymethylene with R = ∞ and a melting point of about 137°C. Polyoxymethylene and polymethylene are highly crystalline materials and consequently do not show elastomeric properties. At intermediate CH$_2$/O ratios there is a minimum in the melting point curve (around R = 4) at which the flexibility of the CH$_2$—O link seems to have its maximum effect. In fact, polymethylene oxides with R = 3–4 are very useful elastomeric macromolecules. A more quantitative estimate of the relative flexibilities of polymethylene oxide chains can be derived (2) from a correlation of intrinsic viscosity [η] with weight average molecular weights $\bar{M}_w$. In this manner it is possible to derive values for the so-called conformation parameter σ. The latter is defined as the root mean square end-to-end distance of the chain in a theta solvent relative to the value which would be obtained if the hindrance to the rotation of successive bonds were absent except for the bond angle restrictions. Values of σ for the linear, unsubstituted polymethylene oxides are shown in Fig. 2 and although the curve resembles the corresponding melting point plot, it is clear that the molecular chain of polyethylene oxide (R = 2) has the least hindrance to internal rotation.

I. THE POLYMERIZABILITY OF CYCLIC ETHERS

Although certain polyethers can be obtained by polymerization of aldehydes and other carbonyl compounds, (see Chapter 5.D) the most convenient method for obtaining high polymers of the type

$$-\!\!\left(-(CH_2)_R\!-\!O\right)_{\!\!n}\!-$$

is by ring-opening polymerization of cyclic ethers. Oxiranes ($R = $ ⌐, have long been known to undergo polymerization by both cationic and anionic reagents (3,5,6). Larger ring cyclic ethers are less reactive and it is only in the last few years that the principles underlying polymerizability of ring compounds have become generally appreciated (4–6).

Conversion of cyclic monomer to linear high polymer is only feasible when the reaction involves a decrease in free energy (i.e., ΔG_{LC}^0 must be negative) and when there is a suitable reaction mechanism to induce the

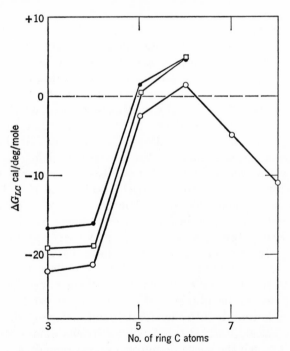

Fig. 3. Free energy of polymerization of cycloalkanes. O, unsubstituted; □, monomethyl; ●, dimethyl derivative.

ring opening process. The thermodynamic factors affecting polymerizability were first rationalized by Dainton, Devlin, and Small (7), who used semiempirical methods to calculate free energy changes for the hypothetical polymerization of pure liquid cycloalkanes. Values of ΔG_{LC}^0 so obtained for cycloalkanes, methylcycloalkanes, and 1,1-dimethylcycloalkanes are indicated in Fig. 3. Small (8) extended the earlier calculations to cyclic ethers and showed that since the bond lengths of C—C and C—O bonds do not differ very much, replacement of a carbon atom in a cycloalkane by a heteroatom such as oxygen, would affect the free energy change only slightly. Data for cycloalkanes (Fig. 3) can therefore be taken to represent roughly, the free energy changes involved in cyclic ether polymerization. The following generalizations emerge:

1. In any one group of derivatives, the thermodynamic feasibility of the ring opening process decreases with increasing ring size in the order 3 > 4 > 5 > 6, at which ring size the free energy change has become positive. With further increase in ring size, ring opening again becomes possible, the free energy change favoring the reaction in the order 8 > 7 > 6.

2. For any one size of ring, the thermodynamic feasibility of ring opening decreases in the order unsubstituted derivative > monomethyl derivate > 1,1-dimethyl derivative.

3. The thermodynamic criterion for a spontaneous process, that the free energy changes must be negative, is of importance only in the case of the six membered rings (unsubstituted, monomethyl-substituted, and 1,1-disubstituted), and substituted five-membered rings. In all these cases the attendant free energy change is positive, thereby precluding their participation in the (hypothetical) ring opening polymerization.

These calculations have been largely supported by experimental observations and the specific compounds will now be considered.

A. Three-Membered Ring Compounds

Ethylene oxide (oxirane) and its simple derivatives readily undergo ring opening polymerization by both anionic and cationic processes. Homopolymers of ethylene oxide, propylene oxide, the isomeric butene oxides, and epichlorohydrin are well known (3,5,6). In addition, more complex epoxides are used extensively as comonomers in epoxy resin manufacture.

$$R—CH\underset{\diagdown\diagup}{\underset{O}{}}CH_2$$

Substituted epoxides give rise to different types of

polymer when polymerized with certain heterogeneous catalysts (9). This follows since monosubstituted epoxides contain an asymmetric carbon atom, e.g., for propylene oxide as shown here. The normal

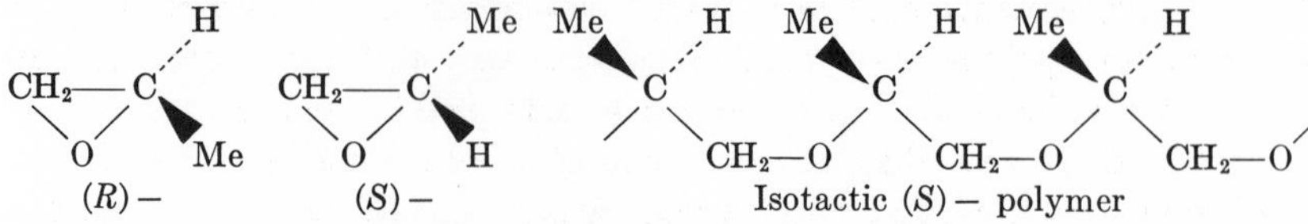

$(R)-$ $(S)-$ Isotactic $(S)-$ polymer

monomer will be a racemic mixture of $(R)-$ and $(S)-$ forms but in polymerization it is possible to control the propagation steps so as to produce an isotactic macromolecule having its asymmetric carbon atoms in long sequences of identical configurations, i.e., all $(R)-$ or all $(S)-$ (10). On the other hand, random polymerization of the racemic monomer gives rise to atactic poly(propylene oxide) with a completely random arrangement of asymmetric carbon atoms. Stereoregularity of the backbone chain has a marked effect on physical properties.

cis- and *trans-*2,3-Epoxy butanes with two asymmetric carbon atoms (11) can in principle yield two diisotactic polymers (meso with $(R)(S)$—$(R)(S)$ carbon sequences and racemic with $(R)(R)$—$(R)(R)$ or $(S)(S)$—$(S)(S)$ carbon sequences) and two disyndiotactic polymers (meso$_1$, with $(R)(S)$—$(S)(R)$ carbon sequences and meso$_2$ with $(R)(R)$—$(S)(S)$ carbon sequences).

B. Four-Membered Ring Compounds

The polymerization of oxacyclobutane (oxetane) and its substituted derivatives occurs very readily and a wide range of polymers have been characterized (4,5,12). However, of the homopolymers, only that (Penton) from 3,3-bischloromethyl oxacyclobutane is of commercial significance. For all the oxetanes, a cationic mechanism is necessary for high polymer formation.

C. Five-Membered Ring Compounds

Tetrahydrofuran was the first medium ring cyclic ether to be homopolymerized (13). Substituted tetrahydrofurans do not homopolymerize under any conditions (14) in accord with the predictions made by Dainton et al. (7), (see Fig. 3) 1,3-dioxolane is readily polymerized to high polymer (5,6) but once again the substituted derivatives are not readily homopolymerized (15).

Substituted five-membered cyclic ethers are polymerizable only when additional ring strain is present, as in the bridged cyclic compound 7-

oxabicyclo[2.2.1]heptane (16):

D. Six-Membered Ring Compounds

It has not been found possible to polymerize tetrahydropyran, 1,3-dioxane, or 1,4-dioxane. The six-membered rings are essentially strain free and there is a positive free energy change on ring opening (see Fig. 3). The polymer which would be obtained from 1,4-dioxane is, of course, identical with that obtained from ethylene oxide. Similarly polyoxymethylene —$(OCH_2)_n$— results from polymerization of formaldehyde $CH_2{=}O$, and its cyclic trimer, trioxane. In this case, the latter can be

$$\begin{array}{c} O \\ \diagup \quad \diagdown \\ CH_2 \qquad CH_2 \\ | \qquad\qquad | \\ O \qquad\qquad O \\ \diagdown \qquad \diagup \\ CH_2 \end{array}$$

Trioxane

used as monomer for the polymerization but it is not yet clear whether breakdown to formaldehyde is a necessary preliminary reaction (5,6).

E. Larger Ring Systems

The homopolymerization of cyclic formals containing seven-, eight-, and eleven-membered rings has been reported (17,18). High polymer corresponding to that which would be obtained from hexamethylene oxide (oxepane) has been prepared by acid-catalyzed condensation of 1,6-hexane diol (19).

II. ELASTOMERIC MATERIALS FROM CYCLIC ETHERS

From the preceding paragraphs it can be deduced that few, if any, homopolymers from cyclic ethers are naturally elastomeric in nature. In particular, polyoxymethylene (polytrioxane), polyethylene oxide, and isotactic polypropylene oxide show the properties of highly crystalline materials at ambient temperatures. On the other hand, copolymerization of mixtures of cyclic ethers readily produces a whole range of elastomers. Any process which can be used to effect homopolymerization of one or other cyclic ether can in principle—and usually also in practice—be used to

TABLE I
Polymerization of Common Cyclic Ethers

Ether	Structure	Polymer mp, °C	Leading references to polymerization[a]
Ethylene oxide (oxirane)	CH_2—CH_2 with O bridging (oxirane ring)	66	3–6,22–29
Propylene oxide (methyl oxirane)	CH_3CH——CH_2 with O bridging	74 (isotactic polymer)	5,6,9,10,28–38
Epichlorohydrin (chloromethyl oxirane)	$ClCH_2CH$——CH_2 with O bridging	119	5,6,20,28,39–41
Oxacyclobutane (oxetane)	CH_2 / CH_2 CH_2 \ O (oxetane ring)	35–40	3–6,12,42
3,3-Bis(chloromethyl) oxacyclobutane [3,3-(Bischloromethyl) oxetane]	CH_2Cl CH_2Cl on C; CH_2 CH_2 \ O / (oxetane ring)	180	3–6,12,43–48
Tetrahydrofuran	CH_2——CH_2 / CH_2 CH_2 \ O / (oxolane ring)	30–40	3–6,13,21,51–54
1,3-Dioxolane	CH_2——CH_2 with O—O and CH_2 (dioxolane ring)	55	5,6,55–58
Trioxane	O; CH_2 CH_2; O O; CH_2 (trioxane ring)	180	5,6,59–66
Oxepane	CH_2——CH_2 / CH_2 CH_2 / CH_2 CH_2 \ O / (oxepane ring)	61	19,67

[a] The references are mainly to specific kinetic and mechanism studies for the particular polymerization.

bring about copolymerization of mixtures.   It is for this reason that this treatise necessarily encompasses references to reagents and mechanisms for homopolymerization to nonelastomeric materials.   Table I lists the most common monomers used in elastomer synthesis.

At the present time there are only three commercially promising rubbers derived from cyclic ether monomers (20,20a,20b).

Polypropylene oxide rubbers are essentially high molecular weight block copolymers of isotactic and atactic segments, incorporating a small amount of unsaturated epoxide to facilitate vulcanization (20a).   These materials (e.g., Dynagen XP-139) show a combination of properties which are not available in other rubbers.   Prominent features are good tensile and tear strength, exceptional low temperature flexibility, good ozone resistance, excellent dynamic properties and resistance to flex fatigue, good heat resistance, and moderate oil resistance.   The constancy of properties over a broad, useful, temperature range is exceptional and better than for most elastomers.   Polypropylene oxide rubbers are white or light colored compounds with escellent ozone resistance and good physical properties.   This combination of characteristics is not available in existing rubbers and should find wide application.   A comparison of the low temperature flexibility and heat resistance of polypropylene oxide and conventional rubbers, is shown in Figs. 4 and 5.

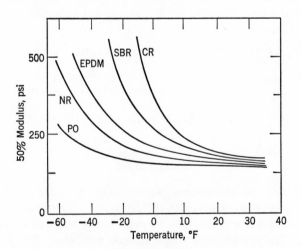

Fig. 4. Low temperature flexibility of elastomers. PO, propylene oxide rubber; NR, natural rubber; SBR, styrene-butadiene rubber; EPDM, ethylene-propylene rubber; CR, chloroprene rubber.   50% modulus is the force in psi required to extend the rubber tensile specimen by 50% (Meyer, 20a).

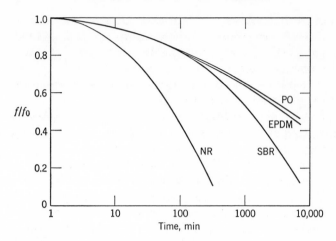

Fig. 5. Continuous stress relaxation of rubber elastomers at 250°F. NR, natural rubber; SBR, styrene-butadiene rubber; EPDM, ethylene-propylene rubber; PO, propylene oxide rubber (Meyer, 20a).

The other commercially promising polyepoxide elastomers show similar characteristics and were developed initially by Hercules Powder Co. (20b). They are a high molecular weight, amorphous, polyepichlorohydrin (CHR) and an even higher molecular weight 1:1 copolymer of epichlorohydrin and ethylene oxide (CHC), as shown in reactions 1 and 2. Both

$$n\mathrm{CH_2}{-}\mathrm{CHCH_2Cl} \rightarrow \quad {-}\!\!\left(\mathrm{CH_2}{-}\mathrm{CH}{-}\mathrm{O}\right)_{\!\!n} \tag{1}$$
$$\underset{\mathrm{O}}{\diagdown\diagup} \qquad\qquad\qquad \underset{\mathrm{CH_2}{-}\mathrm{Cl}}{|}$$

Polyepichlorohydrin (CHR)
$n = $ ca. 5000

$$n\mathrm{CH_2}{-}\mathrm{CHCH_2Cl} + n\mathrm{CH_2}{-}\mathrm{CH_2} \rightarrow \quad {-}\!\!\left(\mathrm{CH_2}{-}\mathrm{CH}{-}\mathrm{O}{-}\mathrm{CH_2}{-}\mathrm{CH_2}{-}\mathrm{O}\right)_{\!\!n} \tag{2}$$
$$\underset{\mathrm{O}}{\diagdown\diagup} \qquad\qquad \underset{\mathrm{O}}{\diagdown\diagup} \qquad\qquad\qquad \underset{\mathrm{CH_2Cl}}{|}$$

Copolymer (CHC)
$n = $ ca. 20,000

CHR and CHC are amorphous, saturated, high molecular weight, aliphatic polyethers with chloromethyl sidegroups. Vulcanization is readily accomplished with a wide variety of reagents reacting difunctionally with the chloromethyl substituent, e.g., diamines, urea, etc. Vulcanized CHR and CHC have equal and excellent oil resistance. CHC is especially interesting since it offers the oil resistance of nitrile rubbers and has low temperature, flame resistance, and rubber properties similar to neoprene

TABLE II
Comparison of Properties of Epichlorohydrin Elastomers with
Neoprene and Nitrile Rubbers (20)[a]

	CHR	CHC	Neoprene[b,f]	Nitrile[c,f]
Unvulcanized				
Building tack	Excellent	Good	Excellent	Good
Extrudability (Garvey die):				
Garvey rating	Excellent (125–200°F)	Good (150–200°F)	Good (200–250°F)	Good (200–250°F)
Surface appearance	Excellent	Good	Good	Good
Die swell, %	0–2	5–10	25	25
Moldability	Excellent	Excellent	Good	Good
Vulcanized				
Brittle point, °F (ASTM D746)	−15	−50	−44	−20
Heat-aging resistance in air, 250°F	Excellent	Very good	Good–fair	Good–fair
Ozone resistance (135 pphm at 100°F)	Excellent[d]	Excellent[d]	Good	Poor
Solvent resistance (vol % swell, 70 hr at 212°F)				
ASTM No. 3 oil	8	9	73	8
Water	10	10	12	11
Abrasion and tear resistance	Good	Good	Good	Good
Flame resistance	Very good	Good	Very good	Poor
Electrical properties	Poor	Poor	Fair	Poor
Oxygen permeability, 73°F cc/cm²/sec/atm/ cm × 10⁹	3.9[e]	21	30	30

[a] Unless otherwise noted, the data are based on unplasticized compounds containing about a 5:1 volume ratio of polymer to carbon black.
[b] General-purpose type, Neoprene W.
[c] High acrylonitrile type, Hycar 1041.
[d] >500 hours.
[e] Butyl rubber, oxygen permeability, 9.0.
[f] Unvulcanized data based on a plasticizer-containing formulation.

(see Table II) (20). CHR has outstanding air retention properties together with building tack comparable to natural rubber.

The cyclic ethers, other than epoxides, can only be homopolymerized or copolymerized by essentially homogeneous cationic mechanisms. Epoxides, on the other hand, can be polymerized by cationic and anionic reagents, and the type of polymer produced depends largely on whether the reaction is homogeneous or heterogeneous. The various reaction mechanisms will now be discussed in detail.

III. HOMOGENEOUS CATIONIC POLYMERIZATION OF CYCLIC ETHERS

Apart from minor variations in the various initiation, propagation, termination, and transfer reactions, all the evidence available suggests that reaction mechanisms in cyclic ether polymerization follow a simple general scheme. Most of the detailed kinetic work has been carried out on the polymerization of tetrahydrofuran (THF) and, for the purposes of this review, all mechanisms are based on this monomer. For other ethers, the formulas would need to be modified only as far as changing the number of methylene groups or oxygen atoms in the ring structures. Major departures from the general scheme are specifically referred to under the appropriate section headings.

The choice of tetrahydrofuran as a model system is also justified by the fact that this was the first cyclic ether to be polymerized to a high polymer by cationic processes (13). Early work with tetrahydrofuran was due almost entirely to Meerwein and his collaborators (21) and, although we now have a much wider range of catalyst types to choose from, the factors governing catalyst reactivity, gegenion stability, together with details of the propagation and ether exchange reactions, were clearly seen by these workers many years ago.

Meerwein proposed that overall polymerization of tetrahydrofuran occurred by an oxonium type of ring opening process, following initial reaction between the monomer and some cation forming species added as initiator e.g., as shown in reactions 3,4, and 5.

$$\text{Initiator} \rightarrow R^+ \tag{3}$$

$$R^+ + THF \rightarrow R{-}\overset{+}{O}\!\!\!\diagdown \rule{1em}{0pt} \tag{4}$$

$$R{-}\overset{+}{O}\!\!\!\diagdown \rule{1em}{0pt} + nTHF \rightarrow R{-}[O(CH_2)_4]_n{-}\overset{+}{O}\!\!\!\diagdown \rule{1em}{0pt}, \text{ etc.} \tag{5}$$

The individual steps involved in this reaction scheme are treated separately in the following sections.

IV. INITIATOR SYSTEMS

Most of the initiating systems described are useful for all cyclic ethers. However, the relative efficiency varies in a nonpredictable way from one ring system to another, and in copolymerization of various mixtures.

A. Oxonium Salts

Initiation of cyclic ether polymerization requires reaction between the ether oxygen atom and an ion pair comprising a highly electrophilic cation (X^+) and a gegenion of low nucleophilicity (Y^-), e.g.,

$$X^+Y^- + O\big\langle\;\;\;\big] \rightleftharpoons X\!-\!\overset{+}{O}\big\langle\;\;\;\big]\;Y^-$$

Propagation follows when the initially formed cyclic oxonium ion undergoes ring opening (by reaction with more monomer) more rapidly than competing side reactions such as inter- or intramolecular hydride ion abstractions, i.e.,

$$X\!-\!\overset{+}{O}\big\langle\;\;\;\big]\;Y^-\xrightarrow{\text{(THF)}} XH + \big[\;\;\;\big\rangle\;Y^-$$
$$\underset{+}{\overset{}{O}}$$

Clearly a simple alkyl cation would be desirable as initiator (X^+). Simple alkyl cations, however, are far too reactive and unstable to be handled as preformed catalysts. Consequently, the simple cation must be used in the form of a trialkyl oxonium ion salt (51,53) or a very reactive quaternary ammonium salt (68).

Trialkyl oxonium salts can be regarded as specifically solvated alkyl cations ($R_2O + R^+ \rightleftharpoons R_3\overset{+}{O}$) and consequently they are very powerful alkylating reagents.

Meerwein showed (21) that stable trialkyl oxonium salts could be prepared by reacting epoxides with Friedel-Crafts halides in diethyl ether as solvent. The epoxide was preferably epichlorohydrin or propylene oxide; e.g., for boron trifluoride etherate, see reaction 6. In a similar manner

$$\begin{array}{c}\text{Et}\\\diagdown\overset{+}{O}\!-\!\overset{-}{B}F_3 + ClCH_2CH\!-\!\!-\!\!CH_2 \rightarrow \\\diagup\qquad\qquad\diagdown\;\diagup\\\text{Et}\qquad\qquad O\end{array}\quad\begin{array}{c}\text{Et}\quad CH_2\\\diagdown\diagup\;\diagdown\\\overset{+}{O}\qquad\;CHCH_2Cl\xrightarrow{Et_2\overset{+}{O}BF_3}\\\diagup\;\diagdown\;\;\diagdown\diagup\\\text{Et}\quad\overset{-}{B}F_3O\end{array}$$

$$\begin{array}{cc} & CH_2Cl\\ & |\\Et_3\overset{+}{O}BF_4{}^- + & EtOCH_2CH\!-\!OBF_2\end{array}\qquad\qquad(6)$$

it was possible to prepare triethyl oxonium salts having $SbCl_6{}^-$, $FeCl_4{}^-$, and $AlCl_4{}^-$ gegenions.

Trialkyl oxonium salts are efficient initiators for cyclic ether polymerization but the actual initiation process involves regeneration of the appropriate dialkyl ether which may act as a transfer reagent (see later sections), i.e.,

$$Et_3\overset{+}{O}BF_4{}^- + THF \rightarrow Et\overset{+}{O}\!\!\diagdown\!\!\boxed{}\; BF_4{}^- + Et_2O$$

Furthermore, although this particular reaction would seem to be the most direct method for forming the cyclic oxonium ion necessary for propagation, kinetic studies have shown that the initiation reaction is significantly slower than the ensuing propagation. Trialkyl oxonium salts are sufficiently stable to be used as preformed initiators but it is often more convenient to add the Friedel-Crafts halide and the reactive epoxide directly to the polymerizable ether. In this way, the oxonium salt is generated *in situ* and recent work has utilized this principle. Particularly successful when used in this manner are combinations of epichlorohydrin or propylene oxide with alkylaluminum derivatives (48,69).

B. Protonic Acids

Initiation of cyclic ether polymerization might be expected as a result of protonation of the oxygen atom such as happens in reaction with strong acids. Strong acids may only be used if the anion generated by the loss of a proton is of such low nucleophilic power that interaction of it with the oxonium ion formed is not possible (21). The possible acids are $HFeCl_4$, HBF_4, $HSbCl_6$, and $HClO_4$, which form a secondary oxonium ion by proton addition (reaction 7). Using these acids only liquid polymers

$$HClO_4 + \boxed{}\!\!\diagup_O \;\rightleftharpoons\; \boxed{}\!\!\diagup_{\underset{H}{\overset{+}{O}}}\; ClO_4{}^- \tag{7}$$

are formed. Sulfuric acid is not an initiator for the polymerization, probably because the bisulfate anion is a stronger nucleophile than the monomer and competes for the oxonium ion or the proton. On the other

$$\boxed{}\!\!\diagup_{\underset{H}{\overset{+}{O}}}\;\overset{-}{O}SO_3H \quad\rightarrow\quad \boxed{}\!\!\diagup_{\underset{H}{O}}\!\!-OSO_3H$$

hand, fluorosulfonic acid (HSO_3F) and chlorosulfonic acid (HSO_3Cl) are quite effective initiators.

Anions of low nucleophilicity are most readily formed by complexing

Friedel-Crafts halides with a suitable proton donor, e.g., $BX_3 + HX \rightleftharpoons$ $H^+BX_4^-$. Consequently, the use of Friedel-Crafts halides in conjunction with proton donors such as H_2O, alcohols, hydrogen halides, etc., provides a wide range of initiator systems most of which are better prepared *in situ*. Many reaction systems are not sufficiently free of hydroxylic impurity (especially water) and it is often found that addition of Friedel-Crafts halide alone is sufficient to induce polymerization (49,70,71). In these cases it is likely that adventitious moisture is a necessary cocatalyst. Rigorously dried THF is unaffected by BF_3Et_2O after standing for many months (72). Similarly in the case of oxetane polymerization, Farthing showed that rigorously dried systems were not polymerized by BF_3 alone (73). Detailed kinetic studies of the polymerization of oxetane by Rose (42) established that, for low concentrations of added water, the polymerization rate was given by the expression

$$\frac{d(\text{polymer})}{dt} = k[\text{Monomer}][BF_3][H_2O]$$

Polymerization of ethylene oxide using BF_3 is also markedly affected by the small amounts of water added as cocatalyst (23).

C. Carboxonium Ion Salts

These initiators are readily formed by interaction (*in situ*) between organic molecules having labile halogen atoms and suitable Friedel-Crafts halides such as $AlCl_3$, BF_3, etc. (21). Particularly useful are the alpha chloro ethers (reaction 8). Alternatively, carboxonium salts can be pre-

$$CH_3OCH_2Cl + AlCl_3 \rightleftharpoons [CH_3\overset{+}{O}{=}CH_2]AlCl_4^- \xrightarrow{\text{THF}} CH_3OCH_2\overset{+}{O}{\diagup} \ \boxed{} \ AlCl_4^- \quad (8)$$

pared by the reaction of Friedel-Crafts halides with ortho esters or acetals, as shown in reaction 9.

$$R{-}C{\diagup}^{OR'}_{\diagdown OR'}{-}OR' + BF_3 \rightarrow R{-}\overset{OR'}{\underset{\underset{+}{OR'}}{C}}{-}\overset{\bar{B}F_3}{O}{-}R' \xrightarrow{BF_3}$$

$$R{-}\overset{R'}{\underset{R'}{C}}{\diagup}^{O}_{\diagdown O}{+} \ BF_4^- + BF_2(OR') \xrightarrow{\text{THF}} R'{-}\overset{+}{O}{\diagup} \ \boxed{} \ BF_4^- + R\overset{O}{\overset{\|}{C}}{-}OR' \quad (9)$$

D. Acylium Ion Salts and Related Molecules

Reactions between acid chlorides and Friedel-Crafts halides give rise to acylium salts and the reaction lends itself readily to initiation of cyclic ether polymerization (21). For example, see reaction 10. Similarly the

$$CH_3COCl + AlCl_3 \rightleftharpoons CH_3CO^+AlCl_4^- \xrightarrow{\text{THF}}$$

$$CH_3CO-\overset{+}{O}\!\!\diagdown\!\!\Big] \; AlCl_4^- \xrightarrow{\text{THF}} CH_3CO-O-(CH_2)_4-\overset{+}{O}\!\!\diagdown\!\!\Big] \; AlCl_4^- \quad (10)$$

acylium cations can be formed *in situ* by reacting strong acids with carboxylic acid anhydrides (21). In particular, the combination of perchloric acid and acetic anhydride has been used extensively (74) (reaction 11). Instead of carboxylic acid anhydride it is possible to use

$$(CH_3CO)_2O + HClO_4 \rightleftharpoons CH_3CO^+ClO_4^- + CH_3COOH$$
$$\downarrow\text{THF}$$

$$CH_3CO-\overset{+}{O}\!\!\diagdown\!\!\Big] \; ClO_4^-, \text{ etc.} \quad (11)$$

thionyl chloride (21), benzene sulfonyl chloride (21), phosphorus oxychloride (21), triphenyl phosphite (75), acetonitrile (76), phenyl isocyanate (76), diketene (76), and β-propiolactone (76) as cocatalyst with Friedel-Crafts halides. The combination of acyl chloride plus Lewis acid has recently been effected by having both components in the same molecule. Thus carboranes of the type $B_{10}H_{10}[C(COCl)_2]$ have been reported as initiators for homopolymerization of THF and propylene oxide (77).

If aromatic acid chlorides are used instead of acyl derivatives then it is quite convenient to pre-form the stable benzoyl salt (e.g., $PhCO^+$ $SbCl_6^-$) (78).

E. Stable Carbonium Ion Salts

Although simple alkyl carbonium ion derivatives are too unstable to permit isolation, the triphenyl methyl cation (Ph_3C^+) and cycloheptatrienyl cation ($C_7H_7^+$) are readily isolated in the form of crystalline salts having stable anions such as BF_4^-, ClO_4^-, $SbCl_6^-$, PF_6^-, etc. The triphenyl methyl cation salts are very good initiators for cyclic ether polymerization and are very convenient to handle (79). Especially useful are triphenyl methyl hexachloroantimonate and the corresponding hexafluorophosphate (80). Cycloheptatrienyl salts are similarly effective but give much slower rates of initiation than the corresponding triphenyl methyl salts (81).

The latters are highly colored crystalline materials having the characteristic absorption spectrum of Ph_3C^+. On introduction into cyclic ethers there is rapid decoloration of the catalyst followed by polymerization of the ether.

Although the initial decoloration of Ph_3C^+ is due to oxonium ion formation (82), it is now clear that a subsequent reaction produces triphenyl methane before polymerization occurs, i.e., as shown in reaction 12.

$$Ph_3C^+SbCl_6^- + THF \rightleftharpoons Ph_3C\!-\!\overset{+}{O}\!\!\diagdown \quad\Big| \quad SbCl_6^- \xrightarrow{\text{(THF)}}$$

$$Ph_3CH + \Big| \overset{\diagup}{\underset{\underset{+}{O}}{}} \Big| SbCl_6^- \quad (12)$$

Evidence for the formation of triphenyl methane was provided initially by attempts to produce dications in the polymerization of tetrahydrofuran. Kuntz (83) showed that the stable dicarbonium ion salt produced

$$\left[Ph_2\overset{+}{C}\!\!\diagdown\!\!\langle\!\!\bigcirc\!\!\rangle\!\!\diagup CH_2CH_2\!\!\diagdown\!\!\langle\!\!\bigcirc\!\!\rangle\!\!\diagup\overset{+}{C}Ph_2 \right](SbCl_6^-)_2$$

polytetrahydrofuran having the same molecular weight as that produced by corresponding concentrations of the mono salt $Ph_3C^+SbCl_6^-$. Later work by Kuntz (84) using NMR techniques showed that triphenyl methane was rapidly formed during initiation. Independent work has shown that triphenyl methane can be detected by gas chromatography during polymerization initiated by $Ph_3C^+SbCl_6^-$ and can actually be isolated from the reaction mixture by chromatography on neutral alumina (81). In a similar fashion the product of hydride ion abstraction, cycloheptatriene, can be isolated from polymerizations of tetrahydrofuran initiated by cycloheptatrienyl salts (81).

F. Aryl Diazonium Salts

Aryl diazonium salts decompose thermally and photochemically by both free radical and ionic mechanisms. It has recently been demonstrated that these reactions can be utilized to effect polymerization of many cyclic ethers. Dreyfuss and Dreyfuss first showed (85) that p-chlorophenyl diazonium hexafluorophosphate was a very effective initiator for tetrahydrofuran polymerization. Extremely high molecular weights were obtained and the system was apparently free from termination reactions. The initiation is thought to occur by hydride abstraction as shown in reaction 13. These authors propose (86) that the cyclic

$$Cl\langle\bigcirc\rangle N_2{}^+PF_6{}^- + THF \rightarrow Cl\langle\bigcirc\rangle\!-\!N\!=\!N\!-\!\overset{+}{O}\!\langle\;\;\rangle\; PF_6{}^- \rightarrow$$

$$Cl\langle\bigcirc\rangle H + N_2 + \left[\underset{\overset{+}{O}}{\bigcup}\right] PF_6{}^-, \text{ etc.} \quad (13)$$

carboxonium so formed reacts further with tetrahydrofuran to give an aldehyde endgroup in the polymer chain, viz:

$$\left[\underset{\overset{+}{O}}{\bigcup}\right] PF_6{}^- + THF \rightarrow H\overset{\overset{O}{\|}}{C}\!-\!(CH_2)_3\!-\!\overset{+}{O}\!\langle\;\;\rangle\; PF_6{}^-$$

Alternatively a cyclic acetal endgroup could arise (see reaction 14).

$$\left[\underset{\overset{+}{O}}{\bigcup}\right] PF_6{}^- + \left[\underset{O}{\bigcup}\right] \rightarrow \left[\underset{O}{\bigcup}\right]\!-\!\overset{+}{O}\!\langle\;\;\rangle\; PF_6{}^- \xrightarrow{\text{THF}}$$

$$\left[\underset{O}{\bigcup}\right]\!-\!O(CH_2)_4\!-\!\overset{+}{O}\!\langle\;\;\rangle\; PF_6{}^- \quad (14)$$

Differentiation between mechanisms by endgroup analysis will be extremely doubtful because of the likelihood that acetal endgroups will copolymerize or act as transfer agent during polymerization.

G. Initiation via Friedel-Crafts Halides

It has been reported that THF and other cyclic ethers can be polymerized using BF_3 (49), SiF_4 (87), PF_5 (70,71), $SbCl_5$ (21,52) and WCl_6 (88). However the experimental conditions employed would not have excluded trace amounts of moisture and consequently (with BF_3 and PF_5) it is more than likely that complex acids are the true initiators. Antimony pentachloride (and perhaps also WCl_6) can apparently function as an initiator without added cocatalyst. This effect was first reported by Meerwein (21) and recently more detailed kinetic studies have been reported (52). In this case, it is probable that the initiation mechanism is consequent on the ability of $SbCl_5$, and similar halides, to undergo reduction to a stable lower valence state. $SbCl_5$ is a well-known chlorinating agent (89) and could function by first chlorinating the ether, after which the resulting alpha chloro ether would function as cocatalyst as indicated earlier. Alternatively, as suggested by Rosenberg (52), the $SbCl_5$–THF complex could react with more $SbCl_5$ to form a growing polymer chain in which the endgroup is —$SbCl_4$, i.e.,

$$THF + SbCl_5 \rightleftharpoons \left[\underset{}{\overset{+}{O}}\!-\!\overset{-}{Sb}Cl_5 \right] \xrightarrow[SbCl_5]{THF} Cl_4Sb\!-\!O(CH_2)_4\!-\!\overset{+}{O} \left[\right] SbCl_6{}^-$$

whichever mechanism operates, it is clear that only one polymer chain results from an initiation reaction involving two molecules of $SbCl_5$.

Use of PF_5 as initiator for cyclic ether polymerization was first reported by Muetterties (70), and a more detailed study has been made by Sims (71). This initiator gives rise to a very high molecular weight polymer (in the case of THF) although, as mentioned above, there is considerable doubt as to whether the simple halide is a true initiator. Polymerizations induced by PF_5 are very similar to those initiated by p-chlorphenyldiazonium hexafluorophosphate (85,86).

H. Solid State Polymerization

Cyclic ethers such as trioxane and bis(3,3-chloromethyl)oxetane can be polymerized in the solid state (90). Mechanisms for these reactions are not yet well understood since high energy γ-rays are necessary for cation formation. More recently it has been shown that certain Friedel-Crafts halides (e.g., $SnCl_4$) can initiate solid state polymerization of the same monomers (91).

V. GENERAL COMMENTS ON INITIATION REACTIONS

The foregoing sections demonstrate that initiation of cyclic ether polymerization follows the well-known pattern of other cationic polymerizations. Most initiator systems are based on the interaction between a Friedel-Crafts halide (or strong Lewis Acid) with a suitable cocatalyst drawn from a wide range of derivatives. Clearly, for a given Friedel-Crafts halide, it is possible to have a constant gegenion and yet have different rates of initiation—according to the type of cocatalyst. This possibility has been realized by Rosenberg and his collaborators (52) who made detailed kinetic studies of tetrahydrofuran polymerization using $Et_3\overset{+}{O}SbCl_6{}^-$, $CH_3CO^+SbCl_6{}^-$, and $SbCl_5$ as initiator systems. As expected, the three initiation systems gave quite different rates of initiation.

Endgroups in polyether chains are, to some extent, controlled by the initiator employed, but transfer reactions can be a more important factor

TABLE III
Endgroups in Polytetrahydrofuran
$$X \rightarrow O(CH_2)_4 \rightarrow_n O(CH_2)_4 - Y$$

Catalyst system	Endgroups	
	X	Y
$R_3O^+BF_4^-$	RO—	—OH
$HAlCl_4$, $HFeCl_4$, H_2SnCl_6	HO—	—Cl
$AlCl_3$—, $FeCl_3$—, $SnCl_4$—, —RCOCl	RCO—	—Cl
$HClO_4$—$(CH_3CO)_2O$	CH_3COO—	—OCOCH$_3$
$ClSO_3H$	HO—	—Cl or —OH

in many cases (see later sections). Examples of the various types of endgroups which can be encountered are shown in Table III.

VI. PROPAGATION AND DEPROPAGATION REACTIONS

The chain growth processes in cyclic ether polymerization are best represented (21) as shown here (gegenions are deliberately omitted in this section). Evidence for the reversibility of these polymerizations is

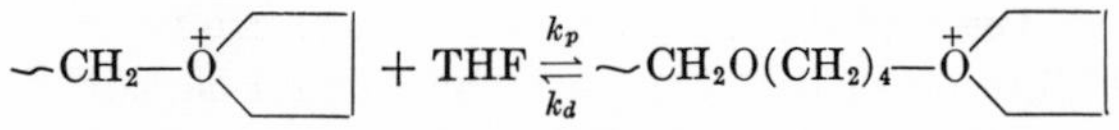

$$\sim CH_2 - \overset{+}{O} \left\langle \quad \right| + THF \underset{k_d}{\overset{k_p}{\rightleftharpoons}} \sim CH_2O(CH_2)_4 - \overset{+}{O} \left\langle \quad \right| \qquad (15)$$

provided by the very rapid depolymerization of polymer to monomer under the influence of, for example, $SbCl_5$ (92). Early work by Meerwein (21) had shown that for a given temperature, there was an upper limit to the percentage conversion of monomeric THF to polymer. The other cyclic ethers behave similarly (4) and this type of behavior is typical of polymerizations exhibiting so-called ceiling temperature phenomena.

The thermodynamics of reversible polymerization processes were first treated (from a kinetic viewpoint) by Dainton and Ivin (93,94) and later (from an equilibrium viewpoint) by Tobolsky and Eisenberg (95). Both treatments lead to essentially the same results and hence the former method will be used here. For a polymerization which is readily reversible the rate of the forward (propagation) reaction R_p is given by

$$R_p = k_p[\text{Monomer}][\text{active centers}]$$

The reverse reaction (depropagation) reforms monomer and is a unimolecular reaction, proceeding at a rate R_d given by

$$R_d = k_d[\text{active centers}]$$

At equilibrium the rates of forward and reverse reaction are equal, i.e.,

$R_p = R_d$ and hence

$$k_p \text{ [Monomer]}_e[\text{active centers}] = k_d[\text{active centers}]$$
$$\therefore k_p[\text{Monomer}]_e = k_d$$

[Monomer]$_e$ represents the concentration of monomer at equilibrium ([M]$_e$) and is readily measured. If values of k_p are known from measurements at much lower conversions, then k_d can be estimated.

In this way it is possible to estimate values of k_p and k_d at various temperatures and hence to calculate values of the enthalpy of activation for both forward ($\Delta H_p^{\ddagger}$) and reverse ($\Delta H_d^{\ddagger}$) reactions. From reaction rate theory it follows that

$$\Delta H_d^{\ddagger} - \Delta H_p^{\ddagger} = \Delta H$$

where ΔH is the thermodynamic enthalpy of conversion liquid monomer to dissolved polymer. The treatment of Dainton and Ivin can then be used to show that

$$\ln [\text{M}]_e = \frac{\Delta H}{RT} - \frac{\Delta S^{\circ}}{R}$$

and hence plots of $\ln [\text{M}]_e$ against $1/T$ can be used to evaluate ΔH and ΔS° for the particular systems. An excellent example of the effect of temperature on the equilibrium conversion for bulk polymerization of tetrahydrofuran is given by Dreyfuss and Dreyfuss (86). Their data are reproduced in Figs. 6 and 7.

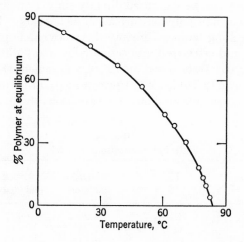

Fig. 6. Ceiling temperature in THF polymerization (86).

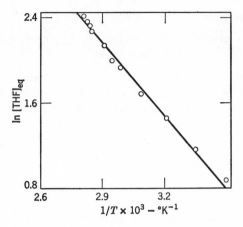

Fig. 7. Equilibrium monomer concentration as a function of temperature (86).

Dainton and Ivin (93) defined ceiling temperature (T_c) by

$$T_c = \frac{\Delta H}{\Delta S}$$

i.e., it is the temperature at which the polymerization exhibits zero change in free energy $(\Delta G = 0)$. Ceiling temperatures can, therefore, be estimated from measurements of $[M]_e$ at different temperatures or, more conveniently, they can be measured directly since T_c represents the temperature at which high polymer formation disappears. For THF polymerization the ceiling temperature and thermodynamic parameters have been measured, and estimated, respectively, by many different groups of workers. Using catalysts based on PF_6^- as gegenion, there is excellent agreement as shown in Table IV. However, with other initiator systems lower values of T_c are indicated but these are probably due to termination

TABLE IV
Bulk Polymerizations of THF

Initiator	T_c, °C	$-\Delta H$, kcal/mole	$-\Delta S$ (cal/deg/mole)	Ref.
PF_5	83.0	4.28	17.0	71
$p\text{-}ClC_6H_4N_2^+PF_6^-$	84.0	4.58	17.7	86
$Ph_3C^+PF_6^-$	84.0	4.81	18.2	81

reactions occurring at the higher temperatures (51,79). There is now little doubt that systems based on PF_6^- are highly stable and give reproducible data. Ivin and Leonard (96) have recently pointed out that nonideal mixing of monomer and polymer should be allowed for when making calculations of this type.

VII. MECHANISM OF THE PROPAGATION AND DEPROPAGATION REACTIONS

The propagation reaction in cyclic ether polymerization may be regarded either as involving nucleophilic attack by the monomeric ether oxygen atom on the alpha carbon of the cyclic oxonium ion, or perhaps more correctly as repetitive alkylation of cyclic ether units by the very reactive (and electrophilic) cyclic oxonium ion. Depropagation similarly can be viewed in two different ways but is best regarded as a good example of a neighboring group effect (97) (Scheme I). Depropagation to monomer is normally the expected reaction sequence because of the favorable ring structures which are likely to occur in the transition state. However, for all polyethers there is the possibility of intramolecular degradation (depolymerization) involving a main chain oxygen atom other than that required for depolymerization to monomer (Scheme I).

In the case of five-membered rings the monomer unit is thermodynamically more stable than dimer, trimer, etc., and the occurrence of other reactions—leading to cyclic polymers of random ring sizes—can be expected only when the system is specially designed to effect this, i.e., very low concentration of active centers and elevated temperatures.

For oxetane polymerization, the conditions referred to above lead to formation of significant amounts of a cyclic tetramer (42)

$$\boxed{-(CH_2)_3-O-}_4$$

The three-membered ring system, particularly ethylene oxide, does not depolymerize to monomer but produces the thermodynamically more stable dimer, 1,4-dioxane (98). For all systems however, the relative rates of propagation and the various depolymerization processes will be affected by the nature of the gegenion. This point is discussed more fully in the next section.

VIII. REDISTRIBUTION REACTIONS

Examination of Scheme I reveals the possibility for reactions between the growing polymeric, cyclic oxonium ion and ether oxygen atoms

The growing polymeric oxonium ion

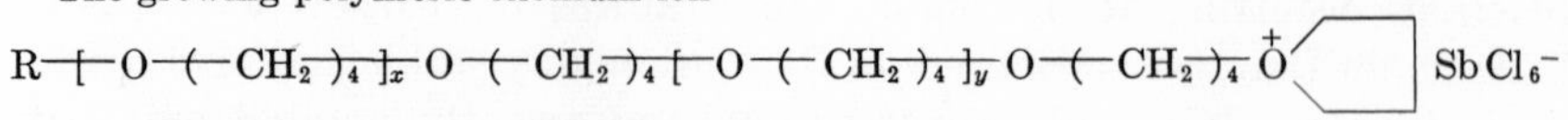

$$R\text{---}[\text{---}O\text{---}(\text{---}CH_2\text{---})_4\text{---}]_z\text{---}O\text{---}(\text{---}CH_2\text{---})_4\text{---}[\text{---}O\text{---}(\text{---}CH_2\text{---})_4\text{---}]_y\text{---}O\text{---}(\text{---}CH_2\text{---})_4\text{---}\overset{+}{O}\!\!\diagup\!\!\boxed{} \quad SbCl_6^-$$

may react without propagating in either of two ways:

(1) To eliminate monomer

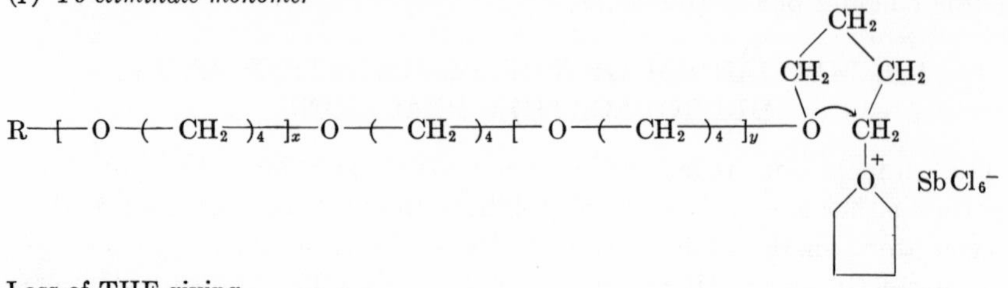

$$R\text{---}[\text{---}O\text{---}(\text{---}CH_2\text{---})_4\text{---}]_z\text{---}O\text{---}(\text{---}CH_2\text{---})_4\text{---}[\text{---}O\text{---}(\text{---}CH_2\text{---})_4\text{---}]_y\text{---}O\text{---}CH_2 \cdots$$

Loss of THF giving

$$R\text{---}[\text{---}O\text{---}(\text{---}CH_2\text{---})_4\text{---}]_z\text{---}O\text{---}(\text{---}CH_2\text{---})_4\text{---}[\text{---}O\text{---}(\text{---}CH_2\text{---})_4\text{---}]_y\text{---}\overset{+}{O}\!\!\diagup\!\!\boxed{} \quad SbCl_6^- + O\!\!\diagup$$

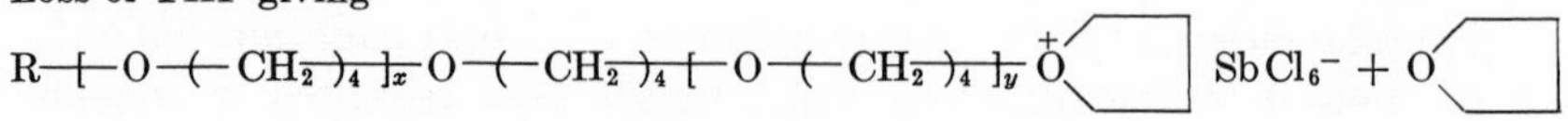

(2) To eliminate a cyclic tetramethyleneoxy polymer

Nucleophilic attack from further back along the chain gives

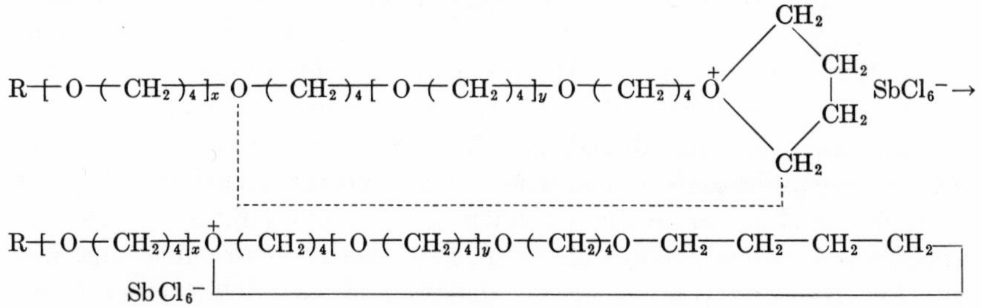

$$R\text{---}[\text{---}O\text{---}(CH_2)_4\text{---}]_x\text{---}O\text{---}(CH_2)_4\text{---}[\text{---}O\text{---}(CH_2)_4\text{---}]_y\text{---}O\text{---}(CH_2)_4\text{---}\overset{+}{O} \quad SbCl_6^- \rightarrow$$

$$R\text{---}[\text{---}O\text{---}(CH_2)_4\text{---}]_x\text{---}\overset{+}{O}\text{---}(CH_2)_4\text{---}[\text{---}O\text{---}(CH_2)_4\text{---}]_y\text{---}O\text{---}(CH_2)_4O\text{---}CH_2\text{---}CH_2\text{---}CH_2\text{---}CH_2\text{---}$$
$$SbCl_6^-$$

which with THF gives

$$R\text{---}[\text{---}O\text{---}(\text{---}CH_2\text{---})_4\text{---}]_x\text{---}\overset{+}{O}\!\!\diagup\!\!\boxed{} \quad SbCl_6^-$$

$$+ \quad O\text{---}(\text{---}CH_2\text{---})_4\text{---}[\text{---}O\text{---}(\text{---}CH_2\text{---})_4\text{---}]_y\text{---}O\text{---}(\text{---}CH_2\text{---})_4\text{---}O\text{---}CH_2\text{---}CH_2\text{---}CH_2\text{---}CH_2\text{---}$$

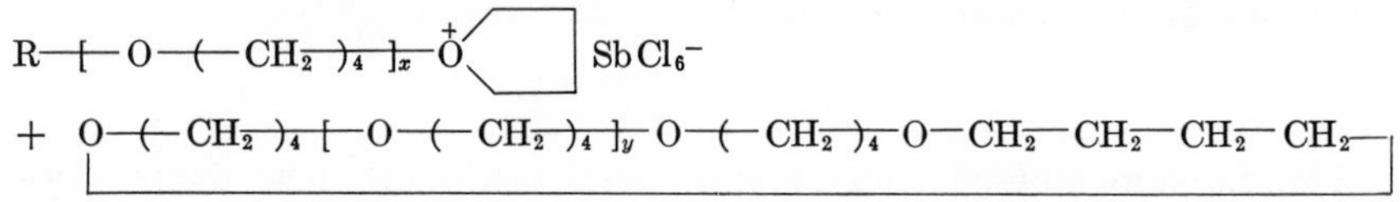

Scheme I

distributed randomly in the system. Scheme I depicts only the intramolecular version of this reaction and clearly, for most systems, it is even more probable that a similar reaction will occur intermolecularly. Before describing the effect of such reactions, it is pertinent to consider the likely environment of the propagating oxonium ions.

Most cyclic ether polymerizations are carried out either in bulk or in diluents of low dielectric constant. Consequently, whatever the nature of the initiating ion pairs, the propagating entities will almost certainly exist at best, as ion pairs. During chain growth it is also highly probable that the ion pairs will agglomerate to larger ionic clusters.

With increase in molecular weight these clusters may even form micelles whereupon chain growth would become a truly heterogeneous reaction. Micelle formation is most likely at high conversions using relatively high catalyst concentrations. At low degrees of conversion and with moderate catalyst concentrations, agglomeration to clusters is the probable behavior of the propagating ion pairs. Degree of agglom-

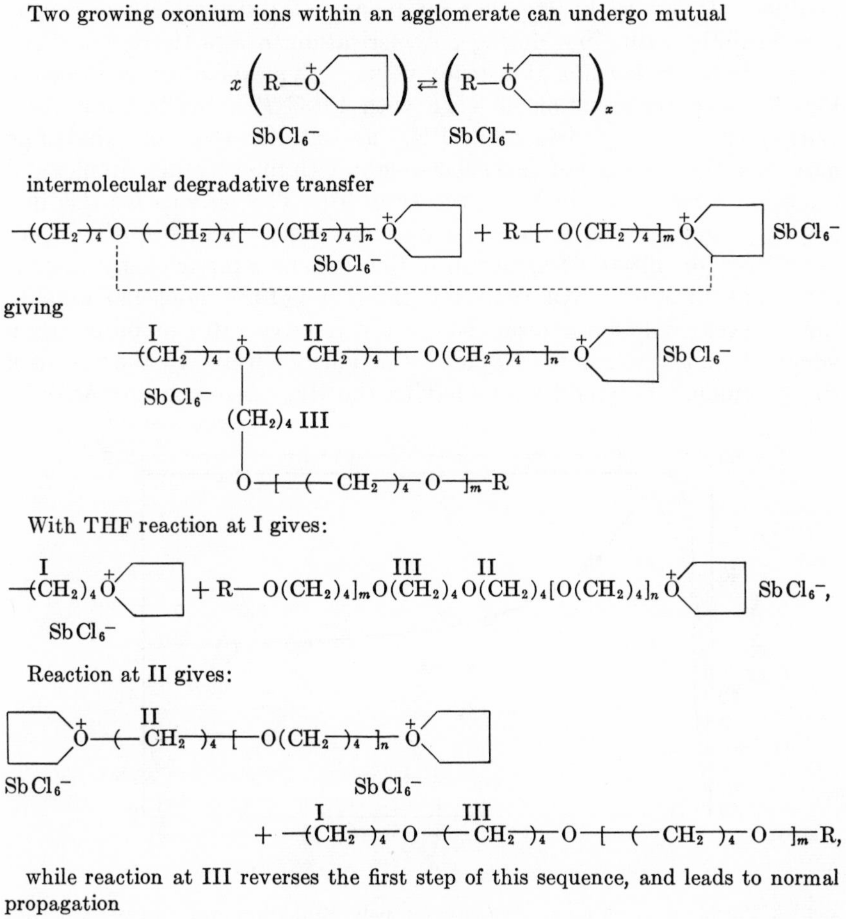

Two growing oxonium ions within an agglomerate can undergo mutual

intermolecular degradative transfer

giving

With THF reaction at I gives:

Reaction at II gives:

while reaction at III reverses the first step of this sequence, and leads to normal propagation

Scheme II

eration will depend markedly on the nature of the negative gegenion as well as on the type of polymeric oxonium ion and hence it is not possible, as yet, to make generalizations about this phenomenon. It is clear, however, that any degree of ion pair cluster formation permits the possibility of intermolecular ether exchange reactions *within a cluster*.

Similar intermolecular reactions can, of course, also occur between "free" ion pairs and "free" polymeric ethers. Scheme II depicts the most likely reactions to be encountered during intermolecular ether exchange processes. Redistribution reactions of this kind would lead to a change in the molecular weight distribution for the particular polymeric product. It is well known that for many catalysts, molecular weight rises initially with time during polymerization, passes through a slight maximum before leveling at a steady value. Typical curves are shown in Fig. 8 for polymerization of THF with $Ph_3C^+PF_6^-$ as initiator (81). With p-chlorophenyl diazonium PF_6^- as initiator (86) and the same monomer, the variation of molecular weight with time is much simpler and there is no maximum in the curve (Fig. 9). The reasons for this discrepancy are not well understood but may reflect presence of transfer impurities (or initiator fragments) in the systems using triphenyl methyl cations as initiators. Alternatively, the shape of the molecular weight–time curves could be determined by the relative rates of propagation versus chain transfer both of which are seriously affected by the nature of the gegenion. Polymerizations having the PF_6^- gegenion appear to be

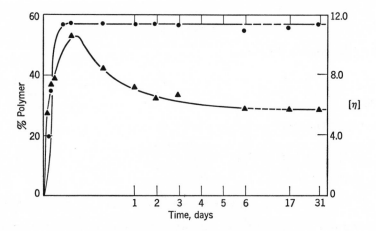

Fig. 8. Variation of intrinsic viscosity for polytetrahydrofuran polymerized with $PH_3C^+PF_6$. ●, % polymer; ▲, $[\eta]$.

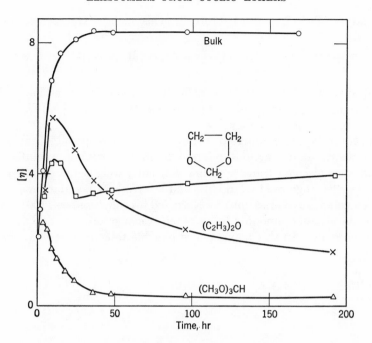

Fig. 9. Comparison of the effect of additives on the intrinsic viscosity of poly(THF) as a function of time (86).

least subject to redistribution and transfer reactions. The initial reaction in redistribution processes involves formation of a trialkyl oxonium ion in which the substitutents are polymeric chains (see reaction 16).

$$\sim \overset{+}{O} \square + O \overset{(CH_2)_4\sim}{\underset{(CH_2)_4\sim}{\big<}} \rightarrow \sim O(CH_2)_4 - \overset{(CH_2)_4\sim}{\underset{+}{O}} \underset{(CH_2)_4\sim}{\big<} \tag{16}$$

Such reactions will occur extensively at equilibrium and would lead to a great increase in viscosity of the polymerizing solution. Rosenberg and his collaborators have reported observation of this phenomenon and have shown that viscosity rapidly falls on addition of small amounts of cation destroying reagents (99).

IX. CHAIN TRANSFER REACTIONS

It is immediately apparent that any cationic polymerization is susceptible to a chain transfer process involving protic reagents, e.g., excess cocatalyst (Eq. 17). When protonic acids such as HSO_3F are used as

$$\sim CH_2 - \overset{+}{O} \diagup \Big| + HY \xrightarrow{\ THF\ } \sim CH_2O(CH_2)_4Y + H - \overset{+}{O} \diagup \Big| \qquad (17)$$

catalyst such transfer reactions are likely to predominate and low molecular weight polymers result. The use of $HClO_4/(CH_3CO)_2O$ systems results in a similar tendency (74).

However, a quite different transfer reaction is possible, similar to the redistribution reactions referred to in the previous section. The significance of chain transfer to low molecular weight acyclic ethers was first clearly recognized by Dreyfuss and Dreyfuss (86), and independently (for trioxane polymerization) by Kern and his collaborators (100). Once again the reaction amounts to an ether exchange process, i.e., as shown in reaction 18. In this manner it is possible to control the equilibrium

$$(18)$$

molecular weight of a given polymerizing system as shown by the curves in Fig. 9. The most efficient chain transfer reagent investigated was trimethylorthoformate which reacts as shown in reaction 19. It follows

$$(19)$$

that polymers prepared in the presence of trimethylorthoformate, will have predominantly methyl endgroups.

X. TERMINATION REACTIONS

The destruction of active centers, rather than transfer of charge to another chain, is not a simple process. Perhaps the most obvious method for chain termination during polymerization would be by anion recombination, i.e.,

$$\sim\overset{+}{O}\diagup\Big|MY^-_{x+1} \rightarrow \sim OCH_2CH_2CH_2CH_2Y + MY_x$$

Conceivably the active center destruction occurring during polymerizations using $Ph_3C^+SbCl_6^-$ results from recombination reactions of this type.

Several groups of workers have made detailed kinetic studies of cyclic ether polymerization (50,51,53,71,72,81). In most cases, however, the reaction kinetics are complicated either by slow initiation reactions and/or competing termination processes. Nevertheless, with certain systems it is possible to demonstrate lack of termination and thus establish a "living" cationic polymerization (50,85,86,107). Polymerizing systems having perfluoro anions such as BF_4^-, PF_6^-, SbF_6^- appear to be the most stable and demonstration of "living" character in the system is easily achieved by waiting until the system reaches equilibrium (or all the monomer is used up) and adding extra monomer, with a consequent increase in molecular weight. Alternatively, addition of a second monomer results in block copolymer formation.

The living ends of growing polyether chains are oxonium ions having a polymeric substituent and will react readily (cf. Et_3O^+) with other cyclic ethers to form block copolymers. However, there is no evidence that trialkyl oxonium ions (R_3O^+) will initiate polymerization of α-olefins and consequently it is not possible to copolymerize cyclic ethers with simple α-olefins. This type of copolymerization can only be expected when the α-olefin contains carbonium ion stabilizing groups such as alkoxy or dialkyl amino (e.g., p-methoxystyrene).

XI. COPOLYMERIZATION OF CYCLIC ETHERS

Many examples of copolymerization of cyclic ethers are known (6, 102,103). In most cases the only evidence for copolymer formation is derived from a comparision of the physical characteristics of the copolymers and the appropriate homopolymers and this may be misleading. Quantitative estimation of relative reactivities in polymerization for cyclic ethers is very difficult because of the complications arising from equilibrium depolymerizations and the likelihood of block copolymer formation. It is possible, however, to estimate the relative basicities of cyclic ethers by several different techniques. Basicity of ethers can be estimated by measurements of equilibrium constants for complex formation with BF_3 (104), i.e.,

$$\overset{+}{Et_2O}-\overset{-}{BF_3} + \stackrel{\displaystyle \frown}{O} \underset{}{\overset{K}{\rightleftharpoons}} \stackrel{\displaystyle \frown}{\overset{+}{O}}-\overset{-}{BF_3} + Et_2O$$

Boron compounds are isoelectronic with carbonium ions and hence measurement of K provides an estimate of the relative reactivities of cyclic ethers toward carbonium ions. Similar information can be obtained from measurements of shift in stretching frequency of the —O—D bond in CH_3OD, when it is hydrogen bonded to the various ethers. These results have been reviewed in detail (102) and show that for cyclic ethers the basicity increases with ring size in the order: 4 > 5 > 6 > 3. For cyclic formals the order is 7 > 5 > 6 and for all types of cyclic ether derivatives, methyl substitution increases the basicity. A comprehensive survey of apparent reactivity ratios for cyclic ether polymerization has been made but the data show only a very crude relationship with the much more reliable basicity values (102).

XII. HOMOGENEOUS ANIONIC POLYMERIZATION OF CYCLIC ETHERS

Reactivity toward hydroxide or alkoxide ions is restricted to epoxides. The overall reaction can be represented, viz (108):

$$RO^- + R'CH\overline{}CH_2 \rightarrow ROCH_2-\overset{\displaystyle R'}{\underset{\displaystyle |}{CH}}-O^-$$

and is induced by a great variety of basic reagents. In solution a repetition of this process provides the first known example of a "living" polymer system. The degree of polymerization increases with time, as expected for a stepwise chain growth reaction, and block copolymers are readily produced by successive additions of the appropriate epoxides. The very nature of the growing species (i.e., alkoxide ions) make it possible to control molecular weights by the introduction of measured amounts of water, alcohols (and other protic reagents) as chain transfer agents. For example,

$$\sim OCH_2CH_2O^- + ROH \rightleftharpoons \sim OCH_2CH_2OH + RO^-$$

The rapid nature of this equilibrium ensures that for homogeneous reactions, any polymer with —OH endgroups is in equilibrium with the growing alkoxide ions and may be regarded as an active center for purposes of computing the expected degree of polymerization (22). In the presence of alcohols uniform molecular weight distributions can be obtained and the molecular weights are low. If hydroxide ion in aqueous media is used as catalyst system, then the resulting polymer has a α,ω-diol structure

$$HO—[CH_2CH_2—O]_n—H$$

Such materials are of special significance as intermediates in the synthesis of polyurethane elastomers. Propylene oxide derivatives are most commonly employed in this manner and a typical polymerization can be represented as in reaction 20. Ring opening reactions of this type do not

$$KOH + CH_3CH{-}{-}CH_2 \xrightarrow{\quad\diagdown\!\!O\!\!\diagup\quad} HO—CH_2—CH—O^-K^+ \xrightarrow{\quad nCH_3CH{-}{-}CH_2 \atop \diagdown\!\!O\!\!\diagup \quad}$$

$$HO{-}\!\!\left[CH_2—\underset{CH_3}{\overset{\overset{\textstyle CH_3}{|}}{CH}}—O\right]_n\!\!—CH_2—\underset{CH_3}{\overset{\overset{\textstyle CH_3}{|}}{CH}}—\bar{O}K^+ \xrightarrow{H_2O}$$

$$HO{-}\!\!\left[CH_2—\underset{CH_3}{\overset{\overset{\textstyle CH_3}{|}}{CH}}—O\right]_n CH_2—\underset{CH_3}{\overset{\overset{\textstyle CH_3}{|}}{CH}}—OH + KOH \quad (20)$$

disturb the configuration of the substituted carbon atom. Thus Price and Osgan (107) were able to polymerize optically active (−)propylene oxide to a crystalline, optically active, isotactic polymer. Homogeneous base catalyzed polymerization proceeds by a stepwise chain growth process, as already indicated. Solutions of KOH in various alcohols or glycols are the most common initiating systems. However, St. Pierre and Price showed (109) that solid KOH was a specific, heterogeneous catalyst for bulk polymerization of propylene oxide. The polymers so obtained were largely unsaturated monohydroxy polyethers with an average molecular weight around 5000, regardless of the monomer to catalyst ratio. KOH, RbOH, and CsOH have been found to be effective catalysts whereas LiOH and NaOH were inactive. Accordingly, it was suggested that this type of anionic polymerization was surface catalyzed, involving a true chain reaction. Chain transfer involved formation of unsaturated endgroups (see reactions 21 and 22). Recent work (32,33,110) suggests

$$n(CH_3CH\!\!-\!\!CH_2)$$

$$KOH(solid) + CH_3CH\!\!-\!\!CH_2 \rightarrow CH_3CH\!\!-\!\!CH_2\!\!-\!\!O^-K^+ \xrightarrow{}$$
$$\underset{O}{\diagdown\diagup} \qquad \underset{OH}{|}$$

$$\overset{OH}{|}$$
$$CH_3CH\!\!-\!\!CH_2\!\!-\!\!(\!\!-\!\!OC_3H_6\!\!-\!\!)_{\overline{n}}\!\!-\!\!O^-K^+ \quad (21)$$

$$ROCH_2\overset{O}{\overset{\diagup\diagdown}{CH}}\quad\overset{\diagdown}{CH_2} \xrightarrow{} \quad\begin{matrix}ROCH_2CH\!\!=\!\!CH_2\\ or\\ ROCH\!\!=\!\!CHCH_3\end{matrix} \quad + \quad \overset{-O}{\underset{HO}{\diagdown\diagup}}\overset{\diagdown}{\underset{CHCH_3}{CH_2}} \quad (22)$$

that these unsaturated endgroups are formed by base catalyzed isomerization of propylene oxide to unsaturated alcohols prior to polymerization. The same unsaturated endgroup then arises during chain initiation as shown in reactions 23, 24, and 25.

$$KOH + CH_3\!\!-\!\!CH\!\!-\!\!CH_2 \rightarrow CH_3\!\!-\!\!CH\!\!-\!\!CH_2\!\!-\!\!\overset{-\,+}{OK} \quad (RO^-K^+) \quad (23)$$
$$\underset{O}{\diagdown\diagup} \qquad \underset{OH}{|}$$

$$RO^-K^+ + CH_3\!\!-\!\!CH\!\!-\!\!CH_2 \rightarrow ROH + CH_2\!\!=\!\!CH\!\!-\!\!CH_2\overset{-\,+}{OK} \quad (24)$$
$$\underset{O}{\diagdown\diagup}$$

$$CH_2\!\!=\!\!CH\!\!-\!\!CH_2\!\!-\!\!\overset{-\,+}{OK} + n\left[CH_3CH\!\!-\!\!CH_2\right] \rightarrow$$
$$\underset{O}{\diagdown\diagup}$$
$$CH_2\!\!=\!\!CH\!\!-\!\!CH_2\!\!-\!\!O\!\!-\!\![CH_2CHO]_{n-1}\!\!-\!\!CH_2\!\!-\!\!CH\!\!-\!\!O^-K^+ \quad (25)$$
$$\underset{CH_3}{|} \qquad\qquad \underset{CH_3}{|}$$

XIII. HETEROGENEOUS POLYMERIZATION OF EPOXIDES

Commercially, the most important homopolymers obtained from epoxides, are produced using heterogeneous catalysts. Catalysts for ethylene oxide polymerization are basic derivatives of the alkaline earth metals, e.g., CaO, SrO, BaO, SrCO$_3$, Ca(NH$_3$)$_6$, etc., and extensive surveys have been made (5,6). In all cases the polyethylene oxide produced is of very high molecular weight (10^4–10^6) and the materials are highly crystalline. Mechanisms for these reactions are not known in detail but clearly the reaction is a special case of conventional homogeneous base catalyzed polymerization—use of the catalyst surface for initiation drastically reducing transfer/termination possibilities.

In the case of substituted ethylene oxides such as propylene oxide and epichlorohydrin, there is an even more extensive range of heterogeneous catalysts available. This latter group is comprised of complex alkoxy derivatives of iron, aluminum, and zinc. There is a significant difference in the polymers produced by the aluminum/zinc catalysts when compared with those produced using iron catalysts and so the two types of catalyst are considered separately.

XIV. IRON CATALYSTS FOR EPOXIDE POLYMERIZATION

Pruitt and Baggett originally disclosed (9) that reaction of anhydrous ferric chloride with four moles of propylene oxide produced a complex material which had a high reactivity for polymerizing propylene oxide to a crystalline, stereoregular product. Later work (111) showed that partial hydrolysis (via adventitious water) was desirable for high catalyst activity and consequently it is now common to partially hydrolyze the complex with controlled amounts of water.

The precise nature and structure of the ferric chloride–propylene oxide catalyst is not yet known but it is clear that partial hydrolysis produces a polymeric iron–oxygen–iron structure, infusible and insoluble, in which there are uncoordinated, highly reactive oxygen atoms in the terminal units of the iron catalyst chains. Evidence for catalyst structure and reaction mechanism has recently been analyzed by Gurgiolo (106) and the most probable reaction path for isotactic polymer formation is shown in Scheme III.

In structure I monomer coordinates with the iron at a vacancy with simultaneous orientation of the unsubstituted ring carbon atom by the abnormally electrophilic oxygen atom. Structure II depicts an incipient bicyclic structure in a concerted reaction leading to a dimer unit, as in III. Translocation of the reactive oxygen atom then occurs, with consequent regeneration of the vacant site necessary for primary coordination of monomer. This process then repeats itself to give polymer as in IV. Catalysts of this type are known to produce large amounts of amorphous polymer concurrently with synthesis of isotactic material. Mechanisms for this process are not well understood but it is thought that a nonconcerted stepwise growth of ferric-alkoxy derivative is responsible.

XV. POLYMERIZATION OF EPOXIDES WITH ZINC AND ALUMINUM ALKYL DERIVATIVES

Although formally similar to the ferric alkoxide catalyzed reaction— now regarded as a coordinated anionic polymerization—the zinc and aluminum systems are even less well defined or understood. Recent

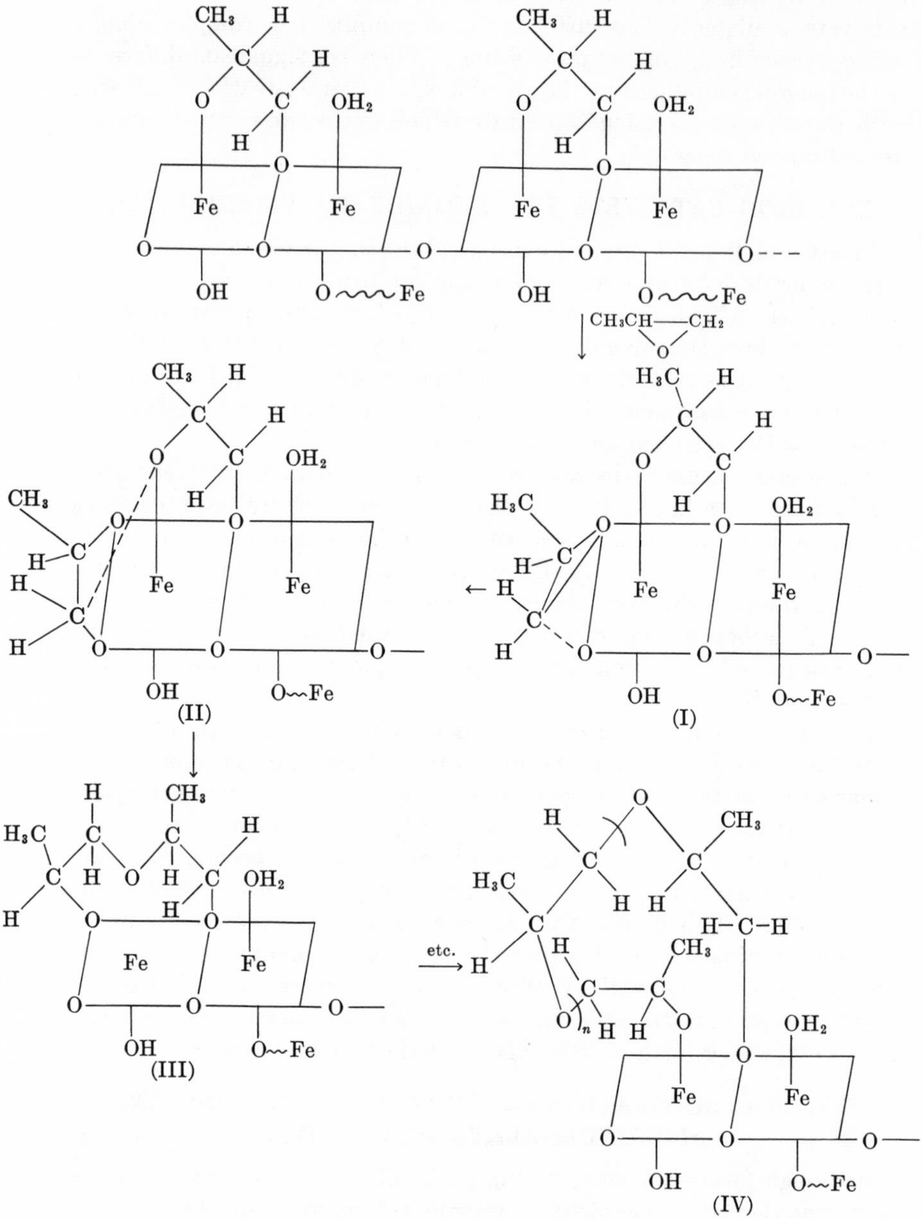

Scheme III

work (30,31,112) suggests that the reactions should be regarded as coordinated cationic polymerizations.

Polymerization of propylene oxide induced by zinc alkyls does not occur in the absence of water (112). The latter causes hydrolysis of zinc–carbon bonds producing condensed zinc–oxygen derivatives. In this

$$ZnR_2 + H_2O \rightarrow RZnOH + RH$$
$$RZnOH + ZnR_2 \rightarrow RZnO-ZnR + RH$$
$$RZnOZnR + H_2O \rightarrow RZnOZnOH + RH$$
$$RZnOZnOH + ZnR_2 \rightarrow RZnOZnOZnR + RH, \text{ etc.}$$

way there will be formed mixtures of polymeric —Zn—O—Zn— derivatives, some of which will be soluble in the reaction media and others (higher molecular weight) completely insoluble. Two types of active centers are known to be formed (112) but details of the polymerization mechanisms are not known. The most probable reaction involves coordination of epoxide with a zinc atom leading to a dipolar oxonium ion (see reaction 26). Propagation would then be formally classified as

$$(26)$$

cationic in nature although many different types of inter- and intramolecular association may be envisaged (106).

Cocatalysis of the zinc alkyl initiated polymerization with alcohols instead of water, produces condensed and complex alkoxy derivatives of zinc which function in a related manner (113). It is interesting (114) that stable coordination compounds can be formed between dimethyl zinc and cyclic ethers such as ethylene oxide, oxetane, tetrahydrofuran, and tetrahydropyran. In these compounds, the stability of the bond between zinc and the ether oxygen atom, increases with increasing ring size. However, the fact that a coordinate bond between ethylene oxide and zinc can be formed at all, without disrupting the ring, is evidence of a specific mode of interaction between the oxygen lone pair electrons and dimethyl zinc.

Aluminum alkyl derivatives are another class of initiating system for epoxide polymerization (5,6). As in the case of zinc alkyls, water is necessary for stereospecific polymerization. Detailed kinetic studies by Colclough and his collaborators (30,31) using trimethyl aluminum, showed that three types of reaction occurred.

1. At 0°C a fast reaction yielding polymer of molecular weight less than 1000. The importance of this reaction falls rapidly after the first few percent conversion.

2. Heating (to 80°C) increased the amount of low polymer formed in *1*.

3. A slow polymerization producing high molecular weight, partly stereoregular polymer.

In the absence of water (*1*) and (*2*) were observed but not (*3*).

Reactions (*1*) and (*2*) were assumed to result from stepwise cationic growth of a dipolar species formed by direct reaction of monomer with aluminum triethyl (Eq. 27). When water was added as cocatalyst,

$$\tfrac{1}{2}Al_2Me_6 + CH_3CH\overset{\diagdown\diagup}{\underset{O}{}}CH_2 \rightarrow CH_3CH \cdots$$

condensation of aluminum trimethyl into polymeric —Al—O—Al— units occurred. Chain growth would then proceed as for the analogous zinc derivatives.

Vandenberg has recently shown (115) that the aluminum alkyl–water catalyst is considerably improved if the aluminum alkyl is first treated with acetylacetone. The resulting chelate, when activated with water then provides a very efficient catalyst for stereoregular polymerization of epoxides, and, in addition, can give new, amorphous polymers of high molecular weight from styrene oxide, epichlorohydrin and other mono-substituted epoxides.

Although there is a formal similarity between the iron and aluminum/zinc–water systems, the former catalyst gives polymers with a very high degree of crystallinity reflecting a high degree of stereoregularity in the backbone chains. Aluminum/zinc–water systems on the other hand, give polymers which are much less crystalline and are therefore more readily usable as elastomers. It has been suggested (30,31) that the latter catalysts produce a block copolymer of stereoregular and amorphous units.

The variation in stereoregularity produced by the various catalyst systems is amply demonstrated by results obtained with optically active systems.

It has already been noted that 1-propylene oxide polymerized to optically active, isotactic polymer using solid KOH or the $FeCl_3$–propylene oxide catalyst (107). Chu and Price (34) have extended the original work and have investigated the polymerization of 1-propylene oxide with both zinc alkyl–water, and ferric chloride catalyst systems. For all catalysts employed the resulting polymer had exactly the same optical rotation, but melting point and crystallinity were much less for the diethylzinc–water catalyst. The same series of metallic components were then converted to optically active catalysts by initial reaction with 1-propylene oxide. These "asymmetric" catalysts were then used to polymerize racemic monomer. It was found that any asymmetry associated with the catalyst was rapidly lost since all the polymers, even at low conversion, had essentially no optical activity.

Independent parallel work by Tsuruta and his collaborators has shown (37) that the diethylzinc catalyst could be made to give higher stereospecificity by using optically active d-borneol as cocatalyst. These workers concluded that the active species producing polypropylene oxide was a polymeric zinc dialkoxide formed *in situ*.

References

1. E. L. Eliel, N. L. Allinger, S. J. Angyal, and G. A. Morrison, *Conformational Analysis*, Interscience, New York, 1965.

1a. J. C. Swallow, *Proc. Roy. Soc. (London)*, **238A**, 1 (1956).

2. K. Yamamoto, A. Teramoto, and H. Fujita, *Polymer*, **7**, 267 (1966).
   K. Yamamoto and H. Fujita, *Polymer*, **7**, 557 (1966).

3. A. M. Eastham, *Fortschr. Hochpolymer Forsch.*, **2**, 1 (1960).

4. J. B. Rose, *The Chemistry of Cationic Polymerization*, P. H. Plesch, Ed., Pergamon, London, 1963, Chapter 11.

5. N. G. Gaylord and H. F. Mark, *Linear and Stereoregular Addition Polymers*, Interscience, New York, 1959; N. G. Gaylord, *Polyethers*, Part I, Interscience, New York, 1963.

6. J. Furukawa and T. Saegusa, *Polymerization of Aldehydes and Oxides*, Interscience, New York, 1963.

7. F. S. Dainton, T. R. E. Devlin, and P. A. Small, *Trans. Faraday Soc.*, **51**, 1710 (1955).

8. P. A. Small, *Trans. Faraday Soc.*, **51**, 1717 (1955).

9. M. E. Pruitt and J. M. Baggett, U.S. Pat. 2,706, 182 (1955) (Dow Chem. Co.).

10. C. E. H. Bawn and A. Ledwith, *Quart. Rev.*, **26**, 361 (1962); M. H. Lehr, *Survey of Progress in Chemistry*, A. F. Scott, Ed., Academic Press, London, 1966, Vol. 3, 184–267.

11. E. J. Vandenberg, *J. Am. Chem. Soc.*, **83**, 3538 (1961); *J. Polymer Sci. B*, **2**, 1085 (1964).

12. E. J. Goethals, *Ind. Chim. Belg.*, **30**, 559 (1965).

13. H. Meerwein, E. Battenberg, H. Gold, E. Pfeil, and G. Willfang, *J. Prakt. Chem.*, **154**, 83 (1939).

14. C. L. Hamermesch and V. E. Hanry, *J. Org. Chem.*, **26**, 4748 (1961); S. C. Temin, *J. Polymer Sci.*, **62**, 591 (1962).

15. M. Okada, Y. Yamashita, and Y. Ishi, *Makromol. Chem.*, **80**, A6 (1964); M. Okada, *Kogyo Kagaku Zasshi*, **65**, 691 (1962); T. Kagiya, M. Hatta, T. Shimizu, and K. Fukui, *Kogyo Kagaku Zasshi*, **66**, 1890 (1963).

16. E. L. Wittbecker, H. K. Hall, and T. W. Campbell, *J. Amer. Chem. Soc.*, **82**, 1218 (1960); H. K. Hall, *J. Am. Chem. Soc.*, **80**, 6412 (1958).

17. A. A. Strepikheev and A. V. Volkhima, *Dokl. Akad. Nauk. SSSR*, **99**, 407 (1954).

18. R. C. Cass, S. E. Fletcher, C. T. Mortimer, H. D. Springall, and T. R. White, *J. Chem. Soc.*, **1958**, 1406.

19. T. P. Hobin, *Polymer*, **7**, 225 (1966); T. P. Hobin, *Polymer*, **7**, 367 (1966); J. Lal and G. S. Trick, *J. Polymer Sci.*, **50**, 13 (1961).

20. E. J. Vandenberg, *Rubber Plastics Age*, **46**, 1139 (1965).

20a. J. G. Hendrickson, A. E. Gurgiolo, and W. E., Prescott, *Ind. Eng. Chem., Prod. Res. Dev.*, **2**, 199 (1963); E. E. Gruber, O. A. Meyer, G. H. Swart, and K. V. Weinstock, *Ind. Eng. Chem., Prod. Res. Dev.*, **3**, 194 (1964); Dunlop Rubber Co. Ltd., Belg. Pat. 623, 510 (1963); Carbide Chemical Corp., U. S. Pat. 3,031,439 (1962); Hercules Powder Co., Brit. Pat., 927,817 (1963); E. J. Vandenberg, U.S. Pat. 3,158,591, (1964); C. C. Price, Brit. Pat., 893, 275 (1962); Dynagen XP-139, Technical Bulletin, The General Tire & Rubber Co., 1963; D. A. Meyer and E. E. Gruber, Intern. Automotive Engineering Congress, Detroit, Mich., Jan. 11–15, 1965, paper 977C.

20b. A. E. Robinson, U.S. Pat., 3,026,270 (1962); A. Kutner, U.S. Pat., 3,058,923 (1962); E. J. Vandenberg, U.S. Pat., 3,135,705 (1964), U.S. Pat. 3,135,706 (1964); U.S. Pat. 3,158,580 (1964); U.S. Pat., 3,158,581, (1964), U.S. Pat., 3,158,591 (1964).

21. H. Meerwein, *Angew. Chem.*, **59**, 168 (1947). H. Meerwein, D. Delfs, and H. Morschel, *Angew. Chem.*, **72**, 927 (1960); K. Hamann, *Angew. Chem.*, **63**, 231 (1951).

22. G. Gee, W. C. E. Higginson, and G. T. Merrall, *J. Chem. Soc.*, **1959**, 1345; G. Gee, W. C. E. Higginson, K. J. Taylor, and M. W. Trenholme, *J. Chem. Soc.*, **1961**, 4298; G. Gee, W. C. E. Higginson, and B. Jackson, *Polymer*, **3**, 231 (1962).

23. D. J. Worsfold and A. M. Eastham, *J. Am. Chem. Soc.*, **79**, 900 (1957); D. J. Worsfold and A. M. Eastham, *J. Am. Chem. Soc.*, **79**, 897 (1957).

24. L. A. Bakalo and B. A. Krentsel, *Usp. Khim*, **31**, 657 (1962).

25. J. Furakawa and T. Saegusa, *Pure Appl. Chem.* **4**, 387 (1962).

26. H. Batzer and W. Fisch, *Kunststoffe*, **17**, 562 (1964).
27. J. Furukawa, T. Tsuruta, T. Saegusa, and G. Kakogawa. *J. Polymer Sci.*, **36**, 541 (1959).
28. E. J. Vandenberg, *J. Polymer Sci.*, **47**, 486 (1960).
29. R. O. Colclough, G. Gee, W. C. E. Higginson, J. B. Jackson, and M. Litt., *J. Polymer Sci.*, **34**, 171 (1959).
30. A. J. Burgess and R. O. Colclough, S. C. I. (London) Monograph No. 20., 1966, p. 41.
31. R. O. Colclough and K. Wilkinson, *J. Polymer Sci.*, *C*, **4**, 311 (1963).
32. D. M. Simonds and J. J. Verbanc, *J. Polymer Sci.*, **44**, 303 (1960).
33. E. C. Steiner, R. R. Pelletrer, and R. O. Trucks, *J. Am. Chem. Soc.*, **86**, 4678 (1964).
34. N. S. Chu and C. C. Price, *J. Polymer Sci.*, *A*, **1**, 1105 (1963).
35. H. Imai, T. Saegusa, and J. Furukawa, *Makromol. Chem.*, **82**, 25 (1965).
36. J. Furukawa, T. Tsuruta, R. Sakata, T. Saegusa, and A. Kawasaki, *Makromol. Chem.*, **32**, 90 (1959).
37. T. Tsuruta, S. Inoue, M. Ishinore, and N. Yoshida, *J. Polymer Sci.*, *C.*, **4**, (1963).
38. E. Booth, W. C. E. Higginson, and E. Powell, *Polymer*, **5**, 479 (1964).
39. L. A. Bakalo, B. A. Krentsel, and A. V. Topchiev, *Vysokomolekul. Soedin.*, **4**, 1361, 1962; *Neftekhimiya*, **3**, 206 (1963).
40. S. Ichida, *Bull. Chem. Soc., Japan*, **33**, 727 (1960).
41. A. Zilkha and M. Weinstein, *J. Appl. Polymer Sci.*, **6**, 643 (1962).
42. J. B. Rose, *J. Chem. Soc.*, 542 (1956).
43. T. Saegusa, H. Imai, and J. Furukawa, *Makromol. Chem.*, 53, 203 (1962).
44. V. A. Kropachev, L. V. Alferova, B. A. Dolgoplosk, *Vysokomolekul. Soedin.*, **5**, 994 (1963).
45. M. Hatano, S. Kambara, *J. Polymer Sci.*, **35**, 275 (1959).
46. I. Penczek, and S. Penczek, *Makromol. Chem.*, **67**, 203 (1963).
47. L. C. Case and C. C. Tood, *J. Polymer Sci.*, **58**, 633 (1962).
48. T. Saegusa, H. Imai, and J. Furukawa, *Makromol. Chem.*, **65**, 60 (1963).
49. R. C. Burrows and B. F. Crowe, *J. Appl. Polymer Sci.*, **6**, 465 (1962).
50. P. R. Johnston, *J. Appl. Polymer Sci.*, **9**, 461 (1965).
51. B. A. Rozenberg, N. V. Makletsova, I. V. Epelbaum, E. B. Lyudvig, and S. S. Medvedev. *Vysokomolekul. Soedin.*, **7**, 70, 1051 (1956); B. A. Rozenberg, E. B. Lyudvig, A. R. Gantmakhers, and S. S. Medvedev, *Vysokomolekul. Soedin.*, **7**, 188 (1965); B. A. Rozenberg, O. M. Chekhuta, E. B. Lyudvig, A. R. Gantmakhers, and S. S. Medvedev, *Vysokomolekul. Soedin.*, **6**, 2030 (1964).
52. E. B. Lyudvig, B. A. Rozenberg, T. M. Zvereva, A. R. Gantmakhers, and S. S. Medvedev. *Vysokomolekul. Soedin.*, **7**, 269 (1965).
53. D. Vofsi and A. V. Tobolsky, *J. Polymer Sci.*, *A*, **3**, 3261 (1965).
54. P. Dreyfuss and M. P. Dreyfuss, *Advan. Polymer. Sci.*, **4**, 457 (1967).
55. M. Kucera and J. Pichler, *Vysokomolekul. Soedin.*, **7**, 3, 10 (1965).
56. J. Kagiya, M. Hatta, T. Shimizu, and K. Fukui, *Kogyo Kagaku Zasshi*, **66**, 1890 (1963).
57. M. Okada, *Kogyo Kagaku Zasshi*, **65**, 691 (1962).
58. M. Okada, Y. Yamashita, and Y. Islu, *Makromol. Chem.*, **80**, 196 (1964).
59. W. Kern, *Chem. Z.*, **88**, 623 (1964); H. Bader, V. Jaaks, and W. Kern, *Makromol. Chem.*, **82** 213 (1965). V. Jaacks and W. Kern, *Makromol. Chem.*, **62**, 1 (1963)

60. L. Leese, M. V. Baumber, *Polymer*, **6**, 269 (1965).
61. T. Higashimura, T. Mika, and S. Okamura, *Bull. Chem. Soc., Japan*, **38**, 2067 (1965).
62. G. V. Rakora, L. M. Romanov, and N. S. Enikolopyan, *Vysokomolekul. Soedin.*, **6**, 2178 (1964).
63. V. Rabauvanov, R. Mateva, and M. Natov, *Compt. Rend. Acad. Bulgare Sci.*, **18**, 821 (1965).
64. M. Baccaredda, E. Butta, and P. Giusta, *J. Polymer Sci., C*, **4**, 953 (1964).
65. M. Kucera and E. Spousta, *Makromol. Chem.*, **82**, 60 (1965). M. Kucera and E. Spousta, *J. Chem. Soc.*, **1965**, 1478.
66. P. F. Onyon and K. J. Taylor, *European Polymer J.* **1**, 133 (1965).
67. J. Lal and G. S. Trick, *J. Polymer Sci.*, **50**. 13 (1961).
68. W. Ziegenbeim and K. H. Hornung, Ger. Pat. 1,159,651 (1965).
69. K. Weissermel and E. Nölken, *Makromol. Chem.*, **68**, 140 (1963).
70. E. L. Muetterties, T. A. Bither, M. W. Farlow, and D. D. Coffman, *J. Inorg. Nucl. Chem.*, **16**, 52 (1960).
71. D. Sims, *J. Chem. Soc.*, **1964**, 864.
72. R. M. Bell, Ph.D. Thesis, Liverpool, 1963.
73. A. C. Farthing and R. J. Reynolds, *J. Polymer Sci.*, **12**, 503 (1954).
74. T. Shono, T. Tsujino, and Y. Hachihamg, *Kogyo Kagaki Zasshi*, **61**, 1357 (1958).
75. I. Yamashita and M. Serizawa, *Bull. Chem. Soc., Japan*, **37**, 1721 (1964).
76. T. Saegusa, H. Imai, and J. Furukawa, *Makromol. Chem.*, **54**, 218 (1962).
77. S. I. Trotz (Olin Matheson Co.), U.S. Pat. 3,219,595 (1965).
78. A. Ledwith and J. Weightman, unpublished work.
79. C. E. H. Bawn, R. M. Bell, and A. Ledwith. Chemical Society Anniversary Meeting, Cardiff, 1963; *Polymer*, **6**, 95 (1965).
80. C. E. H. Bawn, R. M. Bell, C. Fitzsimmonds, and A. Ledwith, *Polymer*, **6**, 661 (1965).
81. C. Fitzsimmonds, Ph.D. Thesis, Liverpool, 1966.
82. G. W. Cowell and A. Ledwith, unpublished work.
83. I. Kuntz, *A. C. S. Polymer Preprints*, 187 (1966 (Jan.)).
84. K. Kuntz, *J. Polymer Sci., B.*, **4**, 427 (1966).
85. M. P. Dreyfuss and P. Dreyfuss, *Polymer*, **6**, 93 (1965).
86. M. P. Dreyfuss and P. Dreyfuss, *J. Polymer Sci., A*, **4**, 2179 (1966).
87. D. B. Miller, *A. C. S. Polymer Preprints*, No. 2, 613 (1966).
88. Y. Takegami, T. Ueno, and R. Hirai, *Bull. Chem. Soc., Japan*, **38**, 1222 (1965).
89. J. Holmes and R. Petitt, *J. Org. Chem.*, **28**, 1695 (1963).
90. S. Okamura, T. Higushimura, and K. Takeda, *Makromol. Chem.*, **51**, 217 (1962); S. Okamura, K. Hayashi, and Y. Kitamishi, *J. Polymer Sci.*, **58**, 925 (1962); A. Chapiro and S. Penczek, *J. Chim. Phys.*, **59**, 696 (1962); S. Nakashio, M. Kondo, H. Tsuchita, and M. Yamada, *Makromol. Chem.*, **52**, 79 (1962); H. Rao and D. S. Ballentine, *J. Polymer Science, A*, **3**, 2579 (1962); K. Hayashi, H. Ochi, and S. Okamura, *J. Polymer Sci., A*, **2**, 2979 (1964).
91. S. Okamura, E. Kobayashi, M. Takeda, T. Higashimura, and K. Tomikawa, *J. Polymer Sci., C*, No. 4, 827 (1964); S. Okamura, E. Kobayashi, and T. Higashimura, *Makromol. Chem.*, **88**, 1 (1965).
92. H. Rausch, *Angew. Chem.*, **72**, 927 (1960).
93. F. S. Dainton and K. J. Ivin, *Quart. Rev.*, **12**, 61 (1958).

94. F. S. Dainton, K. J. Ivin, and D. A. G. Walmesley, *Trans. Faraday Soc.*, **50,** 1784 (1960); F. S. Dainton, F. E. Hoare, and T. P. Melia, *Polymer*, **3,** 263 (1962).
95. A. V. Tobolsky and A. Eisenberg, *J. Colloid Sci.*, **17,** 49 (1962).
96. K. J. Ivin and J. Leonard, *Polymers*, **6,** 621 (1965).
97. B. Capon, *Quart. Rev.*, **28,** 45 (1964).
98. G. A. Latremouille, G. T. Merrall, and A. M. Eastham, *J. Am. Chem. Soc.*, **82,** 120 (1960).
99. B. A. Rosenberg, Y. B. Lyudvig, A. R. Gantmakher, and S. S. Medvedev., *Vysokomolekul. Soedin.*, **7,** 188 (1965).
100. V. Jaacks, H. Baader, and W. Kern, *Makromol. Chem.*, **83,** 56 (1965).
101. E. A. Ofstead, *A. C. S. Polymer Preprints*, **6,** 674 (1965).
102. Y. Yamashita, T. Tsuda, M. Okada, and S. Iwatsuki, *J. Polymer Sci.*, *A*, **4,** 2121 (1966).
103. T. Saegusa, T. Uestima, H. Imai, and J. Furukawa, *Makronol. Chem.*, **79,** 221 (1964); T. Saegusa, H. Imai, and J. Furukawa, *Makromol. Chem.*, **56,** 55 (1962); K. Wessermel, E. Fischer, K. Gutweiler, and H. D. Herman, *Kunststoffe*, **54,** 410 (1964); A. Ishigaki, T. Shono, and Y. Hachihama, *Makromol. Chem.*, **79,** 170 (1964); L. A. Dickenson, *J. Polymer Sci.*, **58,** 857 (1962); M. Okada, N. Takikawa, S. Iwatsuki, Y. Yamashita, and Y. Ishi, *Makromol. Chem.*, **82,** 16 (1965); T. Tsuda, T. Nomura, and Y. Yamashita, *Makromol. Chem.*, **86,** 301 (1965).
104. M. Okada, K. Sugana, and Y. Yamashita, *Tetrahedron Letters*, No. 28, 2329 (1965).
105. S. Iwatsuki, N. Takigawa, M. Okada, Y. Yamashito, and Y. Ishi, *J. Polymer Sci.*, *B*, **2,** 549 (1964); C. Quivoran and J. Neel, *Compt. Rend.*, **259,** 1845 (1964).
106. A. E. Gurgiolo, *Rev. Makromol. Chem.*, **1,** 39 (1966).
107. C. C. Price and M. Osgan, *J. Am. Chem. Soc.*, **78,** 4787 (1956).
108. R. E. Parker and N. S. Isaacs, *Chem. Rev.*, **59,** 737 (1959).
109. L. E. St. Pierre and C. C. Price, *J. Am. Chem. Soc.*, **78,** 3432 (1956).
110. C. C. Price and W. H. Snyder, *J. Am. Chem. Soc.*, **83,** 1773 (1961).
111. A. B. Borkovec, *J. Org. Chem.*, **23,** 828 (1958).
112. C. Booth, *Polymer*, **5,** 479 (1964).
113. R. Sakota, T. Tsuruta, T. Saegusa, and J. Furukawa, *Makromol. Chem.*, **40,** 64 (1960).
114. K. H. Thiele, *Z. Anorg. Allg. Chem.*, **319,** 183 (1962).
115. E. J. Vandenberg, *J. Polymer Sci.*, **47,** 481, 489 (1960).

CHAPTER 5

ELASTOMERS BY CATIONIC MECHANISMS
D. Elastomeric Polyacetals

O. VOGL

*Central Research Department, E. I. du Pont de Nemours & Co.,
Wilmington, Delaware*

Contents

I. INTRODUCTION

Polyacetals are polyaddition products of aldehydes.

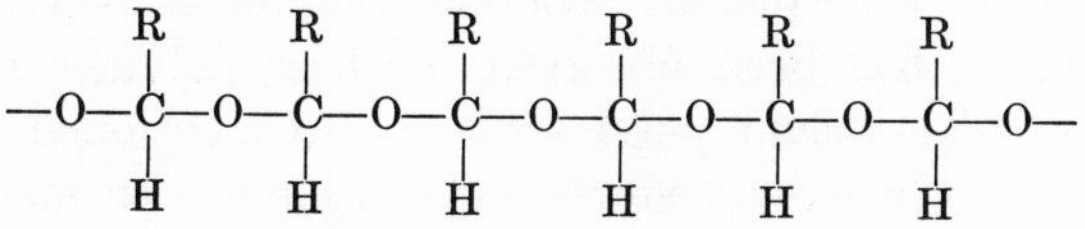

$$-O-\underset{\underset{H}{|}}{\overset{\overset{R}{|}}{C}}-O-\underset{\underset{H}{|}}{\overset{\overset{R}{|}}{C}}-O-\underset{\underset{H}{|}}{\overset{\overset{R}{|}}{C}}-O-\underset{\underset{H}{|}}{\overset{\overset{R}{|}}{C}}-O-\underset{\underset{H}{|}}{\overset{\overset{R}{|}}{C}}-O-\underset{\underset{H}{|}}{\overset{\overset{R}{|}}{C}}-O-$$

R = alkyl

They are still sometimes, improperly, called polyethers.

Polyformaldehyde, the lowest member of the family of aldehyde polymers, is known only as a crystalline polymer. It is a plastic rather than an elastomer. Most higher members of the polyaldehyde family have been prepared in an amorphous and a crystalline form. The crystallinity of the polymers is associated with their stereoregularity. All known stereoregular higher polyaldehydes are isotactic and possess a helical structure with four monomers in the repeat unit. Syndiotactic polyacetaldehydes have not yet been prepared.

This chapter is concerned with the amorphous and elastomeric polyaldehydes and not with stereoregular and crystalline aldehyde polymers. Efforts have been made to segregate the reports on crystalline polyaldehydes from those on amorphous and elastomeric polyaldehydes.

Elastomeric polyacetals, particularly polyacetaldehyde, have become of interest as potential rubber candidates in recent years. There are several reasons for this development: (1) the monomer is competitive in cost with other monomers that are used for the preparation of elastomers, (2) methods have been worked out to polymerize aldehydes rapidly and reproducibly, and (3) substantial advances have been made in the stabilization of polyaldehydes. However, higher polyaldehydes still do not seem to have the stability necessary for commercial development.

Of all polyaldehydes known, atactic polyacetaldehyde has been studied most extensively. Amorphous polyacetaldehyde is soluble in a number of solvents, and when obtained at high molecular weight, it has properties similar to uncured rubber.

The amorphous state and elastic properties of polyacetals are associated with the lack of stereoregularity of the polymers. The tacticity of elastomeric polyacetaldehyde has recently been determined.

Most reports in the literature on aldehyde polymerization are concerned with polyacetaldehyde. Polyacetaldehyde will, therefore, be the major topic of this review.

Elastomeric polymers of aliphatic aldehydes higher than acetaldehyde have been obtained as by-products in attempts to polymerize them to the corresponding crystalline, isotactic polymers. These polyaldehydes are generally of lower molecular weight, often described as sticky, tacky, amorphous products or just as "soluble fraction." Not infrequently, some crystallinity has been observed in these polymer preparations. Very low molecular weight polymers are also mentioned as semisolid masses and oils. No serious efforts seem to have been made, however, to prepare very high molecular weight elastomers of aldehydes higher than acetaldehyde.

It is the purpose of this article to report the polymerization of the aldehydes to their elastomeric polymers, the physical mechanical properties of the polyaldehydes, and to assess the work on polyaldehydes critically. No effort will be made to evaluate the merit of polyaldehydes as potential rubber candidates.

Polyacetone, which is actually a polyketone, will be included here since it has been described as an elastomeric polymer and is formally closely related to acetal polymers.

II. POLYACETALDEHYDE

The availability of acetaldehyde and the substantial knowledge of its polymerization that has been accumulated during the years have made polyacetaldehyde the most widely investigated polyaldehyde.

A. Background

Travers (1) and Letort (2) found independently that a rubbery polymer was obtained when acetaldehyde was frozen and then allowed to melt. Staudinger (3) suggested that this polymer was a polyacetal. This proposal for the polyacetaldehyde structure was later confirmed by an IR study (4), and it was found that the polymer was amorphous (5).

Over the next 20 years, Letort's school (6–11) and other investigators (5,12) studied the condition of this polymerization in detail. It consisted of crystallizing the monomer and allowing it to melt. This technique was later called "crystallization polymerization."

Progress was made during the years in improving the polymer yield by standardizing the polymerization conditions (13,14). It was found that polymerization occurred at the melting point (8) and actually during the process of melting (15). Virtually no polymer was formed in the crystalline phase of acetaldehyde before melting (5) (Fig. 1).

The rate of melting was found to affect the polymerization significantly (16). Whereas initially it was thought that the crystallization polymerization was spontaneous and uncatalyzed (6), later studies showed that several impurities, like oxygen and acetaldehyde peroxides (17), had an accelerating effect. The first reproducible process for preparing polyacetaldehyde consisted of adding a known amount of oxygen to acetaldehyde and initiating with UV light (14). Even so, the yield was rarely more than 40%. In view of the fact that the initiator was prepared by an obviously free radical route, the polymerization was thought to occur by a free radical mechanism.

A few years later, Letort found that acids could be used very effectively

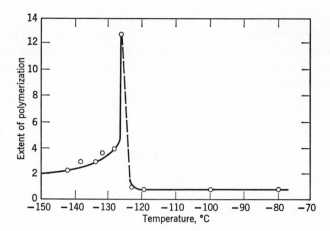

Fig. 1. Effect of temperature on acetaldehyde polymerization.   Extent of polymerization is measured by determining the viscosity of the solution (5).

as initiators in the crystallization polymerization (18) and could give
more reproducible yields.

Apparently unknown to all the investigators interested in acetaldehyde
polymerization because of an incorrect structure assignment, was a patent
issued to Smyers (19) describing a low temperature solution polymerization of acetaldehyde using BF_3 as initiator.   The properties of the
polymer described in this patent were clearly those of Letort's polyacetaldehyde, and not polyvinyl alcohol as Smyers had claimed.

As other methods became known for the polymerization of acetaldehyde to the elastomer, it was recognized that polyacetaldehyde of high
molecular weight is prepared generally by a cationic mechanism (20).
Two major techniques were developed for acetaldehyde polymerization:
bulk polymerization in the presence of insoluble initiators and solution
polymerization with Lewis acid- and Brönsted acid-type initiators.

B. Polymerization with Solid Initiators

One group of cationic initiators that produce polyacetaldehyde of very
high molecular weight are insoluble initiators.   Inorganic oxides, metal
salts, and even glass serve as initiators.

The type of surface and the surface area play an important role in the
polymerization of acetaldehyde.   In some cases it was demonstrated that
doping of the surfaces with acids greatly improves the conversion (21).

First, the crystallization polymerization will be discussed in the light of recent developments in acetaldehyde polymerization. Letort and Mathis (18) had shown that some acid is beneficial for the crystallization polymerization of acetaldehyde. In subsequent work (21) it has been pointed out that the surface of the reaction vessel is of eminent importance for the polymerization. Untreated Pyrex glass or glass washed with acid—and subsequently rinsed with water and dried—gave good yield of polymer while alkali treatment of the same glass reaction vessel completely inhibited polymerization (21). In the case of free radical initiation as in the case of controlled acetaldehyde peroxidation (13,14,17), the cationic initiator was probably formed by a free radical reaction and the polymerization proceeded via a cationic polymerization.

Aluminum oxides were found to be excellent initiators for the acetaldehyde polymerization, even above the melting point, as reported by Furukawa (22,23).

The alumina has to be calcined between 600–1000°C for 15 hr to be active (24) (Table I). Best results were obtained when the acetaldehyde was condensed from the vapor phase onto the alumina which was kept at

TABLE I

Polymerization of Acetaldehyde by Alumina (24)

(Acetaldehyde, 15 ml; alumina, 5 g)

	Alumina				
Origin	Calcination temp., °C	Water content, %	Polym. time, hr[b]	Conversion, %	Inherent visc.[c] [η]
A[a]	200	20.3	43	0	—
A	400	3.6	43	49	7.1
A	600	1.4	43	55	7.2
A	800	0.5	43	91	5.4
A	1000	0.2	43	76	6.6
A	1150	0.0	43	8	0.03
B[d]	—[e]	13.5	65	23	3.9
B	600	1.6	65	66	2.4
B	1000	0.2	43	57	4.3

[a] Prepared by hydrolysis of aluminum isopropoxide.

[b] Polymerization temperature, −70°C.

[c] In 2-butanone at 27°C.

[d] Commercial alumina (8–14 mesh).

[e] Commercial alumina used without calcination.

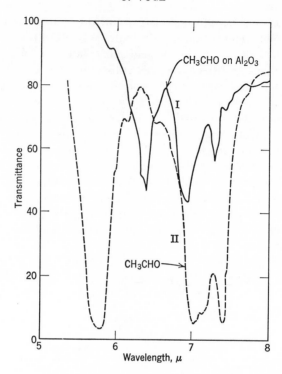

Fig. 2. Initiation of acetaldehyde polymerization by alumina (24).   Infrared spectra of acetaldehyde: I (solid line), adsorbed on alumina; II (dotted line), liquid.

−70°.   Furukawa has shown by infrared spectroscopy that a monomolecular layer of acetaldehyde reacts initially with the alumina to   form   a

$$
\begin{array}{ccc}
— O & & CH_3 \\
& \searrow & \diagup \\
— O — \bar{Al} \leftarrow \overset{+}{O} = C & \\
& \diagup & \diagdown \\
— O & & H
\end{array}
$$

bond (24) (Fig. 2).   This initiator is capable of propagation to form high polymer of acetaldehyde.   Polyacetaldehyde formed on glass surfaces was also obtained in high yields when the monomer was condensed from the vapor phase (21).   Very low yields were noticed when the flask was filled at room temperature with liquid acetaldehyde and subsequently cooled to perform the crystallization polymerization (24).   Similar low yields were obtained by Furukawa in the alumina initiated polymerization of acetaldehyde (22) when liquid acetaldehyde was added to the alumina and then cooled. Both examples suggest that the monomer must be able to penetrate into electrophilic lattice imperfections of the insoluble initiator.

Treatment of alumina with acids produced more effective initiators. HCl (25), trifluoroacetic acid (26), and BF_3 (27) have been reported to improve many solid initiators.

Metal oxides other than Al_2O_3 have been described as initiators for acetaldehyde polymerization. CrO_3, MoO_3 (24,28), and SiO_2 (24,28–31) and silica alumina (31–36) gave high molecular polymers with acetaldehyde (27,29,30) (Table II). A number of other metal oxides gave traces of a sticky, tacky polymer, whereas others were found to be inactive (24,28).

Silica gel (Table III), a mixture of silica gel and Al_2O_3 with polyformaldehyde (Table IV), and other polymers (Table V), activated with BF_3, have been recommended for the preparation of high molecular weight polyacetaldehyde in good yield.

It was found that inorganic salts, particularly sulfates, are also effective initiators for acetaldehyde polymerization (37). These salts were preferably used calcined at 100–230° for 4 hr (Figs. 3, 4). It could

TABLE II
Acetaldehyde Polymerization with Solid Initiators

Solid initiator	Polym. time, hr	Polym. temp. °C	Yield, %	Ref.
Al_2O_3[a]	2	−80	9	28
	5	−50	14	28
	10	−20	3	28
SiO_2	2	−80	11	28
	15–60	−70	45	29
Al_2O_3 techn.	60	−70	4	29
Al_2O_3 pure	30–60	−70	28	39
	16	−70	59	34
	20	−70	64	34

Initiator	DP	Conversion, %	Ref.
Alumina	5300	55–60	36
Silica alumina	3590	40–50	36
Thorium oxide	1410	30–40	36

[a] Solvent: pentane.

TABLE III
Acetaldehyde Polymerizations with
Cationically Activated Solid Initiators[a]
(Polymerization Temperature, −70°) (29)

Solid initiator	Activated with ml BF$_3$ gas	Polym. time, hr	Yield, %	Soln visc., [η]
Silica gel	2	20	46	4.0
Silica gel	1	20	40	4.7
Silica gel	0.5	20	45	5.8
	0.5	8	38	5.7

[a] Dried *in vacuo* for 24 hr at 400–500°.

TABLE IV
Acetaldehyde Polymerization with Cationically
Activated Solid Initiator Mixtures[a] (29)

Wt ratio silica gel:polyoxymethylene	Yield, %	Soln visc., [η]
1:1	51	7.8
1:2	67	4.3
1:6	58	9.5
1:13	44	4.0
1:60	2	—

[a] Silica gel and polyoxymethylene activated with 0.5 ml BF$_3$ gas. Polymerization time, 15 hr.

TABLE V
Acetaldehyde Polymerization with Cationically
Activated Initiator Mixtures
(Silica Gel Modified with Various Organic Polymers)[a] (29)

Additive	Yield, %	Soln visc., [η]
Polyoxymethylene	53	9.7
Polypropylene	61	8.9
Polyvinyl chloride	68	5.6
Polyoxymethylene[b]	42	3.4

[a] Weight ratio of silica gel: polymer, 1:2.5, polymerization temperature, −70°, polymerization time, 15 hr. Activated with 0.5 ml BF$_3$ (gas).
[b] With alumina.

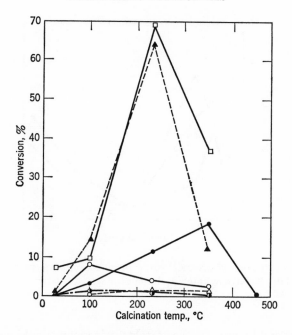

Fig. 3. Polymerization of acetaldehyde with sulfates of divalent metals (37). O, zinc sulfate, ●, nickel sulfate; ◑, manganese sulfate; △, mangesium sulfate; ▲, cupric sulfate, □, ferrous sulfate; ✕, calcium sulfate.   Initiator, 1.0%; polymerization temperature, −78°; polymerization time, 20 hr.

be shown that $CuSO_4$ and $FeSO_4$ (Table VI) gave good conversion to polymer and even $Al_2(SO_4)_3$ and $Fe_2(SO_4)_3$ (Fig. 4) were active initiators, while $CaSO_4$ and $MnSO_4$ were poor.   Takida and Noro were able to relate the acidity of the initiator at optimum calcination temperature to the conversion of acetaldehyde to polymer.   The degree of polymerization (DP) of these polymers ranged between 300 and 1000 (some DP's were as high as 4000); the yield was also dependent on the dispersion of the solid support (38).

Better results were obtained with sulfate–sulfuric acid complexes (21, 39), and polymers were obtained in higher conversions and of higher molecular weights.   Aluminum sulfate–sulfuric acid (21) was extensively studied, but other complexes, such as $CuSO_4$ (40), $SrSO_4$, $MgSO_4$, $ZnSO_4$, $SnSO_4$, $MnSO_4$, and $NiSO_4$, were also active (Table VII).   The polymerization could be initiated with as little as 0.05% of the initiator without affecting conversion and DP.   The polymerization seemed to be essentially complete after 4 hr.

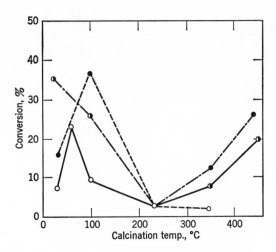

Fig. 4. Polymerization of acetaldehyde with sulfates of trivalent metals (37).   O, aluminum sulfate; ●, ferric sulfate; ◐, chromic sulfate.   Initiator, 1.0%; temperature, −78°C; time, 20 hr.

The effect of diluent upon the acetaldehyde polymerization was studied with metal–sulfate–sulfuric acid complexes.   In the case of magnesium sulfate–sulfuric acid as the initiator, n-hexane (up to 50% hexane) gave a higher conversion to polymer than neat acetaldehyde.   The conversion

TABLE VI
Polymerization of Acetaldehyde with Various Metal Sulfates (37)[a]

Metal salt	Calcination[b] temp,[c] °C	Conversion,[d] %
$ZnSO_4 \cdot 7H_2O$	100	7.9
$NiSO_4 \cdot 7H_2O$	350	18.3
$MnSO_4 \cdot 7H_2O$	100	1.4
$MgSO_4 \cdot 7H_2O$	230	1.3
$CuSO_4 \cdot 5H_2O$	230	64.9
$FeSO_4 \cdot 7H_2O$	230	68.0
$CaSO_4 \cdot 2H_2O$	230	0.9
$Na_2SO_4 \cdot 10H_2O$	up to 350	0

[a] Initiator, 1%; temperature, −78°; time, 20 hr.
[b] Calcination was carried out for 4 hr.
[c] The calcination temperature giving the higher yield of polymers is reported.
[d] The DP of these polymers ranged from 300–800.

TABLE VII
Polymerization of Acetaldehyde with Various Metal
Sulfate–Sulfuric Acid Complexes (40)[a]

Metal salt	Composition of the complex (metal sulfate: $H_2SO_4{}^m : H_2O^n$)		Conversion, %	DP
	m	n		
$CuSO_4$	0.003	0.20	65.7	1750
$SnSO_4$	0.20	0.30	61.4	600
$MgSO_4$	0.66	1.58	66.7	685
$ZnSO_4$	0.05	1.16	63.2	5040
$Al_2(SO_4)_3$	0.81	2.17	75.1	1780
$Fe_2(SO_4)_3$	0.39	1.48	67.9	720
$SnSO_4$	0.34	0.90	46.7	430
$MnSO_4$	0.55	1.02	32.8	580
$NiSO_4$	0.03	1.12	6.4	620

[a] Initiator, 0.5%; temperature, −78°; time, 20 hr.

dropped slightly at higher dilution of acetaldehyde. When toluene was used as diluent, the polymer yield was generally lower.

Aluminum sulfate–sulfuric acid was described in more detail as the initiator (21) for the acetaldehyde polymerization at initial polymerization temperatures of −78°C. In a bulk polymerization a rapid conversion to paraldehyde was noticed. The addition of 10% of toluene gave a mixture of paraldehyde and the elastomeric polyacetaldehyde. When more than 20% of toluene was added to the mixture, the paraldehyde formation was suppressed and only polyacetaldehyde was formed even at −78°C.

Sulfates of organic amines were also described as giving effective initiators for acetaldehyde polymerization. Hydroxylamine sulfate, for example, gave as much as 61% polyacetaldehyde.

It appears that the success of these acetaldehyde polymerizations with inorganic salts depends very much upon the effective acidity of the initiating salt and the relative solubility of these compounds in the polymerization mixture. It has even been suggested that the polymerization proceeds in solution rather than on the solid surface (40).

C. Solution Polymerization of Acetaldehyde

The initiators described in the previous section are insoluble under the conditions of polymerization, namely at temperatures between −78 and

−50°, usually in neat acetaldehyde but sometimes in solution of hydro-carbon solvents. This is not always clear when initiators are apparently liquids at room temperature and completely soluble in the reaction mix-ture. All these initiators will be discussed in the following section. BF₃-etherate, for example, is an excellent initiator for acetaldehyde polymerization under the proper conditions; it is soluble at room tem-perature in pentane but crystallizes when the solution is cooled below −65°. When acetaldehyde polymerization is then carried out, BF₃-ethe-rate could actually be considered an insoluble initiator (20). Very little is known about the solubility of the initiators at the reaction temperature, and the literature does not give a detailed description on this point nor is the miscibility of acetaldehyde with the solvents at reaction temperature clearly stated. Acetaldehyde. for example, is not miscible with all aliphatic hydrocarbons at low temperature; with pentane a phase separation occurs at about −30°. A butyraldehyde solution in pentane separates at about −90°.

It has been pointed out (20,41) that Brönsted and Lewis acids of an acid strength that corresponds to a pH of higher than 2 are initiators for acetaldehyde polymerization. Lewis acids, such as BF_3, $AlCl_3$, AsF_3, $POCl_3$, $TiCl_4$, and protonic acids, such as HCl, HNO_3, CF_3COOH, and CCl_3COOH, are typical initiators. The use of an appropriate cationic initiator does not guarantee the formation of high molecular weight polymer because side reactions can completely eliminate the for-mation of a high molecular weight polymer. The most important side reaction in cationic acetaldehyde polymerization is the formation of the cyclic trimer paraldehyde and the cyclic tetramer metaldehyde.

Paraldehyde is the sole product when acetaldehyde is treated at room temperature with a strong acid. Some metaldehyde is formed in addi-tion to paraldehyde when the initial reaction temperature is lowered to −10 to −30°. Below this temperature, polymerization to high molecular weight is possible. It is not sufficient to start the acetaldehyde poly-merization at a very low temperature. For successful results, the reac-tion mixture must be kept below this temperature throughout the poly-merization. This is not always easy because acetaldehyde polymerization is very rapid and releases a substantial amount of heat (∼7 kcal/mole). A rapid polymerization raises the interior temperature rapidly. If H_2SO_4 or BF_3 was added, for example, to bulk acetaldehyde at −80°, even with efficient outside cooling, the temperature rose rapidly to the boiling point of acetaldehyde. Only paraldehyde, but no polymer, was obtained (21) (Table VIII).

TABLE VIII
Acetaldehyde Polymerization with Very Active Initiators[a]
(Paraldehyde Formation) (21)

Initiator	Solvent	Initiator conc., wt % based on monomer	Monomer: solvent ratio	Remarks[b]
H_2SO_4	Diethyl ether	0.1	2:1	$\oplus$ ($-20°C$)[c]
P_2O_5	Diethyl ether	0.05	2:1	$\oplus$
BF_3- etherate	Diethyl ether	0.1, 0.02	2:1	$\oplus$ ($-10°C$)[c]
$FeCl_3$	Diethyl ether	0.2, 0.05	10:1, 2:1	$\oplus$ (boiled over)[c]
	CH_2Cl_2	0.2	1:1	$\oplus$ ($+10°C$)[c]
	Diethylpentane	0.2, 0.05	1:1	$\oplus$ ($+20°C$)[c]
$SnCl_4$	Diethyl ether	0.1, 0.02	2:1	$\oplus$
$TiCl_4$	Diethyl ether	0.1, 0.02	2:1	$\oplus$
TiF_4	Diethylpentane	0.05	2:1	$\oplus$
$ZrCl_4$	Diethyl ether	0.05	2:1	$\oplus$
PCl_3	Diethyl ether	0.5	2:1	$\oplus$
$SbCl_3$	Diethyl ether	0.5	2:1	$\oplus$

[a] Initial reaction temperature, $-65°C$, reaction time, 5 hr.

[b] $\oplus$ denotes paraldehyde formation; if not noted otherwise, the yield was close to quantitative.

[c] Indicates the temperature to which the system rose during polymerization.

Even when acetaldehyde was diluted with a solvent of relatively high dielectric constant, such as diethyl ether or tetrahydrofuran, the polymerization proceeded very rapidly with active initiators to form paraldehyde exclusively. Lower dielectric constant solvent must be used in order to overcome paraldehyde formation. Ether–pentane combinations gave good polymers with BF_3 as the initiator at an initial temperature of $-78°$ (42).

It was found that ethylene was an excellent solvent for acetaldehyde polymerization. The polymerization proceeded rapidly and in quantitative yield without the formation of paraldehyde as a side product (43). Moderately active initiators gave exclusively polymer even in ether solvents (43) (Table IX).

The rate of acetaldehyde polymerization varies substantially depending upon the reaction conditions but is always very fast. It has not yet been measured for solution polymerizations because the reaction is complete in seconds or minutes. With solid initiators, the polymerization is much slower and reaction times of several hours have been reported. One

TABLE IX
Acetaldehyde Polymerization with Moderately Active Initiators[a]
(Polymer Formation)[b] (21)

Initiator	Solvent	Initiator conc., wt % based on monomer	Monomer:solvent ratio
H_3PO_4	Ether	0.5	10:1
HCl (conc. aqueous)	Ether	0.2	2:1
HNO_3 (conc.)	Ether	0.2	2:1
CF_3COOH	Ether	0.1	2:1
AsF_3	Ether	0.1	2:1
$AsCl_3$	Ether	0.15	2:1, 10:1
SbF_3	Ether	0.05	2:1
$AlCl_3$	Ether	0.1	2:1
Adipoyl chloride	Ether	0.2	2:1
$ZnCl_2$	Ether	0.5	2:1
SbF_3	CH_2Cl_2	0.15	1:1
CCl_3COOH	Ether	0.1	1:1
$ZrCl_4$	Ether	0.015	2:1

[a] Initial reaction temperature, $-65°C$; reaction time, 5 hr.
[b] Polymer was formed in good yield in all cases.

reason why molecular weights are usually higher in polymerizations with solid initiators, with low boiling solvents, or in the crystallization polymerization is that temperature rise and side reactions, such as chain transfer and trimerization, are avoided.

A number of phosphorus-containing compounds have been studied as initiators for acetaldehyde polymerization.

Trialkylphosphines were found to be reasonably good initiators (44,45). It was initially suggested that the basicity of the phosphine is responsible for the effectiveness of these initiators because triphenylphosphine was inactive. Unfortunately, no further study has apparently been undertaken to gain more insight in this interesting type of initiation, but it appears that the initiation is nucleophilic in character.

Japanese workers have studied the initiation of acetaldehyde polymerization using polyphosphoric acid as initiator (46–49). As in the case of trialkylphosphine, the polymers are of low molecular weight.

The authors have studied the acetaldehyde polymerization with polyphosphoric acid alone or in combination with amines, tertiary ammonium salts, urea, amides, and even quinones (46–49). Various metal oxides and metal salts were found useful as carriers for polyphosphoric acid and gave polyacetaldehyde with improved yield. Polyphosphoric acid dispersed on solid carriers, such as glass wool and asbestos, increased the reaction rate and yield, but the molecular weights remained low.

Neutral phosphates, such as sodium hexametaphosphate and potassium metaphosphate (46), are reported to have been activated by phosphoric acid, polyphosphoric acid, and sulfuric acid. Using radioactive phosphoric acid to follow the reaction, the authors concluded that acetaldehyde was first adsorbed on the initiator to form a new C—O—P bond. Polymerization occurred by further addition of acetaldehyde; the polymerization was inhibited by amines.

D. Aluminum Organic Compounds as Initiators

Aluminum organic compounds are good initiators for aldehyde polymerization (50,51). They are, however, reported here very briefly because they generally form substantially isotactic, insoluble fractions, and the rubbery elastomer is obtained in most cases as by-product, called soluble fraction, and is rarely well characterized. It is, however, safe to say that these soluble polymers are of relatively low molecular weight and belong to the sticky and tacky variety.

When Et_3Al was used as initiator for acetaldehyde polymerization, the mixing of initiator with acetaldehyde at the polymerization temperature (Fig. 5) gave a higher proportion of soluble material (Table X)

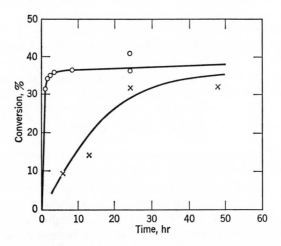

Fig. 5. Rate of acetaldehyde polymerization with $AlEt_3$ as initiator at two different mixing temperatures (52). O, catalyst addition at 20°C; ×, catalyst addition at −78°C; charge, n-hexane 30 cc, acetaldehyde, 0.227 mole, $AlEt_3$, 0.00439 mole.

TABLE X
Effect of Temperature of Initiator Addition upon the
Polymer Composition (51)[a]

Addition temp.	Conversion, % wt	Methanol soluble fraction, % wt
−78°	9.2	74
20°	19.6	54

[a] Initiator (AlEt$_3$) 5%; polymerization temperature, −78°C; time, 20 hr; no solvent.

than mixing the components at room temperature prior to polymerization (52,53).

Recently, Furukawa and coworkers (54,55) have studied the polymerization of acetaldehyde with Et$_2$AlCl, Al$_2$Et$_3$Cl$_3$, and AlEtCl$_2$ and have concluded that they all behave as Lewis acids and induce cationic polymerization of acetaldehyde. AlEt$_3$ and AlEt$_3$–H$_2$O (1:0.5) also produced exclusively the amorphous elastomer when the vapor of acetaldehyde was added to the initiator. The temperature was not allowed to rise substantially above −78°. At higher temperatures, substitution reactions of acetaldehyde with the initiator occurred, and a stereoselective initiator was formed (56). These initiators produced stereoblock polymers of acetaldehyde.

Aldehyde polymerizations initiated with aluminum organic compounds often retain aluminum in the polymer which cannot be removed without destroying the polymer. Such a composite structure of aluminum in the polyaldehyde can substantially change the properties of the polymer. When such polymers are characterized, the aluminum content should always be reported or the absence of aluminum should be mentioned.

E. Radiation Polymerization

Polyacetaldehyde of high molecular weight was obtained by irradiating acetaldehyde with γ-rays at low temperatures in the liquid as well as in the solid state (57,58). Polymers obtained by initiation in the solid state were of substantially higher molecular weight than those prepared in the liquid state. The former were among the highest molecular weight polyacetaldehydes ever obtained. Although benzoquinone seemed to inhibit the polymerization, it was suggested that the x-ray induced polymerization of acetaldehyde was a cationic process (57).

Rubbery polypropionaldhyde was also obtained from propionaldehyde under similar conditions using ionizing radiation.

X-rays induced polymerization of acetaldehyde in the crystalline state (59). No polymerization was obtained, however, when acetaldehyde was liquid or in the glassy state.

F. Mechanism of Acetaldehyde Polymerization

Polymerization of acetaldehyde to the elastomeric atactic polymer is generally considered a cationic reaction. This is based on the experience that Brönsted and Lewis acids, metal oxides with acidic sites, acidic metal salts, and high energy radiation are good initiators for this polymerization (11,21,24). Most of these initiators are also effective in polymerizations of other monomers, such as trioxane and isobutylene, which are known to polymerize only cationically. The usefulness of such compounds in the well-known acid catalyzed trimerization of acetaldehyde to paraldehyde was also used for comparison. In addition to typical cationic initiators, there are a few compounds such as triethylaluminum (51) or triethylphosphine (44), which also initiate acetaldehyde polymerization to give the elastomer. The electrophilic behavior of these compounds in acetaldehyde initiation can be deduced from the fact that they produce the atactic, elastomeric polyacetaldehyde. As will be pointed out in Section II-H, the microstructure of the atactic polyacetaldehyde varies within narrow limits regardless of the cationic initiator used for the polymerization.

Cationic initiation of aldehyde polymerization requires the addition of an electrophile (E) to the carbonyl oxygen (Eq. 1a).

$$\underset{\underset{H}{|}}{\overset{\overset{R}{|}}{C}}=O + E \rightarrow {}^{\oplus}\underset{\underset{H}{|}}{\overset{\overset{R}{|}}{C}}-O-E^{\ominus} \tag{1a}$$

$$\underset{\underset{H}{|}}{\overset{\overset{R}{|}}{C}}=O + A^{\ominus}H^{\oplus} \rightarrow A^{\ominus}\underset{\underset{H}{|}}{\overset{\overset{R}{|}}{C}}=O^{\oplus}-H \tag{1b}$$

The electrophile may be the proton of the protic acid which protonates the carbonyl oxygen to form a hydroxyl endgroup and the corresponding oxonium ion (Eq. 1b). This oxonium ion is capable of being subjected to an electrophilic attack on the terminal carbon atom by the carbonyl oxygen of the incoming monomer thus forming a new oxonium ion. The new oxonium ion is then capable of further propagation. In the past, cationic polymerization of heteroatom- containing monomers such as epoxides have been almost exclusively depicted as a carbonium ion polymerization. It should be emphasized here that it is unlikely that the carbonium ion carries the positive charge of cationic aldehyde polymerization.

In order to have effective propagation, the anion of the ion pair must be of low nucleophilicity.

Initiation with Lewis acids is more difficult to explain. If a cocatalyst, such as water or alcohol, is needed for the initiation, the actual polymerization initiator would be a complex protic acid and the mechanism for the polymerization with protic acids would apply. There is no indication, however, that a cocatalyst is needed in acetaldehyde polymerization.

It is generally assumed, at least in polymerizations with BF_3 initiator, that the first reaction is the direct addition of BF_3 to the carbonyl oxygen to form I

$$BF_3 + O{=}\overset{\overset{\displaystyle R}{|}}{\underset{\underset{\displaystyle H}{|}}{C}} \rightarrow F_3\overset{\ominus}{B}{-}\overset{\oplus}{O}{=}\overset{\overset{\displaystyle R}{|}}{\underset{\underset{\displaystyle H}{|}}{C}} \rightarrow \overset{\oplus}{\sim\sim}O{=}\overset{\overset{\displaystyle R}{|}}{C}{-}H$$

I

Subsequently, I can then polymerize, possibly in its associated form, or the dipolar adduct I can rearrange into other ion pairs capable of propagation.

Since our knowledge about chain termination and chain transfer in cationic acetaldehyde polymerization is still very vague, it will not be elaborated here. It is, however, worth mentioning that unlike anionic aldehyde polymerization, which gives relatively low molecular weight polymers, cationic acetaldehyde polymerization can produce very high molecular weight polymers.

Some attempts have been made to answer the question: Why do typical anionic initiators favor the formation of crystalline-isotactic polyaldehydes and cationic initiators amorphous polymers? It must be pointed out that, unlike most other amorphous vinyl polymers which are about 80% syndiotactic, amorphous polyacetaldehyde is predominantly isotactic; 65–75% acetaldehyde placements are isotactic (in terms of diads). This would mean that during the polymerization of acetaldehyde the addition of monomer to the growing polymer chain is predominantly iso-tactic i.e., *cis* addition takes place. On the average, the incoming mono-mer can rotate in every third to fifth addition and it forms a syndiotactic or *trans* linkage. Occasional rotation seems to be much easier in cationic than in anionic polymerization. This may suggest a greater charge sep-aration and less coordination of the growing ion pair.

Recently, a somewhat different explanation for the formation of atactic polyacetaldehyde has been suggested (67). It was proposed that propa-gation in acetaldehyde polymerization occurs by addition of acetaldehyde as the dimer. It has been known for some time that acetaldehyde is associated in aggregates of 2–5, particularly in solvents of low dielectric constant and at low temperatures. It has been proposed that acetalde-hyde exists in solution as discrete dimers which are in a *meso* arrangement. In acetaldehyde polymerization these dimers add in a syndiotactic manner to give highly heterotactic (in terms of triads) polymer. In addition to this polymerization as associated linear dimers, some mono-meric acetaldehyde is also incorporated, which addition leads to isotactic placement.

While this suggestion appears attractive for acetaldehyde polymeri-zation, it cannot explain the relatively high portion of crystalline isotactic polymer in the cationic polymerization of the more sterically hindered *n*-butyraldehyde and particularly isobutyraldehyde.

Much more work, especially on model compounds, is needed to estab-lish fully the mechanism of cationic aldehyde polymerization.

G. Molecular Weight of Polyacetaldehyde

Polyacetaldehyde is soluble in a number of solvents. Butanone and ethyl acetate solutions (0.05–0.5%) have been used traditionally for the determination of solution viscosities. Elastomeric polyacetaldehyde can be of very high molecular weight, and intrinsic viscosities as high as 8 have been obtained. The molecular weight of polyacetaldehyde has been estimated (60) from the modified Staudinger equation $[\eta] = KM^{\alpha}$ by assuming a value of 0.65 for α and calculating K from the data of Muthana and Mark (61). These authors reported a number average molecular weight of 5.1×10^5 from osmotic pressure measurements on a sample of polyacetaldehyde in butanone, in which the intrinsic viscosity of the polymer was 2.75. From these values K is calculated as 5.36×10^{-4}.

This viscosity–molecular weight relationship has been used subse-quently by all investigators. Weissermel and Schmieder (29) have reinvestigated this relationship (Fig. 6) and found that the molecular weights are too high when this K is used to calculate $\bar{M}_n$.

They also found that unfractionated polyacetaldehyde contains a low molecular weight portion in the range of 4,000–15,000. This low molecular weight portion tends to contribute significantly to the osmotic pressure measurement; the amount of the contribution depends upon the selectivity of the membranes (Table XI).

Fractionated polyacetaldehyde was measured to clarify this problem and a $\bar{M}_n$ of 163,000 was observed on a sample that had an intrinsic viscosity $[\eta]$ of 4.1. These authors propose an equation of $[\eta] = 1.68 \times 10^{-3} \times \bar{M}_n^{0.65}$ as a better representation of the viscosity–molecular weight relationship and suggest that all the data in the literature should be corrected accordingly.

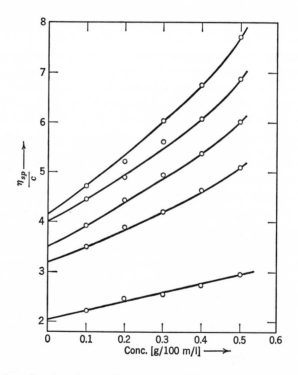

Fig. 6. Solution viscosity of amorphous polyacetaldehyde in dependence on concentration at 25°C (29).

TABLE XI
Solution Viscosities and Osmotic Molecular Weights of Polyacetaldehyde (29)

$[\eta]$, dl/g	$\bar{M}_n$	Sample
2.04	122,000	Unfractionated polymer
3.20	92,000	Unfractionated polymer
3.50	98,000	Unfractionated polymer
4.00	195,000	Unfractionated polymer
4.10	163,000	Fractionated polymer

H. Structure of Elastomeric Polyacetaldehyde

Soon after its discovery, polyacetaldehyde was recognized as a poly-acetal (3). This was later confirmed by an IR study (4). Polyacetaldehyde was also found to be amorphous by x-ray investigation (5). It was recognized that the concept of stereoregularity applies equally well to polyacetaldehyde as it does to polypropylene, since polyacetaldehyde has an ether-oxygen atom in place of the CH_2 group of polypropylene and neither of these groups contribute to the stereochemistry of the polymers. It has been argued by Letort that, since polyacetaldehyde was made by crystallization polymerization, the crystal structure of acetaldehyde should have a determining influence upon the structure of the polymer. A syndiotactic structure was suggested for elastomeric polyacetaldehyde (62–64). Since a syndiotactic structure appeared very unlikely it was desirable to establish definitely the structure of this polymer.

Polyacetaldehyde is one of the simplest examples of a polymer whose stereoregularity can be studied by high resolution NMR. The poly-acetal linkages in polyacetaldehyde in the main chain prevent spin-spin splitting of the signal which in the case of vinyl polymers arises from the continuous arrangements of hydrogen atoms on the C—C backbone of the chain. The side group in polyacetaldehyde can interact only with the single hydrogen atom of the same monomeric unit and vice versa. In addition, the signals of the tertiary H and the methyl group are far apart and can be observed easily.

Independent work in two laboratories (65–67) has shown that elas-tomeric polyacetaldehyde made with a variety of initiators, including Letort's polymers, has essentially the same structure. Using the A–60 Varian NMR spectrometer, it was found that polyacetaldehyde is two-thirds heterotactic and one-third isotactic (Fig. 7) in terms of triads or

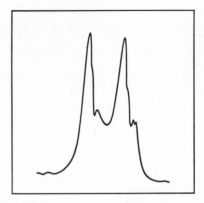

Fig. 7. NMR spectrum of the methyl spectrum of a 10% solution of polyacetaldehyde in perchloroethylene at 93°C (66). Chemical shifts: CH₃ group, 1.31 ppm (CH
group, 5.10 ppm) relative to TMS.

in terms of triads or two-thirds DD and one-third DL in terms of diads.
A small amount of syndiotactic placements was indicated by inspection
of a polyacetaldehyde spectrum obtained from a 100 Mc NMR instrument. An even higher resolution was obtained using the 200 Mc NMR
instrument, and a secondary structure with the new concept of pentads
was suggested for polyacetaldehydes.

I. Degradation

Polyacetaldehyde as a polyacetal is subject to several types of degradation, all of which yield monomer in essentially quantitative yields.

    1. Acid degradation
    2. Base degradation
    3. Radical degradation

Acid degradation can commence at each acetalic oxygen of the polymer
chain because of the nucleophilic character of the oxygen atoms. This
will lead to random chain scission of the polymer. This chain scission
is followed by complete depolymerization of both chain fragments.

$$\begin{array}{ccccccc}
R & & R & & R & & R \\
| & & | & & | & & | \\
-C-O- & C-O- & \rightarrow & -C-O-A & + & {}^{\oplus}C-O- \\
| & | & | & & | & & | \\
H & A & H & & H & & H
\end{array}$$

$$A = H^{\oplus} \text{ or Lewis acid}$$

Base degradation occurs from the chain ends when unstable or hydrolyzable groups are present at the polymer ends. The polyacetal structure when properly end capped, e.g., with methyl endgroups, is base stable.

$$
\begin{array}{cccc}
\text{R} & \text{R} & \text{R} & \text{R} \\
| & | & | & | \\
-\text{C}-\text{O}-\text{C}-\text{O}-\text{X} & \rightarrow & -\text{C}-\text{O}-\text{C}-\text{O}^{\ominus} & \rightarrow \text{degradation} \\
| & | & | & | \\
\text{H} & \text{H} & \text{H} & \text{H}
\end{array}
$$

X = H or acyl

Another type of degradation which is very important for practical purposes is the so-called thermal degradation. It is believed that conditions of true thermal degradation, namely the thermal homolytic dissociation of the —C—O— bond a random, free radical process, are rarely achieved in the case of polyacetaldehyde depolymerization. Normally inactive impurities which have not been completely removed from the rubbery polymer can become active and promote the ionic depolymerization at elevated temperatures but still below the temperature at which homolytic dissociation would occur in the absence of these impurities.

Delzanne and Smets have pioneered the study of polyacetaldehyde degradation in CHCl₃ solution by means of acids (68) and free radicals (69). They were able to show that the degradation proceeds in two steps:

1. Chain cleavage
2. Depolymerization

The chain cleavage results in a drop in molecular weight. Chain cleavage is first order in acid with an overall activation energy of 22.8 kcal/mole. In the case of the degradation by peroxides, the chain cleavage is first order with an overall activation energy of 29 kcal/mole.

The depolymerization is a first order reaction in both cases. The acid depolymerization has an activation energy of 18.8 kcal/mole and the radical depolymerization has an activation energy of 20.8 kcal/mole.

J. Stabilization

Impure polyaldehydes are not very stable and depolymerize even at room temperature. The reason for this instability is the low ceiling temperature of the polymers. The ceiling temperature of polyacetalde-

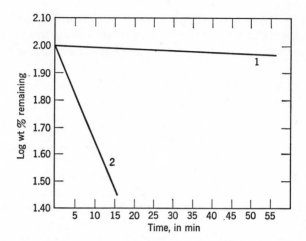

Fig. 8. Thermal stability of uncapped polyacetaldehyde (nitrogen atmosphere). 1, sample contains 0.1% of sym-di(β-naphthyl)-p-phenylenediamine. $k_{111}° = 0.15\%/min$, $\bar{M}_n = 1.48 \times 10^6$; 2, sample contains no additive. $k_{111}° = 9.8\%/min$, $\bar{M}_n = 1.35 \times 10^6$.

hyde is $-18°$. The stability can be improved by: (1) removing all impurities from the polymer, (2) end capping, and (3) addition of antioxidants and thermal stabilizers.

Ionic impurities, particularly cationic initiator residues, are very detrimental to the stability of the polymer and cause random scission of the polymer chain. They are often removed, however, during the subsequent capping reaction. We have shown that crude polymer can be

TABLE XII
Stabilization of Polyacetaldehyde (70)

Additive	Amount	No additive, $k_{111}°$	additive, $k_{111}°$
DNPD[a]	0.2	9.6	0.79
Nylon (66/610/6)	2.0	9.6	0.13
Nylon (66/610/6) ⎱ DNPD ⎰	2.0 0.2	9.6	0.075
Nylon (66/610/6)	2.0	0.89	0.028
Dixylenol butane	0.2	0.89	1.83

[a] DNPD = sym-di(β-naphthyl)-p-phenylenediamine.

stabilized by the addition of aromatic amines (Fig. 8). Better results are obtained, however, when the crude polymer is capped with acetic anhydride using pyridine as the catalyst. This capping has most frequently been carried out in solution (70–73) at room temperature, but higher temperatures, and higher (74) pressures, have been recommended for the acetate capping process.

Table XII shows a comparison of stabilized samples of polyacetaldehyde using various stabilizers (70,72). The stability varies greatly depending upon the stabilizers and the combination of stabilizers used. It may be seen that a secondary amine alone, such as *sym*-di (β-naphthyl)-*p*-phenylene-diamine (70), gave a substantial amount of stabilization. Further stabilization could be effected by the addition of a polyamide. It should be pointed out that even the addition of the polyamide alone improved the stability of polyacetaldehyde. The action of the aromatic amine is not strictly that of an antioxidant. We have shown that when *sym*-di(β-naphthyl)-*p*-phenylenediamine was replaced by a phenolic

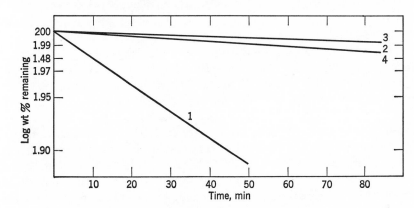

Fig. 9. Thermal stability of acetate-capped polyacetaldehyde (nitrogen atmosphere). Curves 2 and 3 are essentially identical.

Sample no.	Additives	Amount, %	$k_{111°}$	Solution Viscosity [η]	
				Initial	After 90 min at 111°
1	—	—	0.51	1.78	—
2	Nylon (66/610/6)	2	0.026	1.48	1.28
3	Nylon (66/610/6)	2	0.024	1.49	1.37
	DNPD	0.2			
4	Nylon (66/610/6)	2	0.045	1.52	1.28
	Dixylenylbutane	0.2			

TABLE XIII

Stabilization of Acetate-Capped Polyacetaldehyde
with Various Amounts of Nylon Stabilizers (70)

Nylon (66/610/6),[a] %	No additive, $k_{111°}$	With additives	
		$k_{111°}$	$k_{138°}$
0.1	3.23[b]	0.39	—
0.5	3.23	0.030	0.26
1.0	3.23	0.031	0.27
2.0	3.23	0.019	0.17
5.0	3.23	—	0.13

[a] Sample contains 0.2% sym-di (β-naphthyl)-p-phenylenediamine.
[b] Uncapped sample had a $k_{111°}$ of 9.6%/min.

antioxidant (Fig. 9), the stability of polyacetaldehyde was not improved. Even in combination with a polyamide, phenolic antioxidants did not stabilize polyacetaldehyde to any extent.

Table XIII shows a comparison of polyacetaldehyde samples containing various polyamides. It can be seen that the stability varies greatly depending upon the polyamide used. While there is no apparent connection between structure of the polyamide and the degree of stability of the polyacetaldehyde, compatibility of the polyamide with polyacetaldehyde seems to be important. It is interesting that monomeric amides, e.g., adipamide, are not very effective as stabilizers for polyacetaldehyde.

The degradation rate constants $k_{111°}$ and $k_{138°}$ have been used as the measure of stability of polyacetaldehyde. The degradation constant k is derived by plotting the log wt % remaining against the time at constant temperature. After some initial loss of weight, first order rates k are usually observed. When extrapolated to time O, an intersect is sometimes obtained which is often referred to as an "unstable fraction." This unstable fraction does not constitute a significant portion of the polymer and usually does not exceed 5%. In the case of isothermal polyacetaldehyde degradation, toluene (bp 111°) and p-xylene (bp 138°) were used as the bath liquids. The degradation is always carried out in nitrogen atmosphere, and the sole product of polyacetaldehyde degradation is acetaldehyde.

Other authors have used different ways to describe the stability of polyacetaldehyde. Weissermel et al. (29) used the weight loss of poly-

TABLE XIV
Thermal Stability of Polyacetaldehyde (72)

Stabilizer A, 2%	Stabilizer B, 5%	Weight loss after 30 min at 165° in nitrogen, %
β-Naphthylamine	—	92
β-Naphthylamine	P-1[a]	16
β-Naphthylamine	P-2[b]	16
β-Naphthylamine	P-3[c]	21
Diphenylamine	—	96
Diphenylamine	P-1	26
Diphenylamine	P-2	27
Diphenylamine	P-3	19

[a] P-1 = Poly (N-methoxymethyl) caprolactam.
[b] P-2 = Polyamide from sebacic acid and glycol bisaminopropyl ether.
[c] P-3 = Polyamide from oxalic acid and methyl-di-β-aminopropylamine.

acetaldehyde after 30 min. at 165° under nitrogen as their standard for the stability measurement. These authors have also found that both an amine and a polyamide seem to be necessary for effective stabilization of polyacetaldehyde (Table XIV). Neither the amine nor the polyamide alone gave good stability. Weissermel et al. (72) reported that not only aromatic but also aliphatic amines are good stabilizers. However, they found that aromatic amines are more effective than aliphatic ones. As can be seen in Table XV, stearyl amine as well as tributyl amine definitely stabilized polyacetaldehyde when used in combination with polyamides.

TABLE XV
Thermal Stabilization of Polyacetaldehyde (72)

Amine,[a] 2%	Weight loss after 30 min. at 165° in nitrogen, %
Diphenylamine	19
Stearylamine	42
Tributylamine	26

[a] In addition, 5% of a polyamide from oxalic acid and N-methylpropylene was added to the polyacetaldehyde.

K. Properties of Polyacetaldehyde

Samples of polyacetaldehyde were formed into films and bars suitable for measurement of mechanical properties by solvent casting (60) or compression molding (21,29). The mechanical properties of these samples are, generally speaking, similar to those of uncured rubber. There is a sharp drop of elastomeric properties at about −10°, the elongation decreases from 600 to 15%, and the tensile strength increases accordingly (Table XVI).

Torsional modulus of polyacetaldehyde was measured under various experimental conditions. Using Gehman's procedure, the transition temperature of polyacetaldehyde was found to be −7°(60). The transition temperature of natural rubber is −54°. At 0.6 cps, a main amorphous transition of −18° (41) was reported for polyacetaldehyde and a small transition at about −100° (Fig. 10). At 100 Hertz polyacetaldehyde had a transition temperature of 0° (29). Weissermel (29) suggested that in analogy with polypropylene a glass transition temperature at −30 to −40° should be expected for polyacetaldehyde. Measurements according to ASTM D746 gave a T_b of 10° for polyacetaldehyde and −58° for natural rubber.

Nuclear spin lattice relaxation was determined on a sample of polyacetaldehyde (75). A low temperature minimum at −99.5° seems to be associated with the rotation of methyl groups. A higher temperature minimum at 43.5° seems to be associated with a less defined motion of larger sections of the polymer chain. A certain amount of motion was noticeable starting from the glass transition temperature, which was suggested to be at about −40°.

Dielectric studies have been carried out on a sample of polyacetaldehyde by G. Williams and have given an apparent T_g of 30.4° (76).

TABLE XVI
Mechanical Properties of Elastomeric Polyacetaldehyde (21)

Temperature, °C	Tensile strength, psi	Ultimate elongation, %
23	25	580
−10	170	15
−40	990	11

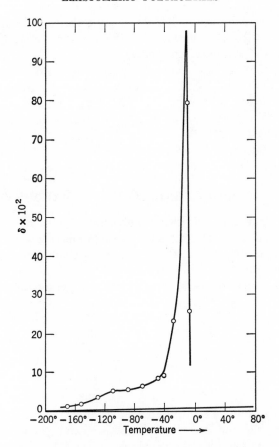

Fig. 10. Internal friction of polyacetaldehyde (21).

III. HIGHER POLYALDEHYDES

Polymers of higher aliphatic aldehydes were first prepared by Bridgeman (77), Conant (78,79), and co-workers by subjecting n-butyraldehyde, isobutyraldehyde, and heptaldehyde to pressures of 10,000 atm. These polyaldehydes have been described as oils or amorphous and soft waxes.

A soluble poly-n-butyraldehyde has been obtained in polymerizations of n-butyraldehyde with Lewis acids, such as $SnCl_4$ and $SnBr_4$, at low temperatures (80). It was also reported that n-butyraldehyde and methoxybutyraldehyde were polymerized (40) to a low molecular weight, soluble polymer by the magnesium sulfate–sulfuric acid complex (38,39).

A polymer of chloroacetaldehyde, completely soluble in chloroform, was obtained with BF$_3$-etherate as the initiator (81); however, the polymer was not characterized further.

It appears that the low molecular weight of higher polyaldehydes is caused by chain transfer to monomer. Chain transfer to monomer plays an important role when the rate of polymerization is decreased due to the steric hindrance of bulky side chains in higher aldehydes. In the case of polyaldehydes obtained by polymerizing higher aldehydes with metal sulfate–sulfuric acid complexes, the lower molecular weight is probably due to protic impurities or excess sulfuric acid in the initiator.

IV. CYCLOPOLYMERIZATION OF DIALDEHYDES

Glutaraldehyde and substituted glutaraldehydes were polymerized with Lewis acid initiators (82–88). The polymers are amorphous and soluble in a number of organic solvents, but insoluble portions were also observed in some cases. Investigation of their structure revealed that a substantial portion of the polymer consists of pyrane units obtained by

cyclopolymerization of glutaraldehyde. Some glutaraldehyde units had, however, polymerized through the carbonyl group in a "normal" aldehyde polymerization. Succinaldehyde has also been polymerized to an amorphous cyclopolymer (89).

All these polymers appear to be of low molecular weight ([η] of up to 0.4), and ultimate physical properties do not seem to have been determined.

V. ALDEHYDE COPOLYMERS

Elastic copolymers of acetaldehyde with formaldehyde have been prepared using Al(iso-Bu)$_3$ (90), amines (91), and a 1:1 complex of diethylaluminum chloride/TiCl$_4$ as the initiators (92). The composition of the copolymer was determined by thermally degrading the polymer and analyzing the monomer composition by gas chromatography. The polymer conposition appeared to be close to the comonomer feed (92).

TABLE XVII

Copolymerization of Acetaldehyde with Various Other Aldehydes (93)

Comonomer	Wt %	BF$_3$ initiator, ppm	Comonomer content of copolymer, %
CH$_2$O	5.0	0.15	7.3
C$_2$H$_5$CHO	1.0	0.30	0.7
C$_2$H$_5$CHO	10.0	2.00	8.3
C$_3$H$_7$CHO	1.0	0.20	1.2
iso-C$_3$H$_7$CHO	2.0	0.60	1.4
C$_6$H$_5$·CH$_2$·CHO	2.0	0.40	1.2
C$_2$H$_5$OCH$_2$OCH$_2$O	2.6	0.15	1.8

| | 2.0 | 1.0 | 1.0 |

The apparent elasticity of the polymer increased with increasing molar ratio of acetaldehyde.

An elastomeric copolymer of acetaldehyde and propionaldehyde was also prepared in a similar way.

Small amounts of aldehydes (1–10% by weight) have been incorporated into polyacetaldehyde as seen in Table XVII (93). BF$_3$ was used as the initiator in all cases. All these copolymers are soluble and apparently elastomeric. In the one case in which a dialdehyde was used, a portion of the copolymer was insoluble and evidently crosslinked. Some of these copolymers still contain reactive double bonds as possible sites for crosslinking reactions.

Copolymers of various higher aldehydes with other aliphatic aldehydes using aluminum alkyls gave partially insoluble polymers (94,95).

In addition to random copolymers of aldehydes, a two-step block copolymerization of acetaldehyde with propylene oxide was carried out (96,97). In the first step, propylene oxide was polymerized with a triethylaluminum–water (1:1) system as the initiator at room temperature. The polymerizing mixture was then chilled to −78°, the appropriate amount of acetaldehyde added, and the reaction allowed to go to completion. From hydrolysis experiments it was concluded that 3–7% of polypropylene oxide was present at the one end of the polyacetaldehyde. Block copolymers of acetaldehyde were also prepared using initial blocks of polyolefins (98). The olefin was first polymerized with a Ziegler-type initiator and the polymerization then completed with acetaldehyde at $-75°$.

Recently, Baker (99) and Takida (100) described, among other block copolymerizations, the preparation of block copolymers of acetaldehyde by initiating acetaldehyde with living polymers of styrene and methyl methacrylate.

Our knowledge of copolymers of aldehydes is very fragmentary. Much work is still required to characterize each individual copolymerization clearly, to establish the uniformity of each copolymer and to prove that copolymers have actually been obtained. Methods are now available to differentiate between block and random copolymers. Reactivity ratios of aldehyde comonomers have not yet been determined and very few copolymer properties have been described.

As in the case of aldehyde homopolymers, great caution must be exercised when aluminum organic compounds are used as initiators for copolymerizations in order to assure aluminum-free polymers.

VI. POLYACETONE

In view of its similarity to polyisobutylene, polyacetone was also expected to be an elastomer. A success in bringing about the polymerization of acetone was considered a challenge for polymer chemists.

Several years ago Kargin reported the successful preparation of polyacetone (101). It was described as a rubbery polymer that depolymerized rapidly at room temperature. Magnesium was used as the initiator of the polymerization which consisted of cocondensation of the initiator with acetone on a cold finger which was held at $-196°$. The initial glassy condensate polymerized upon warmup to a white elastic mass.

Okamura has also reported the preparation of polyacetone in low yield by solid state polymerization using ionizing radiation (102,103).

Recently, Burnop reported that all his attempts to repeat the reported preparation of polyacetone were unsuccessful, and he cast severe doubts on the existence of polyacetone (104).

Calculations have shown (105) that the ΔF of polyacetone formation is unfavorable down to the temperature of liquid nitrogen and unusual reaction conditions would be required for a possible polymerization. We believe it questionable that these requirements have been fulfilled in the reported successful polymerizations of acetone. Until the preparation of polyacetone has been confirmed beyond any reasonable doubt—several laboratories have been unable to repeat the literature reports—and the polymer has been characterized as a polyacetal of reasonable molecular weight, the polymerization of acetone should be considered an unsolved problem.

References

1. M. W. Travers, *Trans. Faraday Soc.*, **32**, 246 (1936).
2. M. Letort, *Compt. Rend.*, **202**, 767 (1936).
3. H. Staudinger, *Trans. Faraday Soc.*, **32**, 249 (1936).
4. G. B. B. M. Sutherland, A. R. Philpotts, and G. H. Twigg, *Nature*, **157**, 257 (1946).
5. H. A. Rigby, C. J. Danby, and C. N. Hinshelwood, *J. Chem. Soc.*, **1948**, 234.
6. M. Letort and X. Duval, *Compt. Rend.*, **216**, 58, 608 (1943).
7. X. Duval and M. Letort, *Bull. Soc. Chim. France*, **1946**, 580.
8. M. Letort, X. Duval, and Y. Rollin, *Compt. Rend.*, **224**, 50 (1947).
9. M. Letort and J. Petry, *Compt. Rend.*, **231**, 519, 545 (1950).
10. M. Letort and A. J. Richard, *Compt. Rend.*, **240**, 86 (1955).
11. M. Letort and P. Mathis, *Compt. Rend.*, **242**, 371 (1956).
12. J. C. Bevington and R. G. W. Norrish, *Proc. Roy. Soc. (London)*, **A196**, 363 (1949).
13. J. Petry and M. Letort, Ger. Pat. 933,785 (1955).
14. M. Letort and J. Petry, Fr. Pat. 1,020,456 (1952).
15. A. J. Richard, Thesis, University of Nancy, France, 1955.
16. M. Letort and J. Petry, *J. Chim. Phys.*, **48**, 594 (1951).
17. M. Letort and A. J. Richard, *J. Chim. Phys.*, **57**, 752 (1960).
18. M. Letort and P. Mathis, *Compt. Rend.*, **241**, 651, 1765 (1955).
19. W. H. Smyers, U.S. Pat. 2,274,749 (1942).
20. O. Vogl, *J. Polymer Sci.*, **46**, 261 (1960).
21. O. Vogl, *J. Polymer Sci.*, **A2**, 4591 (1964).
22. J. Furukawa, T. Saegusa, T. Tsuruta, H. Fujii, and T. Tatano, *J. Polymer Sci.*, **36**, 546 (1959).
23. J. Furukawa, T. Tsuruta, T. Saegusa, H. Fujii, and T. Tatano, Jap. Pat. 22,742 (1961), Appl. Jan. 6, 1959.
24. J. Furukawa, T. Saegusa, T. Tsuruta, H. Fujii, A. Kawasaki, and T. Tatano, *Makromol. Chem.*, **33**, 32 (1960).
25. J. Furukawa, T. Tsuruta, T. Saegusa, H. Fujii, and T. Tatano, Japan. Pat. 22,743 (1961), Appl. Mar. 30, 1959.
26. H. E. Podall, D. Lee, N. Filipescu, and D. H. Rosenblatt, paper presented to the Division of Polymer Chemistry, 142nd Meeting, American Chemical Society, Atlantic City, N.J., September 1962; *Polymer Preprints*, **3**, No. 2, 411 (1962).
27. J. Smidt and J. Sedlmeier, Ger. Pat. 1,106,075 (1961); Appl. Nov. 24, 1958.
28. J. Furukawa, T. Saegusa, T. Tsuruta, H. Fujii, A. Kawasaki, and T. Tatano, *Kogyo Kagaku Zasshi*, **62**, 1925 (1959).
29. K. Weissermel and W. Schmieder, *Makromol. Chem.*, **51**, 39 (1962).
30. K. Weissermel, Ger. Pat. 1,156,223 (1963), Appl. Feb. 26, 1960.
31. J. Furukawa, T. Saegusa, T. Tsuruta, and H. Fujii, Can. Pat. 634,440 (1962); Jap. Pat. Appl. Dec. 16, 1959.
32. Farbwerke Hoechst, Brit. Pat. 966,096 (1964); Ger. Appl. Nov. 14, 1959.

33. S. Artmeyer, F. Broich, W. Franke, and E. Heinrich, Ger. Pat. 1,150,812 (1963); Appl. March 14, 1960.
34. K. Weissermel, K. Küllmer, and H. Schmidt, Ger. Pat. 1,148,738 (1963); Appl. Dec. 17, 1959; Fr. Pat. 1,285,285 (1960).
35. J. Furukawa, T. Saegusa, and H. Fujii, U.S. Pat. 3,094,509 (1964); Japan. Appl. Jan. 6, 1959.
36. A. Lupu and M. Coman, *Rev. Chim. (Bucharest)*, **11**, 298 (1960).
37. H. Takida and K. Noro, *Kobunshi Kagaku*, **21**, 23 (1964).
38. H. Takida and K. Noro, *Kobunshi Kagaku*, **21**, 109 (1964).
39. H. Takida and K. Noro, *Kobunshi Kagaku*, **20**, 705 (1963).
40. H. Takida and K. Noro, *Kobunshi Kagaku*, **20**, 712 (1963).
41. G. Natta, G. Mazzanti, and P. Chini, Brit. Pat. 946,832 (1964); Ital. Appl. Nov. 6, 1959; Ger. Pat. 1,162,563 (1963); Fr. Pat. 1,257,561 (1961).
42. O. Vogl, *Advan. Chem. Ser.*, **52** (1966).
43. O. Vogl, *Macromol. Syn.*, in press.
44. J. N. Koral and B. W. Song, *J. Polymer Sci.*, **54**, S34 (1961).
45. J. N. Koral and B. W. Song, U.S. Pat. 3,122,524 (1964); Appl. Feb. 10, 1961.
46. T. Miyakawa and N. Yamamoto, *Kogyo Kagaku Zasshi*, **65**, 390 (1962).
47. T. Miyakawa, F. Yaku, and N. Yamamoto, *Kogyo Kagaku Zasshi*, **66**, 1694 (1963).
48. T. Miyakawa and N. Yamamoto, *Kogyo Kagaku Zasshi*, **66**, 1697 (1963).
49. T. Miyakawa, F. Yaku, and N. Yamamoto, *Kogyo Kagaku Zasshi*, **66**, 1703 (1963).
50. G. Natta, G. Mazzanti, P. Corradini, and I. W. Bassi, *Makromol. Chem.*, **37**, 156 (1960).
51. S. Ishida, *J. Polymer Sci.*, **62**, 1 (1962).
52. S. Ishida, *Kobunshi Kagaku*, **18**, 187 (1961).
53. Belg. Pats. 603,088 (1961), 603,087 (1961), 603,563 (1961).
54. S. Ohta, T. Saegusa, and J. Furukawa, *Kogyo Kagaku Zasshi*, **67**, 608 (1964).
55. H. Fujii, I. Tsukuma, T. Saegusa, and J. Furukawa, *Makromol. Chem.*, **82**, 32 (1965).
56. H. Fujii, T. Saegusa, and J. Furukawa, *Kogyo Kagaku Zasshi*, **65**, 695 (1962).
57. C. Chachaty, *Compt. Rend.*, **251**, 385 (1960).
58. Houllieres du Bassin du Nord et du Pas-de-Calais, Fr. Pat. 1,270,529 (1961); Appl. May 10, 1960; Ger. Pat. 1,145,357 (1963).
59. V. S. Pshezhetskii, V. A. Kargin, and N. A. Bakh, *Vysokomolekul. Soedin.*, **3**, 925 (1961).
60. F. A. Bovey and R. C. Wands, *J. Polymer Sci.*, **14**, 113 (1953).
61. M. S. Muthana and H. Mark, *J. Polymer Sci.*, **4**, 91 (1949).
62. M. Letort and A. J. Richard, *Compt. Rend.*, **249**, 274 (1959).
63. M. Letort and A. J. Richard, *J. Chim. Phys.* **57**, 752 (1960).
64. M. Letort, *Chim. Ind. (Paris)*, **89**, 155 (1963).
65. J. Brandrup and M. Goodman, *J. Polymer Sci.*, **B2**, 123 (1964).
66. E. G. Brame, R. S. Sudol, and O. Vogl, *J. Polymer Sci.*, **A2**, 5337 (1964).
67. J. Brandrup and M. Goodman, *J. Polymer Sci.*, **A3**, 236 (1965).
68. G. Delzenne and G. Smets, *Makromol. Chem.*, **18/19**, 82 (1956).
69. G. Delzenne and G. Smets, *Makromol. Chem.*, **23/24**, 16 (1957).
70. D. L. Funck and O. Vogl, U.S. Pat. 3,001,966 (1961); Appl. Aug. 18, 1958.

71. O. Vogl, *Chem. Ind. (London)*, **1961**, 748.
72. H. D. Hermann, K. Weissermel, and G. Lohaus, Ger. Pat. 1,131,401 (1964); Appl. March 1, 1960; Can. Pat. 683,847 (1964).
73. Charbonnages de France, Fr. Pat. 1,262,700 (1961); Appl. April 21, 1960; Brit. Pat. 911,959 (1962); Belg. Pat. 602,765 (1961); Ger. Pat. 1,180,938.
74. Charbonnages de France, Fr. Pat. 1,290,180 (1962); Appl. Feb. 4, 1961; Brit. Pat. 933,128 (1963).
75. T. M. Connor, *Polymer*, **5**, 265 (1964).
76. G. Williams, *Trans. Faraday Soc.*, **59**, 1397 (1963).
77. P. W. Bridgeman and J. B. Conant, *Proc. Natl. Acad. Sci. U.S.*, **15**, 680 (1929).
78. J. B. Conant and C. O. Tongberg, *J. Am. Chem. Soc.*, **52**, 1659 (1930).
79. J. B. Conant and W. R. Peterson, *J. Am. Chem. Soc.*, **54**, 628 (1932).
80. O. Vogl, *J. Polymer Sci.*, **A2**, 4609 (1964).
81. T. Iwata, G. Wasai, T. Saegusa, and J. Furukawa, *Makromol. Chem.*, **77**, 229 (1964).
82. C. G. Overberger, S. Ishida, and H. Ringsdorf, *J. Polymer Sci.*, **62**, S1 (1962).
83. W. W. Moyer, Jr. and D. A. Grev, *J. Polymer Sci.*, **B1**, 29 (1963).
84. K. Meyersen, R. C. Schulz, and W. Kern, *Makromol. Chem.*, **58**, 204 (1962).
85. C. Aso and Y. Aito, *Bull. Chem. Soc. Japan*, **35**, 1426 (1962).
86. C. Aso and Y. Aito, *Makromol. Chem.*, **58**, 195 (1962).
87. K. Yokota, Y. Ito, and Y. Ishi, *Kogyo Kagaku Zasshi*, **66**, 1112 (1963).
88. C. Aso and Y. Aito, *Bull. Chem. Soc. Japan*, **37**, 456 (1964).
89. C. Aso, A. Furuta, and Y. Aito, *Makromol. Chem.*, **84**, 126 (1965).
90. K. Jost, Ger. Pat. 1,154,274 (1963); Appl. Sep. 5, 1959.
91. Charbonnages de France, Fr. Pat. 1,302,017 (1962); Appl. June 13, 1961.
92. H. F. Mark and N. Ogata, *J. Polymer Sci.*, **A1**, 3439 (1963).
93. Consortium für Elektrochemie, Brit. Pat. 876,956 (1961); Ger. Appl. 6/30/59; Fr. Pat. 1,262,179 (1961); Belg. Pat. 592,462 (1960).
94. Farbwerke Hoechst, Fr. Pat. 1,299,537 (1962); Ger. Appl. April 16, 1960.
95. B. N. Bastian, Ger. Pat. 1,174,987 (1964); U.S. Appl. Oct. 30, 1961; Fr. Pat. 1,337,682 (1963).
96. H. Fujii, T. Fujii, T. Saegusa, and J. Furukawa, *Kogyo Kagaku Zasshi*, **66**, 846 (1963).
97. H. Fujii, T. Fujii, T. Saegusa, and J. Furukawa, *Makromol. Chem.*, **63**, 147 (1963).
98. S. Ohta, T. Saegusa, and J. Furukawa, *Kogyo Kagaku Zasshi*, **67**, 947 (1964).
99. W. Baker, Fr. Pat. 1,357,936 (1964); Brit. Pat. 997,497 (1965).
100. H. Takida and K. Noro, *Kobunshi Kagaku*, **21**, 467 (1964).
101. V. A. Kargin, P. Kabanov, V. P. Zubov, and I. M. Pabisov, *Dokl. Akad. Nauk SSSR*, **134**, 1098 (1960).
102. S. Okamura, K. Hayashi, and S. Mori, *Doitai To Hoshasen*, **4**, 70 (1961).
103. S. Okamura, K. Hayashi, S. Mori, H. Sobua, Y. Tabata, and H. Shibano, Japan Pat. 16,341 (1963).
104. V. C. E. Burnop, *Polymer*, **6**, 411 (1965).
105. O. Vogl and W. M. D. Bryant, *J. Polymer Sci.*, **A2**, 4633 (1964).

AUTHOR INDEX

A

Abbott, M., 101(19), *122*
Abe, A., 7, 8, *19*
Abell, W. R., 248(134), *252*
Abere, J. F., 266, *272*
Adamek, S., 113(82), *124*
Adams, C. R., 131(12), 179, 280(16), *289*
Adams, H. E., 98(9), *122*
Agius, P., 106(38), *123*
Aito, Y., 448(85, 86, 88, 89), *453*
Alexander, R. L., 110(63), *123*
Alferova, L. V., 384(44), *415*
Alfrey, T., 158(95), *182*, 259(41), *271*
Alimov, A. P., 344(95), 349(114), 350 (114), *374, 375*
Alkhazor, T. G., 132(17), *180*
Allied Chemical Corp., 231(23, 24), *249*
Alliger, G., 97(1–4), 102(20), 117, *122, 124*
Allinger, N. L., 378(1), *413*
Altier, M. W., 232(30), *249*
Amato, F., 100(12), *122*
Ambelang, J. C., 117, *124*
Amberg, L. O., 359(140), *375*
American Cyanamid Co., 261(53), 262, *271, 272*
Amundson, N. R., 178(145–147), *183*
Andersen, D. E., 232(34), *249*
Andersen, H. M., 155(75), *181*
Anderson, J. J., 108(54), *123*
Andrews, E. H., 115, 118, *124, 125*
d'Anghiera, P. M., 22, *88*
Angier, D. J., 114, *124*
Angyal, S. J., 378(1), *413*
Anoshina, N. P., 237(75, 76), *250*
Anosov, V. I., 241(101), *251*, 324(63), *329*
Anspon, H. D., 110(63), *123*
Appenrodt, J., 79(152), *93*
Arbuzova, I. A., 262, *272*

Arcozzi, A., 117(103), *124*
Arganbright, R. P., 131(13), *179*, 231(24), 232(26), *249*
Armstrong, H. E., 43, *90*
Armstrong, W. E., 130(8), 131(12), *179*
Arnett, L. M., 257(31), *271*
Arnold, R. G., 232(34), *249*
Artmeyer, S., 425(33), *452*
Aschan, O., 54, *91*
Ashe, W., 109(57), 110(57), *123*
Ashikari, N., 158(90), *182*
Aso, C., 448(85, 86, 88, 89), *453*
Athey, R. J., 109(58), *123*
Atkinson, A. S., 112(76), *124*
Auerbach, C., 156(76), *181*
Aufdermarsh, C. A., Jr., 229(14), 236 (70), *248, 250*

B

Baader, H., 404(100), *417*
Baccaredda, M., 384(64), *416*
Bacskai, R., 294(25), *328*
Bader, H., 384(59), *415*
Badische Anilin- & Soda-Fabrik, 53(85), 63(111), 64, 65, 68, 78, 85, 86(174), *91–94*
Bähr, K., 80, *93*
Baer, J., 75, *92*
Baggett, J. M., 382(9), 384(9), 409(9), *414*
Bahary, W. S., 103(29), *122*, 171, *182*
Bahr, H., 131(13), *179*
Bailey, J. T., 166(109), 169(109), 171 (109), *182*
Bailey, P. S., 39, *89*
Bajars, L., 131(12, 13, 16), *179, 180*
Bakalo, L. A., 384(24, 39), *414, 415*
Baker, W., 450, *453*

455

472 AUTHOR INDEX

Vogl, O., 422(20, 21), 423(21), 424(21), 427(21), 429(21), 430(20, 21), 431(21, 42, 43), 432(21), 435(21), 438(66), 439(66), 442(70), 443(70, 71), 444(70) 446(21), 447(80), 450(105), *451-453*

Vogt, R. R., 230(15), 242(122), *248, 251*

Vohwinkel, K., 102, *122*

Volkenstein, M. V., 4, 8, *18, 19*

Volkhima, A. V., 383(17), *414*

Vrij, A., 18, *19*

W

Wadelin, C., 117(108), *124*

Wadell, H. H., 320, *328*

Wagner, F. C., 238(86), *250*

Wagner, R. I., 113, *124*

Wake, W. C., 115, *124*

Wakefield, H. F., 346(109), *374*

Wakefield, L. B., 74, *92*

Wakelin, J. H., 260(45), *271*

Walker, H. W., 159(99), *182*, 232(37), 233(47, 51), *249, 250*

Walker, J. F., 82(162), *93*

Wall, F. T., 13(31), *19*, 157(83), 158(83), 177(144), *181, 183*

Wall, L. A., 238(95), *251*

Wallace, A. J., 106(41), *123*

Wallach, O., 35, 36, *89*

Walling, C., 144, 145(41), 149(55a), 157 (41, 85), 158(85), 159(85), 177, *180, 181*, 322(62), *329*

Walmesley, D. A., G., 396(94), *416*

Walton, W. W., 160(100), *182*

Wanless, G. G., 360(141), *375*

Wands, S. C., 437(60), 446(60), *452*

Wasai, G., 257(34), *271*, 448(81), *453*

Watanabe, T., 336(42), 337(42), 347 (42, 112, 113), 348(113), 349(112, 113), 350(42, 112, 113), *373, 375*

Watanabe, W. A., 264(67), *272*

Watson, H. B., 257(28), *271*

Watson, K. M., 129(7), *178*

Watson, W. F., 157(82), *181*

Watterson, J. G., 143(38), *180*

Wattimena, F., 131(13), *179*

Wearsch, N. C., 255(15), *270*

Weber, C. O., 36, 37, *89*

Wei, P. E., 360, *375*

Weiden, H., 232(27), 242(122), *249, 251*

Weightman, J., 392(78), *416*

Weinberger, M. A., 293(12), *327*

Weinstein, A. H., 173(129), *183*

Weinstein, M., 384(41), *415*

Weinstock, K. V., 110(66), *123*, 385 (20a), *414*

Weissbein, L., 269(82), *272*

Weissermel, K., 390(69), 405(103), *416, 417*, 425(29, 30, 34), 437, 440(29), 443(72), 444-446, *451-453*

Weissert, F. C., 97(2-4), 102(20, 23), 105, *122, 123*

Weitz, H. M., 140(29), 142(29), 144(29), *180*

Welch, F. J., 256(24), *271*

Wenisch, W. J., 117(106), *124*

Wenisch, W. T., 145(41), 157(41), *180*

Wenz, A., 81(159), *93*

Wesenberg, 78(150), *93*

Wheelans, M. A., 101(18), *122*

Whitby, G. S., 75(134), *92*, 129(5a, 5b), 132(5c), 139(23a), 144(23b), 148(52), 159(23c), 166(23d), 169(23d), 177 (23e), *178, 180, 181*, 254(3), *270*

White, R. C., 238(92), *251*

White, R. M., 107(48), *123*

White, T. R., 383(18), *414*

Whitman, R. D., 108, *123*, 265(72), *272*

Wich, G. S., 238(88), 239(88), *251*

Wichterle, O., 296(33), *328*

Wicklatz, J. E., 153(67), 154(67), *181*

Wilder, F. N., 232(37), *249*

Wiley, R. H., 13(29), *19*, 263(62, 63), *272*

Wilkinson, K., 384(31), 411-413(31), *415*

Willer, R., 81(161), 82(161), *93*

Willfang, G., 382(13), 384(13), 388(13), *414*

Williams, A., 257(30), *271*

Williams, C. G., 32, *89*

Williams, D. J., 172(125), *183*

Williams, G., 446, *453*

Williams, H. C., 278(19), *289*

Williams, H. L., 153(70), 154(68-70, 73), 156, 158(88, 89, 91, 92, 96, 97), 159 (96), *181, 182*

SUBJECT INDEX

A

Acetal endgroup, 338
Acetaldehyde, block copolymers of, 449
  peroxides of, 421
  preparation of, 188
  solubility of, 430
Acetaldehyde polymerization, initiators
    for, 433–434
  mechanism of, 435–437
  propagation in, 437
  in solution, 429–433
    initiators for, 431–432
    solvents for, 431
  using radiation, 434–435
  using x-rays, 434
  using γ-rays, 434
Acrylate copolymers, 266
Acrylate esters, from acetylene, 254
  from ethylene oxide, 254
  polymerization of, 255–258
  from β-propiolactone, 254
  from propylene, 255
Acrylate monomers, hydrolysis of, 257
  reactivity ratios of, 259–260
Acrylate polymers, isotactic, 258
  syndiotactic, 258
Acrylic acid, 261
Acrylic elastomers, 110, 253–270
  See also individual compounds.
  aging of, 268–270
    effect of carbon black on, 269
  commercially available, 269
  copolymerization of, 259–260
  glass transition temperatures of, 263–
    268
  monomer synthesis for, 254–255
  oil resistance of, 263–268

Acrylic elastomers, synthesis, 255–258
  in bulk, 255
  in emulsion, 256–257
  initiators for, 255–258
  stereoregular, 257–258
  in suspension, 255–256
  reactivity ratios for, 259–260
  vulcanization of, 260–263
    cure systems for, 261
Acrylonitrile, 266
  See also Butadiene-acrylonitrile
    copolymerization, 192–206.
  analysis of, 191
  catalysts for preparation of, 189–190
  as comonomer with butadiene, 73
  copolymer with butadiene, 192–206
  effect on nitrile rubber vulcanizates,
    202
  preparation of, 187–191
  properties of, 191
Acrylonitrile modified natural rubber,
    221–222
  swell resistance of, 222
Acyl peroxides, 157
Acylium ion salts, 392
Adipamide, as stabilizer, 444
Adiprene, 275
Aldehydes, copolymers of, 448–450
  reactivity ratios of, 450
Aldol-α-napthyl amine, 217
Alfin catalysts, derivation of name, 87
  discovery of, 87
Alfin polymers, 103
Aliphatic aldehydes, elastomeric
    polymers of, 420
Alkali metals, as polymerization
    initiators, 78
1-Alkenyl alkyl ethers, 342

477